Cultural Studies of Science Education

Volume 13

The series is unique in focusing on the publication of scholarly works that employ social and cultural perspectives as foundations for research and other scholarly activities in the three fields implied in its title: science education, education, and social studies of science.

The aim of the series is to establish bridges to related fields, such as those concerned with the social studies of science, public understanding of science, science/technology and human values, or science and literacy. *Cultural Studies of Science Education*, the book series explicitly aims at establishing such bridges and at building new communities at the interface of currently distinct discourses. In this way, the current almost exclusive focus on science education on school learning would be expanded becoming instead a focus on science education as a cultural, cross-age, cross-class, and cross-disciplinary phenomenon.

The book series is conceived as a parallel to the journal *Cultural Studies of Science Education*, opening up avenues for publishing works that do not fit into the limited amount of space and topics that can be covered within the same text.

More information about this series at http://www.springer.com/series/8286

Alberto Bellocchi • Cassie Quigley
Kathrin Otrel-Cass
Editors

Exploring Emotions, Aesthetics and Wellbeing in Science Education Research

Springer

Editors
Alberto Bellocchi
Faculty of Education
Queensland University of Technology
Brisbane, QLD, Australia

Cassie Quigley
Department of Teaching and Learning
College of Education
Clemson University
Clemson, SC, USA

Kathrin Otrel-Cass
Department of Learning and Philosophy
Aalborg University
Aalborg, Denmark

ISSN 1879-7229 ISSN 1879-7237 (electronic)
Cultural Studies of Science Education
ISBN 978-3-319-43351-6 ISBN 978-3-319-43353-0 (eBook)
DOI 10.1007/978-3-319-43353-0

Library of Congress Control Number: 2016955693

Printed on acid-free paper

This Springer imprint is published by Springer Nature
The registered company is Springer International Publishing AG Switzerland

Contents

About the Editors and Contributors

Editors

Alberto Bellocchi is a researcher and academic at the Queensland University of Technology, Brisbane, Australia. He is currently a recipient of a 3-year funded research fellowship focusing on the interplay between social bonds and learning science. His broader research programme addresses teaching and learning within university pre-service teacher education classes and high school science classrooms.

Cassie F. Quigley is an associate professor of science education at Clemson University. Her research works to expand the participation in and perspectives of science. In doing so, her work is community- and place-based. To date, she has explored these questions in a variety of settings, focusing attention on specific places to potentially broaden the scope of whose knowledge garners value in scientific and science education communities.

Kathrin Otrel-Cass is professor MSO (with special responsibilities) in visual ethnography at the Department of Learning and Philosophy at Aalborg University. She is an experienced science and technology education researcher primarily in classroom-based settings with a particular interest in the study of interactions and what this means for teaching practices and critical education research.

Contributors

Konstantinos Alexakos is a professor in the School of Education at Brooklyn College (CUNY) and at the Graduate Center at CUNY. His interests include teaching and learning with a focus on emotions, mindfulness and wellness, including emotional bonds and fictive kinships (close personal friendships), coteaching, radical listening, cogenerative dialogues, emotional climate, breathing and mindful practices, with the goals of improving learning and teaching, personal wellness and the emotional climate in the classroom, as well as creating safe spaces for discussing challenging topics, valuing difference and learning from each other.

Steve Alsop is a professor and director of the graduate diploma in environmental sustainability in the Faculty of Education. He teaches courses and supervises graduate students in the fields of education, science and technology studies, environmental sustainability and interdisciplinary studies. Steve has held a series of administrative appointments including associate dean (Research and Professional Development, York University), teaching studies area coordinator (Science, Roehampton Institute, University of Surrey, UK) and head of physics (Haverstock School, London, UK) and a series of honorary positions at the Froebel Institute, Roehampton University and the Universidad Baja California, Mexico.

Per Anderhag has a PhD in science education and is working with school science development at Education Administration, City of Stockholm. His research interest is the difference teaching can make for students' interest and learning in science.

Alberto Bellocchi is a researcher and academic at the Queensland University of Technology, Brisbane, Australia. He is the recipient of a 3-year funded research fellowship focusing on the interplay between social bonds and learning in science. His broader research programme focuses on learning and teaching science at high school science and in university pre-service teacher education classes and the role of emotions in learning.

Gayle Buck is a professor of science education at Indiana University. Her scholarship and teaching foregrounds relationships across learner diversity and science. Her research focuses on student populations traditionally underserved by science education and neglected epistemological assumptions in science teaching and learning.

Tatiana Chemi is an associate professor at Aalborg University. She is the chair of Educational Innovation, where she works in the field of artistic learning and creative processes. She is currently involved in research projects examining artistic creativity, arts-integrated educational designs in schools and the role of emotions in learning. In 2013, Aalborg University Press named her Author of the Year.

Kristin Cook is an assistant professor of science education at Bellarmine University. She received her PhD at Indiana University in curriculum and instruction, specializing in science education and environmental science. Kristin's research focuses on engaging students and pre-service teachers with the community of science through exploration of socio-scientific issues.

James P. Davis is a researcher and lecturer at the Queensland University of Technology, Brisbane, Australia. James has a background in qualitative and quantitative research in health and science education. His current research focuses on analogical reasoning in science teaching and learning, the sociology of emotions in school science, micro-sociology and ethnomethodology.

Niels Bonderup Dohn is an associate professor in science education. His special field of interest is methodological approaches towards measuring interest development in science education.

Al-Karim H. Gangji is a high school physics teacher and an adjunct lecturer in the Physics Department at Queens College of CUNY. He is currently a doctoral student in the Urban Education Program at the CUNY Graduate Center. His research interest involves mindfulness in preparing elementary teachers to teach science and mathematics. He is the recipient of the Queens College President's Award for Excellence in Teaching by Adjunct Faculty (2013), the MIT Inspirational Teacher Award (2009), Educator of Distinction from the Coca-Cola Scholars Foundation (2007) and the Hofstra University/New 12 Long Island Educator of the Month (2003).

Linda Hobbs is a senior lecturer in science education at Deakin University, where she teaches primary science education. She has a range of research projects, which mainly focus on the teacher. Her latest research is in exploring the issue of teachers teaching across subject boundaries and the implications for teacher learning, identity and support needs. Another research agenda of hers is exploring the efficacy of school-based approaches to primary science education.

Leissa Kelly has recently completed a PhD in marine education working with teachers and educators in the marine environment. Leissa has worked as a science educator in a number of formal and non-formal science education settings and facilities, including hospitals, schools, universities, zoological gardens and parks. Leissa is currently project officer on a number of STEM-related projects forging links between education researchers, scientists and school teachers and students.

Liv Kondrup Kristensen is a PhD student at the Department of Health Science and Technology at Aalborg University. Her research is classroom-based, in which she explores different ways of capturing the role of the body in learning processes. At the same time she is an associate professor at the University College of Zealand, where she teaches health related courses in the Social Education program

Renée Lyons is a doctoral candidate in curriculum and instruction at Clemson University. Her research explores how science projects and educational experiences can become third spaces merging the discourse, practices, goals and values of the world of science with the world a person experiences outside of science. The goal of her research is to broaden participation in science by creating new forms of participation in science, forms which present participants with a vision of how participating in science fits in the larger context of their lives.

Anna Malyukova is a doctoral student at the Graduate Center. Anna is very passionate about mindfulness and emotions in the process of learning on the birth-death continuum. She began her education in Russia in 1997 pursuing her degree in engineering, but has since become interested in education and received an AS in early childhood education and a BA in liberal arts from CUNY. In her collaboration with Kenneth Tobin and Konstantinos Alexakos, she became very interested in the role of emotions in the process of learning. She has recently had a publication with Carol Korn-Bursztyn in *Immigrant Children and Youth: Psychological Challenges* (2015).

Jennifer Beers Newlands is a full-time secondary school teacher and a part-time doctoral student at Murdoch University in Western Australia. Her research focuses on the teaching and learning of STEM for students who are at risk of disengaging with school.

Kathrin Otrel-Cass is professor MSO (with special responsibilities) in visual ethnography at the Department of Learning and Philosophy at Aalborg University. She is an experienced science and technology education researcher primarily in classroom-based settings with a particular interest in the study of interactions and what this means for teaching practices and critical education research.

Morten Rask Petersen is an assistant professor in science education. His special field of interest is the theoretical foundation of interest development in science education.

Cassie F. Quigley is an associate professor of science education at Clemson University. Her research works to expand the participation in and perspectives of science. In doing so, her work is community- and place-based. To date, she has explored these questions in a variety of settings, focusing attention on specific places to potentially broaden the scope of whose knowledge garners value in scientific and science education communities.

Stephen M. Ritchie has been interested in the complexities of classroom teaching and learning throughout his research career. His recent research has focused on the emotional engagement of students in science classes and the emotional experiences of beginning science teachers. Although he is now acting provost, his substantive role is dean of education at Murdoch University.

Kenneth Tobin is Presidential Professor of Urban Education at the Graduate Center of CUNY. In 1973, Tobin began a programme of research on teaching and learning that continues to the present day. The current emphasis of his work involves mindfulness, wellness, environmental harmony and the transformative potential of social research. Tobin has published more than 20 books, 200 refereed journal articles and 125 book chapters. He is recipient of numerous awards, including Distinguished Contributions to Science Education through Research Award (2007, National Association for Research in Science Teaching), Mentoring Award as an exemplary scholar and mentor (2008, Division G, American Educational Research Association) and the National Science Foundation Director's Award for Distinguished Teaching Scholars (2004).

Per-Olof Wickman is a professor in science education and director of science education research at Stockholm University, Sweden. His main research interest is in modelling classroom interactions holistically to support teacher and student agency.

Contributors

Konstantinos Alexakos Learning Sciences, Urban Education, The Graduate Center of CUNY, New York, NY, USA

Steve Alsop Faculty of Education and Department of Science and Technology Studies, York University, Toronto, ON, Canada

Per Anderhag Research and Development Unit, Education Administration, City of Stockholm, Stockholm, Sweden

Alberto Bellocchi Faculty of Education, Queensland University of Technology, Brisbane, QLD, Australia

Gayle Buck Indiana University, Bloomington, IN, USA

Tatiana Chemi Department of Learning and Philosophy, Aalborg University, Aalborg, Denmark

Kristin Cook Bellarmine University, Louisville, KY, USA

James P. Davis Faculty of Education, Queensland University of Technology, Brisbane, QLD, Australia

Niels Bonderup Dohn Aarhus University, Aarhus, Denmark

Al-Karim H. Gangji Learning Sciences, Urban Education, The Graduate Center of CUNY, New York, NY, USA

Linda Hobbs Faculty of Arts and Education, Deakin University, Waurn Ponds, Geelong, VIC, Australia

Leissa Kelly Faculty of Arts and Education, Deakin University, Waurn Ponds, Geelong, VIC, Australia

Liv Kondrup Kristensen Centre for Social Educational Studies, University College Zealand, Roskilde, Denmark

Renée Lyons Department of Curriculum and Instruction, Clemson University, Clemson, SC, USA

Anna Malyukova Learning Sciences, Urban Education, The Graduate Center of CUNY, New York, NY, USA

Jennifer Beers Newlands School of Education, Murdoch University, Murdoch, WA, Australia

Kathrin Otrel-Cass Department of Learning and Philosophy, Aalborg University, Aalborg, Denmark

Morten Rask Petersen University of Southern Denmark, Odense, Denmark

Cassie F. Quigley Department of Teaching and Learning, College of Education, Clemson University, Clemson, SC, USA

Stephen M. Ritchie School of Education, Murdoch University, Murdoch, WA, Australia

Kenneth Tobin Learning Sciences, Urban Education, The Graduate Center of CUNY, New York, NY, USA

Per-Olof Wickman Department of Mathematics and Science Education, Stockholm University, Stockholm, Sweden

Chapter 1
Emotions, Aesthetics and Wellbeing in Science Education: Theoretical Foundations

Alberto Bellocchi, Cassie F. Quigley, and Kathrin Otrel-Cass

The idea for this collection emerged during a 3-day exploratory research workshop in Luxembourg associated with the journal *Cultural Studies of Science Education* (CSSE). This workshop was entitled 'Innovation and collaboration in cultural studies of science education: Towards an international research agenda'. The 3-day event held at the University of Luxembourg, between 17 and 19 June 2014, was organized by Professor Christina Siry and had the overarching purpose to create new possibilities for collaboration towards viable, tangible research projects and international networks. This event was funded in part by the Luxembourg National Research Fund (Fonds National de la Recherche Luxembourg), the University of Luxembourg and Springer Publishing. Scholars who were invited had an interest and record of working within sociocultural perspectives. Each scholar was asked to submit a short paper in advance, to be aligned with one of five strands of the workshop. The five strands were equity and social justice, cultural diversity and multilingualism, emotions in teaching and learning, globalization and neoliberalism and representation and voice in the research process.

Amongst the group of researchers who joined the strand *emotions in teaching and learning*, a discussion developed that focused on the challenges and interests in investigating the role of emotions in science teaching and learning. This book captures the essence of the workshop discussions from those original members of the

---

A. Bellocchi (✉)
Faculty of Education, Queensland University of Technology,
Brisbane, QLD, Australia
e-mail: alberto.bellocchi@qut.edu.au

C.F. Quigley
Department of Teaching and Learning, College of Education, Clemson University,
Clemson, SC, USA

K. Otrel-Cass
Department of Learning and Philosophy, Aalborg University, Aalborg, Denmark

A. Bellocchi et al. (eds.), *Exploring Emotions, Aesthetics and Wellbeing in Science Education Research*, Cultural Studies of Science Education 13,
DOI 10.1007/978-3-319-43353-0_1

special interest group and also includes a number of other scholars who were subsequently invited to contribute. These collaborative efforts led to an expansion of topics that went beyond the original focus on emotions to include also aesthetics and wellbeing.

What emerged during the discussion in Luxembourg was that most of us had, at some point, been met with critical reactions to research that considered emotions in relation to teaching and learning in science. In our discussion, we explored the reasons for our common experiences. We concluded that historically, the science education research community had focused on cognitive matters and had only lately considered research underpinned by sociocultural theories. However, even amongst those scholars who have focused on emotion and affect, the diversity of perspectives meant that communication would not always be straightforward. Through our joint contemplations, we realized that perhaps some science education scholars were interested, yet unfamiliar, with the diversity of theoretical constructs and approaches that provide paradigms, conceptual frameworks and analytical lenses for researching emotions, aesthetics and wellbeing. To this end, we agreed that an edited collection in which each author would outline the theoretical and philosophical assumptions underlying their work would be one way of explicating the pluralism evident in these emerging fields of inquiry. In addition, we invited authors to illustrate empirical applications of their research to offer clear examples of how different theories, philosophical positions and methodologies were applied in studies of learning science or learning to be a science teacher.

Subsequent to the Luxembourg workshop, a number of authors presented work from their chapters at the 2015 *European Science Education Research Association* (ESERA) annual conference in Helsinki, Finland. One of our coeditors, Cassie Quigley, organized a symposium called 'Beyond cognition in science education: Considering the role of emotions, wellbeing, and aesthetics'. Our discussant for the symposium, Steve Alsop, kindly agreed to join our crew of authors and to write an afterword for this volume. In this way, this book represents the collective efforts and diverse views of an international collective of scholars. Inviting Steve to write the endnote meant that we as editors were not the only voice to bracket the diverse collection of chapters that form the body of this text. This was important, as one of the goals of this book project was to avoid homogenizing the collection by inviting only those scholars who shared our theoretical or methodological persuasions. The ESERA symposium was well attended, and a separate strand on emotion research also appeared in the conference schedule. We take these to be signs of the increasing interest in these fascinating and emergent disciplines of inquiry. Perhaps our volume has been prepared at a time when interest in emotions, aesthetics and wellbeing in science education is growing. As an international research community that is interested in problems such as student disaffection with school science and how to redress such wicked issues, we feel there is an urgent need to expand on more traditional approaches of investigating science teaching and learning that have been dominant in the past. The approaches shared in this collection are an attempt to provide a platform for expanding science education research and practice further in new and different directions.

1.1 The Need for Emotion, Aesthetics and Wellbeing in Science Education Research

Science teaching and learning has been taught from perspectives that focus on objectivity, rationality and rigor, where emotion, wellbeing and aesthetics have received scant attention in the past and may even be perceived to cloud the outcomes derived from systematic scientific investigations (Fricker 1991). The idea that science is devoid of emotion has been the historical product of the Western world (Ingold 2011) but has been challenged by a number of scholars including those who argued for the need to connect body and mind (Merleau-Ponty 1968; Latour 2004). The uncoupling of emotion from science in teaching and learning and research has been attributed to 'a long-standing Newtonian–Cartesian tradition of separation, prizing apart the mind and body, divorcing and polarizing reason from feeling' (Alsop and Watts 2003). Yet, emotions and wellbeing play an important role in everyone's life, including in educational settings. Research on emotions in education since the 1990s has established the key role they play in 'students' learning, developmental trajectories, psychological health,…teachers' classroom instruction, and professional careers' (Pekrun and Linnenbrink-Garcia 2014; p. ix). Although research on emotion, wellbeing and aesthetics is growing in science education, it is still underrepresented in academic journals (Fortus 2014) potentially because cognitive skills have traditionally been ranked higher than the emotional and expressive aspects of learning and teaching (Ingold 2011). This book seeks to build on work that connects the mind and the body, with a focus on emotion, aesthetics and wellbeing and their relevance for science education research and practice. The perspectives presented herein are necessary and timely, given the long and dominant role that cognition and conceptual understanding have had in science education research.

Constructs such as emotion, wellbeing and aesthetics have a long history in philosophy, psychology and sociology. A diverse range of conceptualizations of the central phenomena they seek to explore and understand constitutes these fields of inquiry. Although this diversity provides a rich intellectual basis for researchers to frame studies in science education, it can pose a challenge when authors from different traditions seek to establish links with and critiques of the works of others or establish their own new ground.

This book presents a selection of work that bridges selected theories and/or philosophical traditions dealing with emotion, wellbeing and aesthetics with the intent of informing science educators who want to engage with this research and scholarship. It also seeks to provide clear discussion of the foundations of some key conceptual traditions underpinning contemporary work in the field. Through this book the authors will provide insights and frameworks that support researchers working in this area. In each chapter, the authors present philosophical and/or theoretical threads that inform and represent their work, or that of others, on teaching and learning in science education. Assumptions and traditions of different research designs are foregrounded to inform future studies and to raise questions about approaches to inquiry on emotion, wellbeing and aesthetics. The goal is to expand views of these constructs and how they can be investigated and understood.

1.1.1 Organization of the Book

The book is divided into three major sections. Part 1 focuses on approaches to the study of aesthetics in science education. We chose to begin with this topic because historically, emotions and wellbeing fell under the branch of aesthetics in philosophy. In Chap. 2, Per-Olof Wickman (Sweden) discusses how different schools of philosophy have dealt with thinking, acting and feeling. By outlining the philosophical roots of these approaches, Wickman outlines the consequences of an aesthetics approach for empirical research in science education, as well as in psychology and neurophysiology. This is followed in Chap. 3 by Per Anderhag's (Sweden) account of a *taste for science*: a construct that he uses to understand learning science as a situated cultural and social practice. Anderhag opens the chapter with a theoretical discussion of taste and how it is relevant for science education. Drawing on empirical examples from classroom-based research, he then outlines how taste can be used to explore norms, aesthetics and cognition as transactional phenomena. Chapter 4, authored by Linda Hobbs and Leissa Kelly (Australia), addresses the affective/emotional dimension of becoming science educators. Hobbs and Kelly investigate how a teacher's passion for science develops, as this is one of the essences of teaching practice that students report to be highly influential in their interest levels with science.

Studies of emotion are the focus of Part 2. Alberto Bellocchi (Australia) outlines the applications of interaction ritual theory and sociology of emotions in science education research in Chap. 5. He then proposes a *microsociology of learning* based on connections between contemporary sociology of emotion and rituals and Emilé Durkheim's social epistemology of knowledge. Steve Ritchie and Jen Beers Newlands (Australia) develop an argument for an *event* as the unit of analysis relevant for studies of emotion in Chap. 6. Drawing on Sewell's sociological analysis of historical events and Žižek's philosophical discussion of events, Ritchie and Beers Newlands provide empirical examples from a science teacher education class to show how this construct provides a meaningful unit of analysis in empirical research that captures transformative classroom practices. This is followed in Chap. 7 by James Davis' (Australia) study of emotion and analogical reasoning informed by ethnomethodology. Davis outlines the ontological and epistemological assumptions in his work that brings together ideas from scholars such as David Hume and Emilé Durkheim to understand classroom analogy as an emotional and embodied practice. In Chap. 8, Kristin Cook and Gayle Buck (United States) detail their applications of interaction ritual theory to the study of out-of-school science. Liv Kondrup Kristensen and Kathrin Otrel-Cass (Denmark) draw on Maurice Merleau-Ponty's philosophical theorization of emotions as publicly embodied enactments that are shaped by behaviours between people in Chap. 9. Adopting this theoretical orientation to emotion, Kristensen and Otrel-Cass then provide an analysis of an eighth grade science class engaged in learning physics. Examples are provided for ways of analyzing bodily stance to interpret emotions in empirical research. Morten Petersen and Niels Dohn (Denmark) discuss psychological constructs of emotion and interest in Chap. 10. In this chapter, Petersen and Dohn tease apart the different constructs of interest asking questions about its relationship to emotion. They focus on examples of topic and epistemic emotions that relate to the content of learning and engagement with the content, respectively. Two cases from upper secondary biology classes are used to illustrate empirical examples of these constructs. Chapter 11 by Tatiana Chemi (Denmark) presents examples of integrated arts and science curriculum. This chapter outlines the way in which artistic experiences related to science can facilitate dedication to science, self-development and learning.

Wellbeing and wellness are the focus of Part 3 of the book. This is an area of research that is still emerging in the science education literature. In Chap. 12, Ken Tobin, Konstantinos Alexakos, Anna Malyukova and Al-Karim Gangji (United States) begin by outlining developments of an approach to authentic inquiry that engages multiple systems of logic and theory. After detailing some of the underlying assumptions of these sociocultural traditions, the authors provide examples of the inclusion of Jin Shin Jyutsu (JSJ), an Eastern system of logic that is also a traditional form of medicine, into their recent research and practice. Through examples from a university research class, Tobin and colleagues demonstrate how the JSJ system can be used for coding episodes of classroom interaction whilst also offering an intervention for ameliorating excessive emotions that are detrimental to learning or interactions. This highly novel approach to inquiry and classroom intervention offers scope for authentic inquiry in science education that may benefit practitioners, researchers and participants. Cassie Quigley and Renée Lyons (United States) introduce wellbeing in the context of environmental education research in Chap. 13. Wellbeing and emotion is presented in the context of an ethic of care. Quigley and Lyons describe experiences of an ethic of care in teaching environmental science within a middle school setting. Their study presents an approach to deconstructing normative rhetoric surrounding environmental care in order for students to reconstruct their understanding with a focus on values and social justice.

The book concludes with an afterword in Chap. 14 by Steve Alsop (Canada). Undaunted by the task of summating the collection, Alsop offers suggestions for what more can and should be known and studied about emotions, aesthetics and wellbeing, indicating directions that were not addressed in the book chapters. After a review of chapters presented in this collection, Alsop offers a forward-looking perspective providing scope for future international collaborations for subsequent editions of this manuscript. He also suggests how the topics addressed in this volume can be and are being studied by other scholars in the field.

Preparing this collection has been a worthwhile experience for us as editors because we found a substantial community of international scholars who share our interests in these topics. The level of scholarship with which we were met as we collected the various contributions was impressive. This book is not only an attempt to collect theories, philosophical traditions and empirical examples of inquiry on emotion, aesthetics and wellbeing but also an attempt to build a community of scholars and to foster international collaborations and discourse on topics that will have ever-increasing importance for science education and education more generally for decades to come. To the readers of this book, we hope to have provided you with inspiration to join us in these exciting lines of inquiry.

We thank our authors for their thoughtful contributions and the work that went into achieving this international collaboration. We are also indebted to Adeyanju Odutola and Heidi Cian for their assistance with collating chapters and author information.

The Editors
Alberto Bellocchi, Cassie Quigley and Kathrin Otrel-Cass
February, 2016

References

Alsop, S., & Watts, M. (2003). Science education and affect. *International Journal of Science Education, 25*, 1043–1047. doi:10.1080/0950069032000052180

Fortus, D. (2014). Attending to affect in science education. *Journal of Research in Science Teaching, 51*, 821–835. doi:10.1002/tea.21155

Fricker, M. (1991). Reason and emotion. *Radical Philosophy, 57*, 14–19.

Ingold, T. (2011). *Being alive: Essays on movement, knowledge and description*. Milton Park/Abingdon/Oxon: Routledge.

Latour, B. (2004). How to talk about the body? The normative dimension of science studies. *Body & Society, 10*, 205–329. doi:10.1177/1357034X04042943.

Merleau-Ponty, M. (1968). *The visible and the invisible: Followed by working notes*. Evanston: Northwestern University Press.

Pekrun, R., & Linnenbrink-Garcia, L. (2014). Conclusions and future directions. In R. Pekrun & L. Linnenbrink-Garcia (Eds.), *International handbook of emotions in education* (pp. 659–675). New York: Routledge.

Alberto Bellocchi is a researcher and academic at the Queensland University of Technology, Brisbane, Australia. He is currently the recipient of a 3-year funded research fellowship focusing on the interplay between social bonds and learning science. His broader research program addresses teaching and learning within university pre-service teacher education classes and high school science classrooms.

Cassie F. Quigley is an associate professor of science education at Clemson University. Her research works to expand the participation in and perspectives of science. In doing so, her work is community and place-based. To date, she has explored these questions in a variety of settings. Focusing attention on specific places potentially broadens the scope of whose knowledge garners value in scientific and science education communities.

Kathrin Otrel-Cass is Professor MSO (with special responsibilities) in visual ethnography at the Department of Learning and Philosophy at Aalborg University. She is an experienced science and technology education researcher primarily in classroom-based settings with a particular interest in the study of interactions and what this means for teaching practices and critical education research.

Part I
Aesthetics in Science Education

Chapter 2
Back to the Drawing Board: Examining the Philosophical Foundations of Educational Research on Aesthetics and Emotions

Per-Olof Wickman

Thinking is an act. Feeling is a fact.

Clarice Lispector (1977/2014, p. 3)

2.1 Introduction

There is much to contemplate in the short quotation from Clarice Lispector's novel *Hour of the Star*. This humorous statement is perplexing in a wonderful way, mostly, I think, because it unites what is typically separated. To be prosaic, it concerns the three basic elements of how Western thought since antiquity has divided the human mind into faculties. In Lispector's rendering these three faculties are thinking, acting, and feeling. In dealing with emotions, wellbeing, and aesthetics, this book orients itself primarily toward "feeling" in Lispector's terms. In this chapter I will examine how various schools of philosophy have approached this tripartite division of human faculties and the consequences their basic assumptions have for science education research on aesthetics and emotion.

In modern philosophy the three faculties are usually divided into *cognition*, *morals*, and *aesthetics*. An example, to take a philosophical oeuvre seminal to subsequent understanding of our faculties, is the three "critiques" by the German philosopher Immanuel Kant (1724–1804): *Critique of Pure Reason* (Kant 1781/1998), *Critique of Practical Reason* (Kant 1788/2004), and *Critique of Judgment* (Kant 1790/1987). His first critique concerns the problem of certainty and the grounds for justified true belief. It is about the basic human faculty of cognition, that is, the rational and empirical grounds for knowing that something is the case (a fact). The second critique treats morals or the grounds for human conduct, that is, the proper

P.-O. Wickman (✉)
Department of Mathematics and Science Education, Stockholm University,
10691 Stockholm, Sweden
e-mail: per-olof.wickman@mnd.su.se

A. Bellocchi et al. (eds.), *Exploring Emotions, Aesthetics and Wellbeing in Science Education Research*, Cultural Studies of Science Education 13,
DOI 10.1007/978-3-319-43353-0_2

acts or behavior in various situations. The third deals with grounds for aesthetic judgments concerning what we find agreeable/disagreeable or beautiful/ugly. This tripartite account is still today widely embraced, even beyond philosophy, psychology, neuroscience, and education, where we can find evidence of divisions of the human mind into cognitive (e.g., knowledge), motor (e.g., skills), and affective (e.g., emotions) *functions*. Sometimes this tripartite division, stemming from linguistic and cultural distinctions, also refers to biological *structures* in the brain.

Of interest also for the theme of this book is how *aesthetics* as introduced by Immanuel Kant did not only deal with emotions as inner states of mind (feelings of agreeableness or disagreeableness, i.e., emotion) but also with judgments of taste about qualities of objects in the world, such as beauty or ugliness (aesthetics in the restricted sense). To Kant, and in philosophy more generally, both kinds of judgments are aesthetic. What is beautiful is often also agreeable, and what is ugly is typically disagreeable (Kant 1790/1987). I will here use this classic philosophical account of aesthetics as dealing also with emotions and wellbeing, to avoid restricting myself by definition to a review of the third faculty about taste and feeling as only being about what is covert inside us, and which we can only know something about by asking people, or in some camps even better, by scanning their brains. Kant's project was to ground aesthetic judgment metaphysically—just as he had set out to do with cognition and morals—particularly regarding beauty, as such a judgment is almost like a statement of objective fact, asking for other people's assent (Kant 1790/1987). The factual grounding of feelings, to paraphrase Clarice Lispector, is something scholars have repeatedly returned to with different outcomes depending on their theoretical perspectives.

2.1.1 Analytic and Synthetic Traditions

As an initial guiding simplification, historically there are basically two ways to approach human faculties and the role of aesthetics for meaning, namely, the *analytic* and *synthetic* traditions. Analytic implies that meaning is based on nature or God-given distinctions, usually dichotomies (e.g., mind-world, inner-outer, value-fact; analysis: dividing into elements). Synthetic, on the other hand, implies that meaning is springing from human experience in transacting with the world, where analytic distinctions are secondary and gain meaning through human practices and their purposes (synthesis: combining elements into a whole).

Analytic approaches constitute the mainstream Western tradition in philosophy, where questions of truth and certainty of propositions are the foundational project and where questions of meaning, as well as of emotions and aesthetics, are investigated with regard to their ability as language propositions to say something true about the world. The basic characteristic of such propositions is their representational nature; it is asked whether the proposition corresponds to something that actually exists in the world, an object independent of the beholder. Meaning is nothing but a cognitive and epistemological matter; only that which is objectively the case in the world has meaning. Values such as emotions and aesthetics are meaningless because of their presumed subjective nature. Learning, according to an analytic understanding, thus means to gain correct representations of the world, of that which is the case. Meaning needs to be independent of values of aesthetics or morals, what we like or not or what *ought* to be the case (Putnam 2002).

Synthetic approaches are here defined as all the disparate philosophical schools critical of analytic philosophy and its propositional and representational approach to meaning (cf. Burke 1994). Synthetic critique originated mainly during the late nineteenth century and came to flourish and affect research during the twentieth century. Here we find American pragmatism, later Wittgenstein and Continental philosophy (e.g., phenomenology), and educational research influenced by them such as critical research, post-structuralism, or progressivism. Situated approaches within sociocultural research also belong here, as, for example, activity theory or situated cognition. Within a synthetic approach, meaning is something that happens through historically[1] situated human transaction. Although there is considerable variation as to how the various schools understand the contribution of human faculties to meaning, and their ambitions in relation to metaphysics, of interest here are those schools that approach meaning naturalistically, as an empirical problem of how faculties come into use in various situations, through action in relation to purpose. It is not assumed a priori that truth and certainty of propositional statements are the basic entities giving meaning to human undertakings.

All the same, it needs to be pointed out that although synthetic approaches have produced vast amounts of research on situated cognition, much less empirical research exists on how aesthetic values and concomitant emotions are situationally transacted, for example, in science classrooms (Kwah et al. 2016). Here synthetic research is only in its infancy. It should be noted that much educational research does not explicitly account for its philosophical assumptions and therefore cannot easily be put into one camp or the other, which means that my philosophical categorization of studies of others to a large extent is based upon the units of analysis chosen, the methodologies applied, and the relationships taken for granted between entities examined. These studies rarely see themselves as one coherent field of research; that is my construction designed to illuminate possible alternative takes on research on aesthetic experience. The distinction between analytic and synthetic research should not be understood as a classification system for the studies cited, but rather as an analytic heuristic to illuminate what typically tacit assumptions allow researchers to examine and what questions may have been left out and that could benefit from a synthetic critique and approach.

---

[1] Here *historical* should not be understood as designating something that occurred in the past. Historical means that events are approached as history, that is, as *largely* contingent, indeterminate, unique, and not repeatable as wholes (Rudwick 2005). Repeatable patterns in history need to be extracted from numerous cases to see what is common to them. And to find out how these patterns could be used to help people of certain practices to cope with other historical events they need to be mangled further. The term *transaction* was introduced by Dewey and Bentley (1949/1975) to emphasize that knowledge or experiences, generally, do not come ready-made, but need to be continuously transformed as shared action. Pickering (1995) referred to this historicity as the *mangle of practice*.

2.2 Analytically Oriented Research on Aesthetic Experience

This section examines the mainstream assumptions about aesthetic experience and emotion in science education research as well as in education, psychology, and neurophysiology more generally. I demonstrate how certain conceptual distinctions may be taken for granted and the consequences these distinctions have for methodologies and what can be investigated and discerned through research in these fields. I illustrate how much of this research relies on a conception of learning as change of mental states and as chains of causal law-bound processes. Typically this research searches for structural factors causing emotions within individuals, and these factors need to be changed to change behavior. Emotions are studied as affective inner mental structures, which are defined by nature rather than by learning. It is a reward or arousal system with distinctions that come ready-made by nature. Aesthetics as distinctions about the beautiful and ugly are rarely studied except as a topic specific for the fine arts. In analytically oriented research on emotion, there are often reductionist assumptions, where neuroscience is fundamental for psychology, which in turn is fundamental to education. It is therefore relevant to first ask what the assumptions are of neuroscience.

2.2.1 Neuroscience, Cognitive Science, and Psychology

Neuroscience, or sometimes neurobiology, is the science of the nervous system. In neuroscience faculties are typically defined in terms of inner/subjective and outer/objective. Cognition through our senses can be examined in all its objectivity as represented through well-defined outer stimuli. Emotion is more problematic as it is something subjective inside us, apparently not representing any outer object in a straightforward manner. In a basic textbook on neuroscience, emotions are introduced as follows:

> In this chapter, we explore the neural basis of emotion. But how can we study something as ephemeral as one's feelings? If you are studying a sensory system, you can present a stimulus and seek neurons that respond to it. You can manipulate the stimulus to determine the stimulus attributes (e.g. light intensity, sound frequency) that are best for evoking a response. It is not as straightforward to study emotions in animals that cannot tell us their subjective feelings. What we observe are the behavioral manifestations of the internal emotions. Therefore, we must carefully distinguish between emotional *experience* and emotional *expressions*. (Bear et al. 2007, p. 564, original emphases).

Hence, a first assumption of neuroscience is that emotions are basically inner as opposed to cognition, which can be defined by its outer, physically distinct stimuli. It is also suggested that for researchers to know them epistemologically as emotional experience, researchers have to rely on secondary emotional expressions, which necessarily do not have a one-to-one relationship with the experience, like the alleged exact correspondence between stimulus and response as between light and light receptors. There is no specific physically defined stimulus that corresponds to specific emotional receptors. Emotional research, in lacking objective reference, to a high degree depends on every-day linguistic categories about emotions, relying on how people describe their feelings in interviews or how neuroscientists tend to associate outer emotional expressions with certain ordinary linguistic categories such as happiness and sorrow. Only such common linguistic categories, after the emotional experiential fact, can be matched with inner emotional experiences using contemporary research methods of neuroscience that make these "inner" and subjective phenomena "outer" and objective through, for instance, human brain imaging techniques like magnetic resonance imaging (MRI). A corollary of this assumption is that experience is an inner and private phenomenon or state.

The second assumption of neuroscience research is that the nervous system can be examined as causal chains of stimulus and response in relation to the function of different structures in this chain. Research within this field regularly asks where different kinds of inner emotional experiences are located, either in the brain or in the physiological responses of the body. In extension some emotions are also significant as intermediary causes in cognitive and behavioral learning by becoming associated in terms of reward or punishment with certain cognitive stimuli or certain behavior. This field of research concerns motivation for individuals for performing certain behavior. Along these lines clinical neuroscience research ascribes certain behavioral morbidities as due to malfunctions involving emotional causal chains, sometimes successfully so, to alleviate human suffering (Rockwell 2005).

However, this is where textbook presentation of emotions in neuroscience usually stops, and little is known about what emotion is as normally functioning circuits of the brain. Neuroscience research on emotion does not have a central position, because of its ill-defined nature in physical terms. The circularity of reifying the social categories of ordinary speech as biological brain structures and then reusing the brain structures as facts to explain the functioning in terms of ordinary linguistic categories is problematic epistemologically (cf. Lundegård and Hamza 2014), as has also been pointed out by neuroscientists: "the pitfalls all good neuroscientists try to avoid: unconsciously shoehorning neuroimaging results into social categories that may not have any biological meaning" (Zimmer 2004), or, "…our psychological theories pretty much use folk psychology terminology…We talk about emotion, planning, memory, and attention, but just because we use these words, it does not mean that these are psychological discrete entities, or even exist as coherent mechanisms." (Harley 2012). Perhaps this is the reason that definitions of *emotion* are hard to find in this field. In a classic textbook on psychology (16th edition), Susan Nolen-Hoeksema and cowriters (2014) in the glossary define *emotion* as "A complex condition that arises in response to certain affectively toned experiences." *Affect* in turn is defined in the same glossary simply as "Emotions and feelings," completing the circularity.

Mainstream cognitive science (Eysenck and Keane 2015) and psychology (Nolen-Hoeksema et al. 2014) often share these two assumptions of neuroscience. Furthermore, generalizations are often made about how emotions may be structured in terms of discrete, essential (statistically independent) states, either categorized into basic distinct emotions such as happiness, anger, fear, disgust, surprise, and sadness or along dimensions of positive and negative affect. The division into basic emotions has been questioned by neuroscientists on the same grounds as the earlier mentioned critique about applying categories of ordinary language to physiological states or brain structures (see Damasio 2012 and Harley 2012). The multitude of divisions suggests that they are pragmatic rather than natural categories and thereby ad hoc for solving specific problems. They are also chauvinistic in the sense that they are not based on actual comparisons of languages of different cultures and their situated use and thus tend to naturalize distinctions of the English language as given beforehand.[2]

To sum up, neuroscience, cognitive science, and mainstream psychology generally tend to approach emotion as inner *states* and understand them as structures functioning in *causal* stimulus-response chains. Knowledge about these structures and functions are the *ultimate facts*, and culture and language are seen as secondary effects of these biological structures, and it is unnecessary to know anything about them before conducting research on emotion. Rather, it is assumed that neuroscience is foundational to cognitive science, psychology, education, and sociology. A blind spot of this field is thus how emotion also is a fact constituted through social and externally material processes (Harley 2012). It overlooks, for example, how human emotion is learned and may change in terms of taste[3] depending on the people and artifacts encountered when growing up, as, for example, food shared in the culture or topics discussed and valued at home when growing up. This has to do with their blindness for the outer qualities of aesthetics in the restricted sense, that is, how people's sense of beauty *develops* through social and material transactions, a taste that typically is highly contingent upon nurture rather than nature (e.g., Bourdieu 1984). To find the term *aesthetics* in these fields, one has to look to specialized literature on how people experience the beauty of faces, bodies, shapes, or art (e.g., Chatterjee 2014). There is a shift toward understanding emotion as significant for cognition and a shift away from seeing emotion as nonsense and merely

[2] See, for example, the studies by Levenson et al. (1992) and Waller et al. (2008). For the interested reader, the linguist Guy Deutscher (2010) in this regard offers a revealing analysis of how different cultures and their languages, for instance, talk about colors and the consequences this has for experience. His discussion does not concern aesthetics, but questions cognitive distinctions as given, for instance, how *colors* may not be understood as a straightforward relationship between the physical stimuli of wavelengths and the response of neurons in the eye.

[3] Taste is here preferences which are aesthetic and thus from an analytic perspective just a question of what you happen to like or dislike. However, as is discussed at length in Chapter 3, your taste, for instance, regarding music is more than a personal aesthetic bias of what you like or not. Aesthetics is one component. A taste is strongly connected to cognition and action. You need knowledge to appreciate certain music and talking about music in certain ways. Taste in this way develops through socialization or enculturation (Bourdieu 1984).

subjective and not important for cognition. But still, emotions remain inner of which researchers only register secondary expressions.

2.2.2 Science Education

The field in science education studying the aesthetic realm of human experience is vast, and considering the total number of publications, the area is one of the most well researched (e.g., Fortus 2014). However, the constructs used to study this realm vary considerably, although with certain emphases. Table 2.1 is the result of searching the ERIC database for articles in science education in relation to the various constructs used.[4] Although this search is not to be considered as representing the distribution of publications in any exact manner,[5] in qualitative terms it gives a general idea about how this area is typically approached. As can be seen from Table 2.1, the most common constructs used to study aesthetic experience in science education are (1) attitude, (2) motivation, and (3) interest. Far less common are the constructs (4) emotion, (5) aesthetic, (6) affect, and (7) taste that together constitute less than 5 % of the peer-reviewed publications.

The most reasonable explanation for this pattern is that the first three constructs concern aesthetic relationships of a person purportedly to mentally external objects, what students or teachers feel about these, and are thus something that can be seemingly easily asked about. They are about attitudes *toward* (e.g., van Aalderen-Smeets and Walma van der Molen 2015), interest *in* (e.g., Areepattamannil 2012) or motivation *to* (e.g., Vedder-Weiss and Fortus 2011), and so forth. Common objects asked about are *science* generally, or aspects of science (experiments, inquiry, etc.) and

[4]The search was performed August 12, 2015, using the string "("science education" OR"science teaching") AND title:attitude" (*attitude* was of course exchanged depending on construct searched for). When searching, for example, the ERIC (Institute of Education Sciences, http://eric.ed.gov) database collection for titles with "emotion," it also brings titles with suffixes, i.e., "emotions" and "emotional." For each search the first 60 hits were scrutinized to decide the proportion that actually concerned science education and the emotional or aesthetic aspects of the construct in question. The first hits are the most recent ones added to the database. The search was carried out between August 12 and 20, 2015. For example, *affect* first returned 1009 hits of which only 5 of 60 were relevant. 46 concerned the verb affect, 4 were of other school subjects, and 5 were generic articles on affect. The number relevant was calculated as 1009 × 5/60 = 84, which is given in Table 2.1. The original numbers for the other constructs were *attitude* 7275, *motivation* 2310, *interest* 1647, *emotion* 645, *aesthetic* 171, and *taste* 72. The decision to use the first 60 hits was based on the few original hits for taste and as ERIC in the default mode presents 15 hits per page.

[5]There are several other constructs that could be examined, e.g., general ones like *engagement*, *feelings*, and *mood*; overlapping ones like *health*, *well-being*, and *illness*; or specific ones like *fun*, *joy*, *anxiety*, *like*, etc. However, they were either considered to fall under the constructs searched for (e.g., anxiety as an emotion) or of wider more general scope (e.g., value which usually concerns morals). The constructs chosen represent a reasonable selection for the argument of this article. The ones chosen represent the most common constructs and some alternatives to illuminate the analytic-synthetic distinction of this chapter. Other chapters in this book are focusing the close relationship between aesthetic experience and well-being.

Table 2.1 The number of peer-reviewed articles in science education as distributed according to emotional or aesthetic constructs in their titles from a search in the ERIC database

Construct in title	Total number	Percentage
Attitude(s)	6911	62
Motivation	2079	19
Interest(s)	1647	15
Emotion(–s, –al)	204	2
Aesthetic(s)	171	1,5
Affect	84	0,8
Taste	5	0,04
Total	**11,101**	**100**

whether students are interested or feel competent in the subject, often using questionnaires before and after an intervention. *Emotion* and *affect*, on the other hand, conceptually are by definition not connected to any particular object and so seemingly more elusive in the objective sense of differentiating between inner experience and outer expression. Research in science education using these constructs tends to focus on how various instances in the classroom affect the emotional state of students and so their wellbeing (e.g., Liu and Chiang 2014). The construct *aesthetics* is typically used by articles in the database as dealing with art and less so with science. When applied to science education, it often concerns arguments for introducing feelings of wonder, awe, or the sublime (e.g., Milne 2010). *Taste*, for similar historical reasons of use, is not traditionally associated with objective science, but more so with subjective realms like entertainment, design, or food. There are a few publications in science education that involve relationships between students' social backgrounds/aspirations and taste for studying science (e.g., Hilmer and Hilmer 2012).

Although the distribution of articles over constructs is not easily explained in terms of the analytic-synthetic distinction, the methodologies applied together with the units of analysis chosen and the relationships between entities examined make the analytic bias all the more obvious. The methodology most commonly applied in the articles examined in the search of the ERIC database is self-reported data collected usually through questionnaires, but sometimes also through interviews (Table 2.2).[6] Only 6 % of the articles are in situ studies, for instance, ethnographies

[6] Of the first 60 article hits using ERIC, those relevant for each construct (number of articles given as N in the table; see footnote 1) were divided into those that use primary data (*self-report* and in situ studies) and those that use secondary data and argumentation or were suggestions for teaching (*theoretical or review*). *Self-report* uses various kinds of prompts or other ways to reconstruct what occurred earlier. *Questionnaire* includes different kinds of survey instruments, also those with open more qualitative questions, although the vast majority is quantitative. *Interview* includes, for example, focus groups, stimulated recall, and introspection. *Combinations* include studies using both questionnaires and interviews. In situ entails studies of aesthetic experience as it can be directly registered in situated activity, usually in the classrooms as ethnographies. However, a large proportion of the in situ studies using the construct interest is children's questions to scientists as part of, for example, TV shows. *As state* indicates that the study registered the state of children's aesthetic experiences, i.e., whether they seemed interested, motivated, or happy and what they found pleasure in or that gave them anxiety. *As constituted* includes studies interested in following

Table 2.2 The percentages of peer-reviewed articles in the search of Table 2.1 in relation to the methodology used to gain data regarding the various constructs

	Attitude	Motivation	Interest	Emotion	Aesthetic	Affect	Taste	Total
	$N = 57$	$N = 54$	$N = 59$	$N = 19$	$N = 14$	$N = 5$	$N = 4$	$N = 212$
Self-report	92	92	70	63	28	60	50	87
Questionnaire	74	63	42	37	7	40	25	65
Interview	2	9	14	26	7	0	25	6
Combinations	16	20	14	0	14	20	0	16
In situ	2	4	19	27	21	0	25	6
As states	2	2	17	16	14	0	0	5
As constituted	0	2	2	11	7	0	25	1
Theoretical or review	7	4	12	11	50	40	25	8

including records from classroom events (e.g., Mumba et al. 2015), although there are five articles in the sample that analyze voluntary questions young people submit to different media fora, for example, "ask a scientist" (e.g., Baram-Tsabari and Yarden 2009). Discounting the latter makes the proportion studying aesthetic experience in situ even smaller (less than 4 %). Only 1 % of the total number were studies of aesthetic learning, that is, how people's sense of beauty and emotional relationship to objects are transformed bit by bit through the transactions occurring in situ (e.g., Flávia et al. 2003).

All the same, and perhaps somewhat paradoxical, the ambition of the majority of the studies using self-reports or assessments of aesthetic states rather than its constitution is to examine how students' or teachers' interest, motivation, or attitudes may be changed through various interventions. Here several nested assumptions with analytic inclinations are implied to allow such inferences without actually studying the processes.

The first assumption is that as these processes are inner, they are not possible to study as they occur without perhaps advanced brain-scanning equipment, only their outcomes or expressions. Remember the credo from neuroscience: "What we observe are behavioral manifestations of the internal emotions. Therefore, we must carefully distinguish between emotional experience and emotional expressions" (Bear et al. 2007, p. 564). Although this assumption is rarely explicitly formulated in science education research, it sometimes becomes evident in the dependence in some works on neurological foundations for arguments (e.g., Liu and Chiang 2014) or on theoretical frameworks based on internal emotional states as part of causal chains effecting behavior.

The latter kinds of explanations bring us to the second assumption which is that theory about inner causal stimuli-response chains can be applied to infer processes

the transformation of emotions or aesthetic experiences, i.e., how people learn new emotional relationships and aesthetic values in relation to what occurs in transactions. As the first 60 hits are the most recent added to the ERIC database, they cannot be seen as representative of the whole database, but are more reflective of the current situation.

from states and that the states of these chains are biological entities and not pragmatically chosen distinctions to solve specific problems. It is not necessary to study what it entails to learn an interest or attitude socially and materially. Rather ordinary speech categories about emotions and the objects of these emotions (e.g., "I enjoy science") used in, for instance, questionnaires are treated as natural categories which can be made objective through causal chain analysis. One such often used model is the one van Aalderen-Smeets et al. (2012) developed to understand primary teachers' attitudes toward science, depicting "attitude toward teaching science" as constituted by the states "cognitive beliefs," "affective states," (which in turn can be divided into "enjoyment" and "anxiety") and "perceived control," which in turn causes "behavioral intention," which again causes "behavior." These components of the causal chains are put in boxes with arrows representing causation. This model in turn is built on the similar more general and very influential model—theory of planned behavior—developed by Ajzen and Fishbein (1980) to explain how attitudes affect behavior. This is just one example among many causal models to explain behavior (another example is the influential self-efficacy theory of Bandura 1977).

A consequence of these assumptions is that human conduct is approached as behavior[7] caused by what lies behind, rather than as action tinkered bit by bit in relation to purpose in an activity and life. What lies behind human conduct is the mental structures, which depending on the constructs used may be attitudes, interest, or motivation, and to change behavior researchers seek an educational method that will change the attitudes (as, e.g., inner emotions and experience) and so the behavior (outer expressions) to better fit the norms (e.g., more students following science careers). It is almost as if the educational method is seen as a medication, a pill that causes certain physiological processes in the body, resulting in predictable behavior of the patient (the student). The intelligent teacher making decisions along the way together with students is easily lost in this approach to aesthetic experience or experience at large for that matter.

Internal critique of this approach to aesthetic experience mainly concerns how well the constructs are actually measured, where room for interpretation, lack of strict definitions, or consensus is usually seen as a weakness, as there is only one reality out there to know for certain. Measurements need to better correspond to natural entities although voices are also raised for a better conceptual fit of the constructs to the interest of the field of the research (Reid 2006). Science education may need somewhat different constructs and methods to study attitudes than those of, for instance, sociology or psychology.

---

[7] Commonly a difference is made between behavior and action. Scientifically, *behavior* is a biological concept defined in terms of physiological processes as muscle movements or gland secretions resulting from *causes preceding* the motor responses (cf. behaviorism). *Action* is a concept of the social sciences and humanities, where human conduct is given *meaning* in relation to purpose, what is to be achieved, rather than from what causes it (Elster 1988). The teleological explanations of the social sciences and humanities are typically banned or eschewed in the natural sciences (Dennett 1995). Often in research with analytic tendencies, behavior is the preferred term, although without necessarily making an overt distinction to action.

2.3 Basic Assumptions of Analytic Approaches

Mainstream studies in science education on aesthetic experience do not only echo the typical assumptions of neuroscience, cognitive science, and much of psychology. As alluded to in the introduction, roots of these shared assumptions go deeper and amount to old Western traditions of accounting for the human faculties, which are often collected under the generic term analytic philosophy (e.g., Shusterman 2000). These assumptions are rarely laid open and examined by the researchers themselves within these fields, but, as well, become evident, typically by synthetic researchers questioning these assumptions. Inside critique is mostly not about changing assumptions, but rather about giving up realism and replacing it with skepticism (e.g., Chalmers 2013).

Analytic philosophy can be characterized as a pervasive quest for certainty, where theories of *meaning* are conflated with theories of *truth*. It adopts a representational and propositional account of meaning as well as a core of a priori oppositions such as the subject-object, inner-outer, and value-fact dichotomies to explain the operation of human faculties. It is a search for a certain foundation beyond human purpose, giving explanations about what is behind or the causes for what is in the world. In doing this, analytic philosophy construes language as representing the world. The foundation for accomplishing true correspondence between mind, world, and language is the all-encompassing problem to be solved. A meaningful proposition is only one that gets reality right (Rorty 1991). "Snow is white" is meaningful only if "Snow is white." That someone believes in their *mind* that "Snow is white" has no meaning until that *language* proposition "Snow is white" with certainty is established to correspond to a fact in the *world*, namely, with certainty that "Snow is white." This view of meaning entails approaching learning as a sole individual constructing mental representations (concepts) of the world through language. This analytic approach to knowledge as *justified true belief* goes back to Plato. Analytic philosophy also owes Plato for its search for essences, that is, divisions of the world which are natural and represent nature as it is, beyond human interest, purpose, and practice. The tripartite division of human faculties is one such analytic split taken as out there rather than as pragmatic for purpose (Dewey 1929/1958). Another is the division between the mind, language, and world (Rorty 1980).

Aesthetic experience concerns values, which analytic philosophy traditionally considers merely as subjective opinions and so as opposed to facts and justified true belief (Putnam 2002). Thus, the only way to give values meaning is to find what they correspond to in the world (Östman and Almqvist 2011). As we have seen, in educational research their correspondence to the world is usually salvaged by reference to biological structures and functions. Emotion is given meaning as objective states of the brain, and these biologically fixed states are involved in causal chains affecting behavior and so the learning of the *biological* body. Apart from this alleged objective role as biological intermediaries, emotion as culturally situated and constituted has no meaning as merely being a matter of taste. Emotion has no objective existence other than as structures with fixed functional roles in inner causal chains. Language on emotion needs to be explained from its biological essence to say something true of the world and so become a fact with meaning. The prevailing reduction of educational research on aesthetic experience to neuroscience in this light is not surprising.

For emotions or aesthetic experiences to pass as facts and so be meaningful in a scientific endeavor, they cannot be taken as they are experienced jointly together through action and discourse, but have to be construed as objective mental states rather than as values. Experience is a priori assumed to be inner through the dichotomy experience-expression, and expression (e.g., language)—like the shadows in Plato's allegory of the cave—is just a dim representation of the experience that is inside us (cf. Garrison 2003). A feeling is a fact only through a theory about how the brain operates via sensory stimuli to outer behavior and not as a part of lived life. For researchers the continuity between aesthetic experience and life is a question of causal mental chains. The questions returned to are "How can we know with certainty what a feeling corresponds to in the world?" and "How can it be established as a fact which does not beget subjective interpretation?" Here analytic approaches need a theoretical proxy linking outer physical stimuli, inner experience as a biological phenomenon and outer expression as biological behavior. This is where theories about basic emotions and their correspondence to certain facial, gestural, or linguistic expressions or brain areas come in. Theoretical causal models explaining the correspondence between attitudes or interest as behavioral responses on questionnaires also belong here. The caveat of some neuroscientists that emotion cannot be studied in a straightforward way, as it does not correspond to a specific physical stimuli, also reflects analytic epistemology and concomitant distinctions resulting in skepticism. The analytic project is to save the correspondence between aesthetic experiences as reported through language and facts as objectively existing in the world, beyond human purpose and interpretations.

This focus on causation, biology, and behavior is also related to the tradition in science education of rendering science as ultimately being about *explaining* the world as the operation of underlying factors, variables, and causes and leave out of the account the historically contingent and the purposes of the undertaking. There is a risk here of subscribing to an Aristotelian Cosmos, where everything can be explained in terms of inherent causes. Galileo Galilei encouraged us to ask *how* rather than *why* the Cosmos works, and to examine the actual processes so that we better can cope with them, and not be content with constructing hypothetical ultimate causes merely to explain them. Explanations come after we know how the Cosmos works, not before the fact (Schwab 1978). A focus on explanations as causal relationships easily also overlooks the discovery of *history* at the turn of the nineteenth century, where the Cosmos began to be understood not only as law bound and predetermined, but as contingent and depending on what is historically unique at a specific place in time (Rudwick 2005). Besides, analytic philosophy usually dismisses synthetic thinkers such as John Dewey (Burke 1994), who emphasized meaning as occurring through shared human practice and that there is always a human interest involved in sensemaking of the Cosmos. This interest is not *just* inner and private, or biological, but continually shared and transacted.

Finally, the analytic division of aesthetics and cognition as separate faculties of mind did not only result in philosophy approaching science as a pure quest for truth but also that philosophers tended to miss that aesthetic experience is a necessary component of doing science (Dewey 1929/1958; Fischer 1999; Root-Bernstein 1996). In the nineteenth century, aesthetic experience came to be understood as detached from reality and so the opposite of cognitive experience. Aesthetic experience was seen to reach its highest levels in fine art, and still fine art is considered to be qualitatively different from science (e.g., Welsch 1997). This understanding of aesthetic experience in science and in fine art very much prevails today, which is evident from the earlier literature review, where aesthetics is usually studied as fine art or as elevated emotion detached from the regular transactions of the science classroom.

2.4 Emotion and Aesthetics in Synthetic Philosophy

To begin with, of interest here are exceptions in neuroscience which try to go beyond its basic assumptions. I mention this example not as a foundation for synthetic approaches, but rather to make clear that there is no consensual neurological foundation for understanding human conduct and aesthetic experience. Here the pioneering research of Antonio Damasio (2006, 2012) can be mentioned as an example. In his research, based on neurological patients with disorders both of decision making and of emotion, he demonstrates how emotion is necessary for reason and not opposed to it (Damasio refers the opposition to Descartes and his separation of mind and body, cf. also Alsop 2005). Damasio argues that neurologically, the brain circuits of reason and emotion are not separated, but emotion is in the loop of reason, and so involved in moral reasoning and in aesthetic experience. From this Damasio (2012) builds a hypothetical theory of how the brain-body nexus works, very much on the grounds of traditional causal chains of stimuli-response, which is beyond the educational theme of this chapter. All the same, this line of neurological research has implications for a synthetic point of view, as it suggests how cognition, moral decision making, and aesthetic experience are diffusely organized structurally in the brain. Hence, the language of aesthetic experience should be mainly seen as socially and culturally constituted, with little neurological structural basis. In my interpretation this means that to understand the role of aesthetic experience and emotion for people's lives and for learning, as in educational settings, it is necessary to study them in action, as social transactions, in historically constituted settings, without an analytic bias that the brain is primary, whereas communication, culture, and social life are secondary (Dewey 1922; Garrison 2003).

2.4.1 Wittgenstein and the Meaning of Language

So what do alternative synthetic assumptions look like and what consequences may these have for the methodologies used? What else is learning and meaning about if they are not about constructing a correct conceptual representation of the world? Ludwig Wittgenstein (1967) in his work *Philosophical Investigations* asked himself how we learn the meaning of a word. He started out in the idea that we learn its meaning by learning what it represents in the world. Adopting this line of reasoning, we may, for example, teach the child the meaning of the word *fly* by pointing at one and say "fly."[8] In examining this analytic account of language, through numerous examples he investigated if learning language in this way is possible at all, and what an alternative route might look like. In the vein of Wittgenstein we may thus ask: How can the child know what the adult is pointing at with regard to the fly? Is it its wings that is pointed at and called *fly*? Or is the word *fly* representing insects or animals generally or any animal with wings? How can the child know that the adult, already mastering language, has a particular group of insects in mind? How is this possible without telepathy? One solution also investigated by Wittgenstein would be to explain in words to the child what we intend with the word, but this would of course not work if the child still had a very primitive or nonexisting language. And in the end words would just be pointing at other words without reference to the world. It was made obvious by Wittgenstein that we do not learn the meaning of language in this way.

Wittgenstein (1967) demonstrated how language does not acquire meaning mainly by learning what the words represent in the world and of acquiring the corresponding mental concepts to that of the adult. Rather, he argued that we learn the meaning of words by using them in all the regular tasks that a child shares with adults and other children (Williams 2002). The important reference for meaning between the adult and the child is the activity which they share and how language helps them to carry it on. It is not necessary to have the same mental conceptualizations in mind to cope with the world together. Language is situated action, and therefore public, suggesting that it is possible and necessary to study how language uses or discourse changes as changed habits situated in practice (Wickman and Östman 2002). Meaning is something that happens or may not happen to language as we are involved in life with others. The intent here is not to explain how a child first learns language, but, rather, to see how Wittgenstein's investigations of language could help us see aesthetic experience not as representation, but as enmeshed in life and activities where it makes sense and is transformed in certain ways.

Wittgenstein (1967) called the situated contexts where meaning happens *language games* to emphasize that language is action and part of shared activities such as buying clothes, traveling on a bus, or for that matter taking part in science class. Apart from shared *activity*, the metaphor emphasizes that it is *as if* there are rules

[8] This is not Wittgenstein's example, but mine. This is to give an example with reference to science education.

which needs to be learned and followed to take part (Williams 2002). There are ways to use language that work in a specific activity, whereas other usages do not. An important part of learning is thus not only to use the word *fly* for specific organisms, but to know how the word can be used in furthering certain activities together with various other words and actions. The language games can be seen as habits, customs, and institutions through which meaning happens. Wittgenstein demonstrated that there are no ultimate structural causes for action which gives it meaning. To educational science, Wittgenstein's investigations of meaning offer a possibility to leave causal explanations and instead look at how meaning happens habitually and institutionally. We do not need to *explain why* human behavior is structured in certain ways, for instance, because of biological or cultural causes, but rather focus on how people learn to take purposefully part in them. We do not just *behave* mechanically as a result of stimuli through given structures, but need to be given *agency* and *act* in relation to purpose to be deemed to have learned anything of *value* by others and ourselves. If we better know *how* meaning is made, we as teachers may be able to better support certain meanings in the classroom.

The same kind of reasoning can be applied to the meaning of aesthetic language and its relationship to aesthetic experience and to experience generally (cf. Wittgenstein 1966). If the meaning of language is not learned by pointing, definitions and explicit rules alone, but critically through the consequences of its use in activity, it cannot be assumed a priori that feelings are meaningless because they by prior definition are opposed to facts (Langsdorf 2002). Instead we need to examine *how* the various feelings, as in a science education class, are transacted in furthering action and so giving them certain meanings. Only when we know how aesthetic experiences are transacted through the practice of teaching may we say how other practices and their language uses, such as that of brain scanning, questionnaires, or interviews, could be used to better support the meaning made in class. Meaning cannot without further ado be reduced to other practices than those were it happens.

The meanings of aesthetic experience and emotion may thus not be wholly a private internal affair based on ready-made biological categories, which are the same irrespective of where and when they occur. The emotional words and the aesthetic judgments we use are already imbued with shared meanings as part of human institutions, customs, and prior transactions. Biology does not come before culture when the child learns (Dewey 1922). Biology and culture are transacted in parallel and are inseparable in the process except analytically. At the same time, meaning does not come complete either from biology or culture, but is transacted and a result of the bit-by-bit consequences furthering the activity taken part in. In transactions with the world, every situation has certain unique characteristics, and what is happening cannot be fully anticipated. This is the historical dimension of meaning (Rudwick 2005). The discovery of history meant understanding how the Cosmos does not develop in a completely deterministic law-bound way, but to a large degree also in contingent and unique ways varying with local circumstances. Such *historicity* is apparent in human transactions, also with nature and in science (Pickering 1995). Only empirical analysis can disentangle the more regular patterns in such historical occurrences and identify how such regular patterns can be conceptualized to become useful for supporting teachers. It would be unfortunate to decide on purely rational grounds that such patterns are mediated merely by mental causal chains and that this is the best way to cope with these habits (Wickman 2012).

Taking Wittgenstein seriously suggests that researchers need to look at how aesthetic experiences of, for example, beauty or joy come into play in human transaction to know better how they are related to learning to participate and wellbeing in science class. Hence, studying aesthetic experience largely through self-report, which is the prevailing practice, means transposing aesthetic experience from one language-game context to another. It entails that emotional categories decided by the researcher before the inquiry, and which transforms ordinary words into neurological theories, are replacing the meanings given to aesthetic experience and emotion through the situated classroom transactions of students and teachers. This practice means that we as researchers do not know if these experiences play other roles or are changed through other routes than those envisaged through theoretical causal chains, because this is not investigated. It also often entails that predetermined objects for these emotions are given to students and teachers through questionnaires or that they are extracted in interviews outside where they happen. It may be that students as engaged in classroom activities do not actually transact the objects envisaged in the questionnaires or interviews, as, for example, "Do you enjoy *science*?" or "Do you like doing *experiments*?" (cf. Chap. 3 this volume). There may be other components that are more important for students' meanings and participation than the object categories *science* or *experiments*. Nonetheless these are the objects that students are repeatedly asked to give their opinion about. What if students during classroom work aesthetically evaluate other objects than science or experiments? To know this better, we also need in situ studies on how the constitution of aesthetic experience happens as part of class experience as a whole.

2.4.2 Dewey on Inquiry and the Principle of Continuity

John Dewey argued for a more action-oriented approach and a less representational one to experience. In Dewey's (1929/1958) writing *experience* is not equated with inner sensation, as it is typically used today especially by neuroscientists and cognitivists. Experience is used more in its every-day meaning as not distinguishing the cognitive from the moral or aesthetic sides of an experience. It is also about *living through* an experience, of active participation in events and how the person is transformed through them and so becoming experienced. Dewey used the concept to describe an undivided whole of living and not just as personal, but also as shared and transacted with other persons (Glassman 2004). Experience can be divided conceptually, but such conceptualizations can always be questioned on a pragmatic basis, in asking how they help us with our undertakings. Experience is thus always in the flux of life and transacted. Dewey (1929/1958) asked how experience could be ameliorated. In answering this question, two observations made by

Dewey are critical. These concern the *principle of continuity* and the *empirical method of inquiry*.

Dewey suggested that learning entails the transformation of complex habits (Dewey 1922). Epistemology thus amounts not only to conceptual change, but to the transformation of the person through its already functional habits. Richard Rorty (1991, p. 1) put this as "by an antirepresentationalist account [of epistemology] I mean one which does not view knowledge as a matter of getting reality right, but rather as a matter of acquiring habits of action for coping with reality." A habit is here understood as a complex co-working set of behaviors which jointly are used for specific purposes, like walking. Walking can be used for a number of different purposes, but may need to be adapted in various ways to accomplish certain purposes better. Habits can also be habitual ways of using language, concepts, or aesthetic expectations. According to Dewey (1922), habits are typically transformed slowly and bit by bit by drawing on prior experience in purposefully coping with current experience. In giving an account for how that occurs, he noted how people go about habitually in making sense of experience. This basic habit he named the "principle of continuity of experience which means that every experience both takes up something from those which have gone before and modifies in some way the quality of those which come after" (Dewey 1938/1997, p. 35). Experience may be divided actively but also by custom and habit for a purpose to further people's undertakings. For a researcher or a teacher, the principle means that by studying interaction, it can be seen how students of a class habitually draw on prior experience in a certain activity and also examine the consequences such habits and decisions may have for how they can cope with what occurs in the practice. It is an empirical problem to learn more about *how* such exchanges occur and *how* experience and habits are changed through them. When people share experience to get things done, it gives researchers and teachers an opportunity to study and assess the process. These observations of Dewey suggest that experience transforms us as whole embodied persons. Learning is not only something where stimuli produce conceptual representations and emotional reactions in the brain but also something which can be seen to transform students habitually in moral, aesthetic, and cognitive ways.

The second critical observation of John Dewey concerned the process of *inquiry*, which he also named the *empirical method* (Dewey 1929/1958). Again, his empirical method is not to be considered just as a philosophical, rational, or normative suggestion of his, as a theory of what the scientific method amounts to, but as an observation of how people go about in successfully transforming habits by making experience continuous, and thereby learning. Dewey (1929/1958) noticed that inquiry always starts in already existing practice with its purposes and also with its multitude of prior experiences, habits, and language usages. When we encounter a problem, an inquiry ensues where old habits of action and language use need to be questioned. In meliorating these habits, they cannot be questioned as wholes, but we need to rebuild the boat while sailing. New concepts or methods are tried out to see how their consequences help the current practice and its habits to be transformed bit by bit so we better can accomplish our aims and purposes. In this process new habits often feel awkward to begin with, and a teacher may be necessary to scaffold this process. In the process of transforming habits, our values (e.g., how it feels) play just as an important role as our cognition (e.g., rational arguments or empirical findings). As knowledge is action and learning the embodied transformation of habits, they also have physical and emotional consequences and so consequences for our taste and wellbeing.

There is thus not one way to divide the world in terms of what it's objectively like out there. There are a multitude of possible divisions depending on who is the knower and the practice in which the knower is taking part. Acknowledging that the words of language are not merely objective representations, but that they also mediate what and who is included and excluded from practice. If we want to cope with the world in new ways with new people, we also need to transform the ways we talk about the world. We need to ask: How may extant divisions be functional and have limitations? What other divisions may be appropriate for our purposes, values, and practices? How may distinctions be made continuous in use? The question is not "How can we know for sure what a feeling corresponds to in the world once and for all?" "How can it be established as a fact which does not beget subjective interpretation?" We rather understand that different situations and practices may need that educational science develops different conceptual constructs. Divisions of emotion made on biological or medical grounds by neuroscientists may not be functional or even applicable to cope with what is happening in classrooms. Science education may need somewhat different constructs and methods to study attitudes than those of, for instance, sociology or psychology.

We here see a critical or therapeutic side of pragmatism (Cherryholmes 1988). We need to ask how the concepts we have borrowed from other fields help us with the problems that we encounter in science education and how they may be needed to be revised or mangled further (Pickering 1995). In answering question like this, we need to discuss the consequences that our conceptualizations have for our ability to make tangible operationalizations which can be dealt and coped with. Maybe other constructs than *attitude* or *motivation* are needed. What, for instance, are those distinctions that teachers can observe in classrooms so that they can cope in ways which have consequences for students' aesthetic experience and its continuity with their learning of science and their moral, political, or physical development?

Concepts are not just inner representations of what is out there, but organize common activities through their shared transactions. Asking ourselves how we could adopt and use concepts to discern in new ways is an important side of what science is. The transformation of every-day concepts to new usages was an important step in the development of physics (Schwab 1978). Still physics students struggle with how concepts like *object* could be productively used to talk about how material interactions change motion. Habits need to be transformed so that students could talk also about the *earth* as an *object* in such inquiries (Sensevy et al. 2008).

The centrality of distinctions in coping with the world has also been investigated by post-structuralist scholars in terms of *discourse* (Cherryholmes 1988). Through socialization we learn to make discursive distinctions valued in our culture. A critical interest means that we as teachers or researchers ask whether these habitual distinctions are the only distinctions possible and value them in relation to our purposes. A critical knowledge interest entails inquiring into alternative accounts of what occurs and so opens up a possibility for people to make active choices in line with certain values. A critical interest is liberating and empowering in helping us to find ways to cope more fruitfully with the world. As far as aesthetic experience goes, the idea is not to produce theories corresponding to the functioning of the brain from which best practice could be deduced. The aim is (1) to study *how* aesthetic experience is transacted in a multitude of ways through various classroom practices, (2) to see the *consequences* these various transactions may have for the experiences afforded through them, and (3) to give accounts of these processes and their various consequences conceptually in ways which make them possible to discern and be dealt with by teachers involved in planning, carrying out, and assessing lessons. This is the daunting task for a research program recognizing that variation, multiplicity, and interpretation are not opposed to objectivity, but are facts which any educational reform need to consider seriously to make a difference.

2.5 Science Education Research with Synthetic Tendencies

To give some idea of what a synthetic approach to aesthetic experience in science education may offer, I here give examples from the work we have been engaged in at Stockholm University and its backdrop. I use our research as examples as they have explicitly been developed from synthetic assumptions, especially those of pragmatism and Wittgenstein in relation to sociocultural frameworks. There is other research with related interest (cf. Olitsky and Milne 2012).

Our research on aesthetic experience builds on a general attempt to study the relationship between *what* is learned and *how* it is learned as *practical epistemologies* (Wickman and Östman 2002; Wickman 2004). Knowledge about this relationship is central to the teacher profession in planning, carrying out, and evaluating teaching. A practical epistemology is an empirical description of how a certain practice—in this case the classroom practice—allows students to acquire "habits of action for coping with reality" (Rorty 1991, p. 1). The approach means describing how certain ways of proceeding as habits are transformed through what occurs in class. In this extended sense, epistemology is understood as the transformation of the person as a whole, including also the transformation of aesthetic experience through what occurs and as habits.

The focus on action and habits makes it possible to describe events in the ways teachers encounter them through their joint work with students and how the teacher can see learning of a certain content to occur in these encounters. It allows studying the transformation of the habitual ways students can take part in the unfolding classroom practices and how these habits also may be functional in everyday or in other out of school practices (Wickman 2004). A conception or misconception is thus analyzed as contingent for the situation or as habitual ways of reasoning in specific situations, not as hidden mental states expressing themselves through action (Hamza and Wickman 2008). In a similar way it may be asked how aesthetic experience could be examined as the transformation of the person as an embodied whole through changed habits. The question of the teacher is thus not what lies behind the habit, as a mental state, a conception, or a motivational or emotional state, and how old states could be removed and replaced by new causal mental structures. Rather it is asked: How are habits transformed through the various transactions possible in the classroom and how could knowledge about how such transformations may occur be used by teachers to better cope with what is happening in their classrooms? It is an empirical question about processes and how a teacher may influence them by changing the bit-by-bit encounters offered through teaching. Situated action, activity, and habits are the units of analysis in relation to what students encounter through these activities (Wickman 2006).

To describe and analyze the relationship between how students' acts are transformed by the encounters they experience, the methodological framework *practical epistemology analysis* (*PEA*) was developed (Wickman 2004; Wickman and Östman 2002). In line with synthetic approaches, PEA can be used to examine how what can be seen to meet through student discourse and action influences the direction student activities take. By *direction* is meant what content can be seen to be unfolding and not unfolding through what students are doing and talking about. This analysis was inspired by other microanalytic studies such as discourse analysis (e.g., Gee and Green 1998), but with a focus on the pragmatic transformation of content as habits. Through PEA it can be examined how the different kinds of encounters the students experience (e.g., with the teacher, other students, prior experience, with the physical world) contribute to the meaning made in the classroom and so what students are afforded and not afforded to learn through these encounters. This analytical model should not be seen as a general theory of how learning occurs, but as operationalizing the entities observable and possible to influence by the teacher. The researcher's PEA is making evident (1) what the students find relevant to inquire into, proceed with, or learn more about; (2) what entities the student and teacher already take as given in experience in these inquiries, operationalized as words used or actions performed without hesitation or without asking about their meaning; (3) how those entities of experience are related to each other through discourse and action to better cope with and further the inquiry; and (4) how the setting in terms of the encounters occurring through transactions influences the direction taken in terms of (1), (2), and (3).

As in ethnomethodology practical epistemology analysis starts from an emic first-person perspective, understood as asking what the shared purposes are of the people taking part in the activity studied (Lynch 1993). This is typically not the same purpose as that of the researcher. The interlocutors' purpose should not be understood as the internal mental "behavioral intention" that they hold or the final goal to be achieved with the activity. In Erving Goffman's (1974, p. 25) rendering, it is rather about starting the analysis by asking "What is it that's going on here?" This question is about the shared purpose through action; and the answer means realizing how to proceed through the activity in ways accepted by the others taking part. Such purposes of classrooms could be to find out the concentration of a substance in chemistry or discuss why we have tires on our cars in a physics classroom.

However, the analysis does not stop with the initial naturalist first-person analysis, but is followed by a third-person analysis of how the discourse and action patterns found may have influenced the content learned. Of interest from a third-person perspective, as that of a researcher or teacher, may, for instance, be how the purposes and encounters occurring actually allow students to proceed in the direction necessary to transform their habits into those ascribed in the curriculum.

So how may aesthetic experience be studied synthetically in a similar way? Important sources of inspiration for furthering the practical epistemology framework and analysis to also include aesthetic experience were the studies of Bloom (1992), Säljö and Bergqvist (1997), and Szybek (1999). Jeffrey Bloom (1992) developed what he called a context of meaning perspective to study children's meaning making and learning. "A context of meaning point of view questions the static, semantic, or propositional nature of knowledge and understanding" (p. 405). One of his interests was in how emotion-values-aesthetics (EVAs) may be continuous with cognition. For this purpose he interviewed children to see how they reasoned while observing and trying to make sense of earthworms. Through these interviews it became evident how children's learning is not a simple representational activity, but how they draw on various contingent experiences developing through the process, and among these EVAs played a significant role for the sense children made.[9] Roger Säljö and Kerstin Bergqvist (1997) made recordings of lower secondary students' transactions with artifacts, with the teacher, and with each other during an optic practical. Although their analysis did not specifically focus emotion or aesthetic experience, it demonstrated how aesthetic transactions were overt and possible to analyze as an important content of the moral the lesson taught the students about their possibilities to participate and how they felt about the lesson. Piotr Szybek (1999) analyzed the aesthetic meaning making in a class with 14-year-old students dissecting a squid. He showed how the meaning of the squid was transformed from being a "disgusting fish" to a laboratory practical of making a "nice arrangement" and so making the squid possible to touch.

These studies made it evident that aesthetic experience and its significance for learning may be analyzed from what students do and say as part of settings. Utterances or actions about how people feel and about what they like or dislike are often analyzed in philosophy as *aesthetic judgments* or *judgments of taste* (cf. Bourdieu 1984; Kant 1790/1987). Aesthetic judgments in terms of utterances, such as "what a *disgusting* fish"; facial expressions like smiles; various sounds as, for instance, screams; or bodily positions such as holding something far away from you are all possible to analyze as aesthetic experiences by looking at the situation in which these judgments are made and the consequences they have for the further meanings construed (cf. Wittgenstein 1966). Studies of how aesthetic judgments contribute to meaning making through classroom transactions and their role in the transformation of students as persons thus offer an avenue to identify ways to discern entities that could be used by teachers to cope with the aesthetic experiences in

[9] Bloom's (1992) study is thus an example of how also interviews can be analyzed as situated transaction.

the classroom (Wickman 2006). The aim of research is not to uncover what the individual students *really* feel *inside*. In practical epistemology analysis, aesthetic judgments are approached as an important avenue, for teachers, to inform and assess what is aesthetically and emotionally unfolding in the classroom and its wider significance for the growth of students. The critical interest is thus to analyze the meaning and role of such aesthetic judgments in the development of students' aesthetic experience and their consequences for the meaning made. How can teachers use aesthetic judgments in regular classroom work to support students? How are such judgments made continuous with classroom transaction experiences generally and students' possibilities to proceed with science lessons, classes, careers, or life in general?

We here present some examples from our research to illustrate what a synthetic approach may offer. Our studies of aesthetic experience now cover recordings from preschool (Caiman and Lundegård 2013) and elementary school (e.g., Jakobson and Wickman 2008) through secondary school (e.g., Lundegård 2008) to university (e.g., Wickman 2006). Common to all of these studies are that aesthetic judgments typically are frequent in classrooms, and they are used by both students and teachers.

One salient result was that neither students nor teachers ever asked each other about the meaning of aesthetic judgments. Neither teachers nor students had to explain what they meant by the aesthetic words used. Although there sometimes was disagreement about whether an experience was wonderful or not, which obviously may vary from person to person, the meaning of judgments such as "wonderful" or "nice" never needed further explanation to make sense (Wickman 2006). Aesthetic judgments thus had *immediacy* in communication which cognitive concepts, such as *ocelli*, did not always have as they were novelties to learn about (Wickman and Östman 2002). *Feelings* and taste in this way were treated as *facts* in ordinary speech. Students and teachers trusted each other that their aesthetic judgments were sincere and not faked. This basic observation supports that aesthetic judgments already are useful in class to access aesthetic experience and can be adopted to further examine how it can be used to understand the meanings made and their significance for the transformation of the persons involved.

Another observation of aesthetic judgments is that they when uttered are highly situated and purposeful. To make sense they need to be situated in the activity of which they are part and its purposes. When one university teacher during a chemistry practical on inorganic analysis said "Let's see, you've got many … chlorides that are *fun* here" to a group of students (Wickman 2006, p. 82), the teacher's aesthetic judgment can only be made sense of as part of their undertaking as a whole. The teacher did not mean that there are specific chlorides that always are fun. His utterance need to be made sense of as part of the purpose of the practical, where students were to find out the chemical substances dissolved in a number of test tubes. To anyone who knows a little chemistry, the statement makes sense as the teacher proceeded "Could you find out the silver nitrate?" and a student's answer "We could do that." Chlorides were fun in this situation as they here helped these students to identify the test tube with silver nitrate and thus because the chlorides would create a precipitation in that test tube. And that would likely be *fun*, as it meant that students succeeded with the practical. Here the important thing is not whether the *teacher* really had a feeling of "fun-ness" but rather the consequences for the students, namely, that the anticipation the teacher gave later was experienced as fun by the students when using the chlorides. In this sense aesthetic judgments are treated as facts in conversations: it had to be proven that what the teacher aesthetically promises of the experience is also fulfilled in the transformation of the aesthetic experience of the students. Some teachers use aesthetic judgments to follow up student experiences to see that aesthetic experiences of fulfillment actually happen as students proceed with their tasks (Anderhag et al. 2014).

There are numerous situations of various aesthetic transformations in the material. In a practical in elementary school where students were studying earthworms, an important part of the practical was for the students to learn to handle them, something which they first found disgusting (Jakobson and Wickman 2008). Eventually the children invented ways to handle the earthworms, and after a while they even found them cute and in need of their care. This helped them proceed with the task of studying the earthworms and so constituted both important learning and transformations of aesthetic experience. At the same time, the children later when asked about their cute earthworms by the teacher "Why don't people have worms as pets?" recognized "Maybe they're a little disgusting." This can be compared with the situational solution of students in coping with the disgusting squid in Szybek's study (1999).

Generally, as suggested already by John Dewey (1934/1980), aesthetic judgments are parts in a rhythm of anticipation and fulfillment. This rhythm is a fundamental *habit* also found in the classrooms studied (Wickman 2006). Moreover, in this rhythm positive aesthetic judgments like the one about "chlorides that are fun" concern anticipation about what would lead the activity forward to fulfillment according to purpose. Positive aesthetic judgments encompass objects, events, and actions that should be included to succeed, and they may be evaluated repeatedly along the way in how they contribute to the aim. Upon completion and fulfillment of a task, the whole experience is often summarized by shared positive aesthetic judgments such as "Great!" (Wickman 2006, p. 89).

Negative aesthetic judgments show somewhat more complex uses as parts of the rhythm between anticipation and fulfillment. Negative aesthetic judgments are used in opposite ways to positive ones, in that negative ones are dealing with what should be excluded so as to carry the activity to fulfillment. For example, the university students used positive judgments about pinned insect specimens displayed in ways permitting correct observations and negative ones about those that made observations problematic. When successful, students through such negative experiences learned how to avoid them. However, negative aesthetic judgments sometimes concerned objects, events, and actions that students could not exclude. This may be because they needed to do something to proceed with the purpose of the lesson, but they did not like what they were doing. If students keep judging something as a negative experience and still have to include it to proceed, students may experience a conflict regarding the activity as a whole resulting in final negative aesthetic experiences. Such discordant negative experiences dealt with what the students realized they had to do to fulfill the task, but with which they could not cope with aesthetically.

There is no sharp boundary between these two kinds of negative aesthetic experiences. What is important is that they need to be carefully followed in the classroom as they develop in the rhythm between anticipation and fulfillment. It is clear from our recordings that learning necessarily involved also negative aesthetic experiences and thus suffering for the students to varying degrees (Wickman 2006). In the practical on insect morphology, the students needed, for example, to learn what characterized intact specimens of the various species. If students included tattered and worn ones, they ended up with the wrong conclusions about how the insects were built. In learning such inclusions and exclusions, students also learned what an entomologist meant by a beautiful specimen. In the earthworm example, students needed to overcome their first feelings of disgust. If such negative experiences linger and the student is not able to cope with them, transform them, or exclude them successfully in ways conducive to classroom purposes, the student may feel excluded as a person with possible detrimental consequences for their wellbeing (Arvola Orlander and Wickman 2011). At the same time, an activity that offers no resistance or negative aesthetic experiences, which students need to learn how to exclude, may be one where everything is gray and boring and where there is no room for feeling the pride of successful achievement (Bellocchi and Ritchie 2015) or the joy of final fulfillment. However, how to strike this balance in different situations is an empirical question that needs to be further modeled conceptually to help teachers to deal with such transactions as they appear in the classroom.

Through the basic habit of inclusions and exclusions, aesthetic experiences have discursive and normative roles. When students successfully learn cognitively how to make the inclusions and exclusions of a practice, they may also come to embrace certain norms for action as desirable and come to emotionally like to adhere to the norms. These transformations occur together and are continuous in action, although they may be discerned as separate faculties in relation to the purpose of a third-person analysis. From a synthetic perspective, the purpose of such an analysis is not to demonstrate their well-defined existence, but to examine how they are continuous in relation to certain purposes. One way to talk about the simultaneous transformation of the person as a whole, including all three faculties, is as a transformation of *taste* and to see *how* the taste developing makes it possible for the person to take part in and be accepted as a member of a certain practice or culture. Both John Dewey (1913) and Pierre Bourdieu (1984) have employed the notion of taste to emphasize the continuity of all three faculties for learning as the transformation of habits (*habitus* of Bourdieu) of making certain distinctions (cognitively, normatively, and aesthetically) about what objects, events, and actions should be included and excluded. From this we have developed a methodology to study how a *taste for science* in all its various shades is constituted through classroom transactions (Anderhag et al. 2015) and used it to model what a teacher may do to support the development of such a taste (Anderhag et al. 2014). This research is reviewed in Chap. 3.

Our studies suggest that aesthetic experience as met in the classroom cannot easily be accommodated only in a few emotional categories. For one thing aesthetic judgments are also aesthetic in the restricted sense (i.e., about what is beautiful or ugly) and so not just emotional. But also a rich language is used which in all its nuances makes immediate sense as situated in use to those participating. The situated nature of aesthetic experience does not make it meaningless. Interestingly aesthetic judgments can be successfully used to speak about what is not yet fully known to students and, at the same time, how students feel about this not yet fully known coming up and how they feel about the experience as a whole in moments of closure and consummation of tasks. The situated nature of the meaning of aesthetic judgments implies that it is possible to examine students' aesthetic experiences in relation to the specific objects, events, and actions that they encounter and handle as they proceed with tasks. Such aesthetic experiences can be analyzed in all their details as they develop and change in the classroom. There is little point educationally in separating a core of inner essence as more real in this multitude of linguistic distinctions springing from human culture and practice.

2.6 Conclusions

In this review I have tried to make salient the consequences for educational research of analytic and synthetic assumptions regarding aesthetic experience and emotion. It should again be emphasized that often these assumptions are not made explicit, especially in research with analytic tendencies. I do not here maintain that analytic assumptions are always embraced by researchers involved in research here categorized as analytic, but rather that they are seen through what is included and excluded in their research in comparison with a synthetic approach. Hence, analytic and synthetic are my analytic categories for the purpose of seeing what synthetic approaches could contribute to mainstream research on aesthetic experience. In analytic research feeling is often a problematic fact, and action is ultimately explained by structures of thinking. In synthetic research on the other hand, feeling is a fact both as encountered from an emic point of view and also as taken seriously as an analytic unit of situated activity. Thinking and its transformation as learning is thus approached primarily as an act and as the transformation of habits.

What becomes evident if feeling is treated as a fact of situated action? The review has given some suggestions. However, ultimately their significance should be sought in the consequences they may have in supporting teachers in making fruitful choices. The models of how aesthetic experience works in science classrooms need further revision with teachers before we can better judge the difference they may make, if any, to analytic approaches.

I have not tried to make a review of all research in science education with either analytic or synthetic tendencies. The intention here has been to isolate two main avenues along which Western thought has approached human faculties and so also aesthetic experience. It is left to the readers to judge for themselves the values of either tradition for various situations and how their own research may be enriched by them.

Acknowledgments I am grateful to the editors and to Anthony Burden for comments on various drafts of this chapter.

References

Ajzen, I., & Fishbein, M. (1980). *Understanding attitudes and predicting social behavior*. Englewood Cliffs: Prentice Hall.

Alsop, S. (Ed.). (2005). *Beyond Cartesian dualism: Encountering affect in the teaching and learning of science*. Dordrecht: Springer.

Anderhag, P., Hamza, K. M., & Wickman, P.-O. (2014). What can a teacher do to support students' interest in science? A study of the constitution of taste in a science classroom. *Research in Science Education, 45*, 749–784.

Anderhag, P., Wickman, P.-O., & Hamza, K. M. (2015). Signs of taste for science: A methodology for studying the constitution of interest in the science classroom. *Cultural Studies of Science Education, 10*, 339–368.

Areepattamannil, S. (2012). Effects of inquiry-based science instruction on science achievement and interest in science: Evidence from Qatar. *The Journal of Educational Research, 105*, 134–146.

Arvola Orlander, A., & Wickman, P.-O. (2011). Bodily experiences in secondary school biology. *Cultural Studies of Science Education, 6*, 569–594.

Bandura, A. (1977). Self-efficacy: Toward a unifying theory of behavioral change. *Psychological Review, 84*, 191–215.

Baram-Tsabari, A., & Yarden, A. (2009). Identifying meta-clusters of students' interest in science and their change with age. *Journal of Research in Science Teaching, 46*, 999–1022.

Bear, M. F., Connors, B. W., & Paradiso, M. A. (2007). *Neuroscience: Exploring the brain* (3rd ed.). Baltimore: Lippincott Williams & Wilkins.

Bellocchi, A., & Ritchie, S. M. (2015). "I was proud of myself that I didn't give up and I did it": Experiences of pride and triumph in learning science. *Science Education, 94*, 638–668. doi: 10.1002/sce.21159

Bloom, J. D. (1992). The development of scientific knowledge in elementary school children: A context of meaning perspective. *Science Education, 76*, 399–413.

Bourdieu, P. (1984). *Distinction: A social critique of the judgement of taste*. London: Routledge.

Burke, T. (1994). *Dewey's new logic: A reply to Russell*. Chicago: Chicago University Press.

Caiman, C., & Lundegård, I. (2013). Pre-school children's agency in learning for sustainable development. *Environmental Education Research, 20*, 437–459.

Chalmers, A. F. (2013). *What is this thing called science?* (4th ed.). Maidenhead: Open University Press.

Chatterjee, A. (2014). *The aesthetic brain: How we evolved to desire beauty and enjoy art*. Oxford: Oxford University Press.

Cherryholmes, C. H. (1988). *Power and criticism*. New York: Teachers College Press.

Damasio, A. R. (2006). *Descartes' error: Emotion, reason, and the human brain*. London: Vintage.

Damasio, A. (2012). *Self comes to mind: Constructing the conscious brain*. London: Vintage.

Dennett, D. C. (1995). *Darwin's dangerous idea. Evolution and the meanings of life*. London: Penguin.

Deutscher, G. (2010). *Through the language glass: Why the world looks different in other languages*. New York: Picador.

Dewey, J. (1913). *Interest and effort in education*. Boston: Houghton Mifflin.

Dewey, J. (1922). *Human nature and conduct: An introduction to social psychology*. New York: Touchstone.

Dewey, J. (1929/1958). *Experience and nature* (2nd ed.). New York: Dover.

Dewey, J. (1934/1980). *Art as experience*. New York: Perigee Books.
Dewey, J. (1938/1997). *Experience and education*. New York: Touchstone.
Dewey, J., & Bentley, A. F. (1949/1975). *Knowing and the known*. Westport: Greenwood Press.
Elster, J. (1988). *Vetenskapliga förklaringar*. Göteborg: Bokförlaget Korpen.
Eysenck, M. W., & Keane, M. T. (2015). *Cognitive psychology: A student's handbook* (7th ed.). London: Psychology Press.
Fischer, E. P. (1999). *Beauty and the beast. The aesthetic moment in science*. New York: Plenum Trade.
Flávia, M., dos Santos, T., & Mortimer, E. F. (2003). How emotions shape the relationship between a chemistry teacher and her high school students. *International Journal of Science Education, 25*, 1095–1110.
Fortus, D. (2014). Attending to affect. *Journal of Research in Science Teaching, 51*, 821–835.
Garrison, J. (2003). Dewey's theory of emotions: The unity of thought and emotion in naturalistic function "co-ordination" of behavior. *Transactions of the Charles S Peirce Society, 39*, 405–443.
Gee, J. P., & Green, J. L. (1998). Discourse analysis, learning, and social practice: A methodological study. *Review of Research in Education, 23*, 119–169.
Glassman, M. (2004). Running in circles: Chasing Dewey. *Educational Theory, 54*, 315–341.
Goffman, E. (1974). *Frame analysis: An essay on the organization of experience*. Boston: Northeastern University Press.
Hamza, K. M., & Wickman, P.-O. (2008). Describing and analyzing learning in action: An empirical study of the importance of misconceptions in learning science. *Science Education, 92*, 141–164.
Harley, T. (2012). Why the earth is almost flat: Imaging and the death of cognitive psychology. *Cortex, 48*, 1371–1372.
Hilmer, M. J., & Hilmer, C. E. (2012). On the relationship between student tastes and motivations, higher education decisions, and annual earnings. *Economics of Education Review, 31*, 66–75.
Jakobson, B., & Wickman, P.-O. (2008). The roles of aesthetic experience in elementary school science. *Research in Science Education, 38*, 45–65.
Kant, I. (1781/1998). *Critique of pure reason*. Cambridge: Cambridge University Press.
Kant, I. (1788/2004). *Critique of practical reason*. Mineola: Dover Publications.
Kant, I. (1790/1987). *Critique of judgement*. Indianapolis: Hacket Publishing Company.
Kwah, H., Milne, C., Tsai, T., Goldman, R. & Plass, J. L. (2016). Emotional engagement, social interactions, and the development of an afterschool game design curriculum. *Cultural Studies of Science Education, 11*, 713–740. doi:10.1007/s11422-014-9621-0
Langsdorf, L. (2002). Reconstructing the fourth dimension. A Deweyan critique of Habermas's conception of communicative action. In M. Aboulafia, M. Bookman, & C. Kemp (Eds.), *Habermas and pragmatism* (pp. 141–164). London: Routledge.
Levenson, R. W., Ekman, P., Heider, K., & Friesen, W. V. (1992). Emotion and autonomic nervous system activity in the Minangkabau of West Sumatra. *Journal of Personality and Social Psychology, 62*, 972–988.
Lispector, C. (1977/2014). *Hour of the star.* London: Penguin Classics.
Liu, C.-J., & Chiang, W.-W. (2014). Scale of academic emotion in science education: Development and validation. *International Journal of Science Education, 36*, 908–928.
Lundegård, I. (2008). Self, values and the world – Young people in dialogue on sustainable development. In J. Öhman (Ed.), *Values and democracy in education for sustainable development – Contributions from Swedish Research* (pp. 123–144). Stockholm: Liber.
Lundegård, I., & Hamza, K. M. (2014). Putting the cart before the horse: The creation of essence out of processes in science education research. *Science Education, 98*, 127–142.
Lynch, M. (1993). *Scientific practice and ordinary action. Ethnomethodology and social studies of science*. Cambridge: Cambridge University Press.
Milne, I. (2010). A sense of wonder, arising from aesthetic experiences, should be the starting point for inquiry in primary science. *Science Education International, 21*, 102–115.

Mumba, F., Mbewe, S., & Cabalengula, V. M. (2015). Elementary school teachers' familiarity, conceptual knowledge, and interest in light. *International Journal of Science Education, 37*, 185–209.

Nolen-Hoeksema, S., Fredrickson, B. L., Loftus, G. R., & Lutz, C. (2014). *Atkinson & Hilgard's Introduction to psychology* (16th ed.). Hampshire: Cengage Learning.

Olitsky, S., & Milne, C. (2012). Understanding engagement in science education: The psychological and the social. In B. Fraser, K. Tobin, & C. McRobbie (Eds.), *Second international handbook of science education* (pp. 19–33). Dordrecht: Springer.

Östman, L., & Almqvist, J. (2011). What do values and norms have to do with scientific literacy? In C. Linder, L. Östman, D. A. Roberts, P.-O. Wickman, G. Erickson, & A. MacKinnon (Eds.), *Exploring the landscape of scientific literacy* (pp. 160–175). New York: Routledge.

Pickering, A. (1995). *The mangle of practice: Time, agency, and science*. Chicago: University of Chicago Press.

Putnam, H. (2002). *The collapse of the fact/value dichotomy and other essays*. Cambridge: Harvard University Press.

Reid, R. (2006). Thoughts on attitude measurements. *Research in Science & Technological Education, 24*, 3–27.

Rockwell, W. T. (2005). *Neither brain nor ghost: A nondualist alternative to the mind-brain identity theory*. Cambridge, MA: MIT Press.

Root-Bernstein, R. S. (1996). The sciences and the arts share a common creative aesthetics. In A. I. Tauber (Ed.), *The elusive synthesis: Aesthetics and science* (pp. 49–82). Dordrecht: Kluwer.

Rorty, R. (1980). *Philosophy and the mirror of nature*. Oxford: Blackwell.

Rorty, R. (1991). *Objectivity, relativism, and truth. Philosophical papers volume I*. Cambridge: Cambridge University Press.

Rudwick, M. J. S. (2005). *Bursting the limits of time: The reconstruction of geohistory in the age of revolution*. Chicago: University of Chicago Press.

Säljö, R., & Bergqvist, K. (1997). Seeing the light: Discourse and practice in the optics lab. In L. B. Resnick, R. Säljö, C. Pontecorvo, & B. Burge (Eds.), *Discourse, tools, and reasoning: Essays on situated cognition* (pp. 385–405). Berlin: Springer.

Schwab, J. (1978). *Science, curriculum, and liberal education. Selected essays*. Chicago: Chicago University Press.

Sensevy, G., Tiberghien, A., Santini, J., Laubé, S., & Griggs, P. (2008). An epistemological approach to modeling: Cases studies and implications for science teaching. *Science Education, 92*, 424–446.

Shusterman, R. (2000). *Pragmatist aesthetics: Living beauty, rethinking art* (2nd ed.). Lanham: Rowman & Littlefield.

Szybek, P. (1999). Touched by a disgusting fish. Dissecting squid in biology lessons in a comprehensive school. In J. Leach & A. C. Paulsen (Eds.), *Practical work in science education-recent research studies* (pp. 96–112). Fredriksberg: Roskilde University Press.

van Aalderen-Smeets, S. I., & Walma van der Molen, J. H. (2015). Improving primary teachers' attitudes toward science by attitude-focused professional development. *Journal of Research in Science Teaching, 52*, 710–734.

van Aalderen-Smeets, S. I., Walma van der Molen, J. H., & Asma, L. J. F. (2012). Primary teachers' attitudes toward science: A new theoretical framework. *Science Education, 96*, 158–182.

Vedder-Weiss, D., & Fortus, D. (2011). Adolescents' declining motivation to learn science: Inevitable or not? *Journal of Research in Science Teaching, 48*, 199–216.

Waller, B. M., Cray, J. J., & Burrows, A. M. (2008). Selection for universal facial emotion. *Emotion, 8*, 435–439.

Welsch, W. (1997). *Undoing aesthetics*. London: Sage.

Wickman, P.-O. (2004). The practical epistemologies of the classroom: A study of laboratory work. *Science Education, 88*, 325–344.

Wickman, P.-O. (2006). *Aesthetic experience in science education: Learning and meaning-making as situated talk and action*. Mahwah: Lawrence Earlbaum.

Wickman, P.-O. (2012). How can conceptual schemes change teaching? *Cultural Studies of Science Education, 7*, 129–136.
Wickman, P.-O., & Östman, L. (2002). Learning as discourse change: A sociocultural mechanism. *Science Education, 86*, 601–623.
Williams, M. (2002). *Wittgenstein, mind and meaning. Towards a social conception of mind.* London: Routledge.
Wittgenstein, L. (1966). *Lectures and conversations on aesthetics, psychology and religious belief.* Oxford: Blackwell.
Wittgenstein, L. (1967). *Philosophical investigations* (3rd ed.). Oxford: Blackwell.
Zimmer, C. (2004). A distant mirror for the brain. *Science, 303*, 43–44.

Per-Olof Wickman is professor in Science Education and director of Science Education research at Stockholm University, Sweden. His main research interest is in modelling classroom interactions holistically to support teacher and student agency.

Chapter 3
Taste for Science: A Bourdieu-Pragmatism Approach to Interest, Aesthetics and Learning

Per Anderhag

It has been suggested that the tradition of approaching human conduct as essentially separated into various dualisms, such as social-mental, emotion-cognition, fact-value and body-mind, may explain why the role of aesthetics for learning science has received relatively little attention from the science education research field (Fortus 2014). However, not only have various authors questioned and empirically argued against the description of science as completely rational and emotionally detached (Wickman 2006) but also against Cartesian dualism and the impact this perspective has had for science education research in general (Alsop 2005). One unfortunate consequence is, for example, that emotions, norms and values are treated usually as separate from cognitive aspects of learning (Wickman 2006).

This chapter addresses the separation of emotions, norms and values from cognition by discussing an alternative perspective on how interest and learning can be explored through science education research. More specifically, I will discuss the Bourdieu-pragmatism concept of *taste* as a tool for studying learning as a socially situated process in which facts, norms and values are transacted simultaneously. The concept of taste for science was originally operationalized for studying how students' interest in science comes into being in classroom action (Anderhag et al. 2015a), thereby serving as an alternative to more cognitive-psychological-oriented approaches. To have taste for something, such as soccer or French wines, does not mean merely that you report that you are interested in the subject but entails that you can engage in and take part in social practices by making distinctions on what is, for example, a good game, an interesting player or bad sportsmanship. Hence, taste is here defined as the way in which people's actions (including speech) make distinctions about what kind of language, objects and people belong and do not belong in

P. Anderhag (✉)
Research and Development Unit, Education Administration, City of Stockholm, SE-104 22 Stockholm, Sweden
e-mail: per.anderhag@stockholm.se

A. Bellocchi et al. (eds.), *Exploring Emotions, Aesthetics and Wellbeing in Science Education Research*, Cultural Studies of Science Education 13,
DOI 10.1007/978-3-319-43353-0_3

a certain practice. In this view, taste for science is not only about knowledge, emotions and feelings towards a subject content, but also about will and capability to participate in the social practice of science.

The theoretical and methodological rationale for approaching interest as taste is discussed in this chapter, and classroom data from previous studies are used to illustrate how teaching may support students in developing a taste for science.

3.1 Students' Interest in Science

3.1.1 Attitudes and Identity

Interest in science is framed typically as students' attitudes towards content, areas, subject, methods and so forth, but also as their attitudes towards science as social practices with which some groups of students can identify at various levels. Due to its impact on learning, students' interest in science is well studied, and it is generally accepted that students' interest towards school science is initially positive but declines in subsequent school levels (Tytler et al. 2008). This decline is more consistent among students coming from homes with lower socioeconomic status (The Royal Society 2008). It has also been shown that at the same time as students may do well in science and even report that they like the subject, they may describe themselves as excluded from science as a practice (Archer et al. 2010). Even if students report positive attitudes towards science, it does not necessarily mean that they also perceive science as being relevant to their lives.

Norms and values are essential parts in identity formation and are therefore important aspects of whether students view themselves as being part of the social practice of science (Aikenhead 1996). The extent to which students may develop an interest in science is not only a question of whether they come to enjoy science as content but also about whether students perceive science as something that they can relate to and want to participate in. As argued by Jay Lemke (1990; p. 1), learning science is about "learning to communicate in the language of science and act as a member of the community of people who do so". Becoming a member of this community is likely to be easier for students who are accustomed with and can relate to the rules encompassing this specific way of acting and talking. Such competence is about ways of communicating what is valued as good/bad, important/unimportant, interesting/uninteresting and included/excluded in the practice of science.

The development of an enduring interest that persists beyond a specific classroom practice, hence, entails the learning of how actions, utterances and artefacts are distinguished and valued by members of the science community. This means learning to talk and act science in a way that will be recognized by others or to develop a taste for science.

3.1.2 *Interest as Representation*

Interest is usually conceptualized as an attitudinal construct that is developing in the socially situated and self-governed interplay between the individual and the object of interest (Hidi et al. 2004). The action-oriented dimension of interest is thus recognized by the attitudinal research field. Thomas Koballa and Shawn Glynn (2007, p. 88), for example, suggest that a student who is interested in a science topic "has a readiness to pursue it". The contextual dimension of interest is also usually acknowledged; Hidi et al. (2004) describe interest as a situated outcome of the interaction between the individual and the object in a particular environment.

However, because the social, procedural and situated nature of interest is difficult to explore through questionnaires, which is the data-collecting method that predominates the attitudinal research field (Potvin and Hasni 2014), situated dimensions of interest are often not recognized. The procedural and situated aspects of interest have a tendency of becoming overlooked when interest is explored through students' secondary reports of what is interesting content, teaching, pedagogy and so forth. Rather than being a process that is constituted by and developed in a specific learning situation, students' interest in science tends to transform to an internal, decontextualized, state of mind that can be extracted afterwards.

When interest is operationalized as a mental entity, it becomes necessary to make assumptions – providing its relation to learning is the focus – of how this entity affects, and is affected, by external action. That is, how does interest affect the transformation of knowledge, or vice versa, how does knowledge transform interest? When knowledge and interest are operationalized as personal and mental, but manifested socially and externally, a translation process is needed. This procedure may be a source of ambiguity; it is simply not possible to gain certainty of *how* a reported interest explains, or is explained by, for example, scores on a test, the use of a spectacular experiment or the implementation of a new pedagogy.

By using more precise questions (Bybee and McCrae 2011) or by cross-checking different data (Ramsden 1998), uncertainties regarding relationships between interest, cognition and other attitudinal constructs, such as motivation, can be clarified. Another alternative is to approach interest simply as being constituted in action. By doing so, the process of translating how feelings and emotions relate to observed action, or vice versa, may be sidestepped. The action-oriented nature of interest, and how it relates to a pragmatic understanding of taste and learning science, is discussed in the next section.

3.2 An Action-Oriented Perspective on Students' Interest in Science

In their works, both John Dewey (1925/1996) and Pierre Bourdieu (2000) recurrently question dichotomies rooted in the notion of an internal subjective mind and an external objective reality. Both authors argue that this dualistic world view has its origin in the historical and economic conditions in which some social groupings had the privilege of positioning themselves in a purely theoretical and detached position towards human existence. In principle, this position, which Bourdieu (1998) and Dewey (1925/1996) argue is evident in different forms of philosophical traditions, resulted in a general detachment from human conduct, and the doings of people were primarily understood as the workings of isolated minds in an objective reality. When the meaning of utterances and actions in everyday and ordinary activities is explored without reference to the social situation in which they occurred, Bourdieu (1998) argues that practical reason collapses into theoretical reason which tends to overlook the embedded and embodied logic of practice. A serious fallacy following this detachment from practice is the failure in seeing that aesthetic judgements, such as what is distinguished as beautiful, ugly or interesting, are rooted in a historical and social context (Bourdieu 2000). For example, students who report an interest in experiments may find quite different aspects of such activities enjoyable; the smoke and sound of a chemistry experiment may be interesting for some students, some may appreciate the distraction it offers from regular instruction, whereas yet another group of students may be interested in the mechanisms behind the reaction observed. It may thus be misleading to assume that the *object of interest*, such as an interest in experiments, has some fixed characteristics that are universally shared and can be implemented in classrooms for making science interesting (Anderhag et al. 2016).

For Bourdieu, human action is understood as the habitual doings of the subject in relation to objective structures, and these structuring structures – that is, the habitus – have in turn been internalized through upbringing and education. Although the concept of habitus often has been interpreted as something fixed and deterministic, Bourdieu stressed that peoples' dispositions are open and constantly subjected to new experiences; they are therefore durable in a particular time and space but they are not fixed (Bourdieu and Wacquant 1992). The social, active and habitual dimension of human conduct is also central in Dewey's writings, and Bourdieu's habitus has thus noticeable similarities with Dewey's concept of *habits*. Dewey (1922/1996, p.31) argues that habits are:

> that kind of human activity which is influenced by prior activity and in that sense acquired; which contains within itself a certain ordering or systematization of minor elements of action; which is projective, dynamic in quality, ready for overt manifestation; and which is operative in some subdued subordinate form even when not obviously dominating activity

For Dewey (1922/1996) habits designate a special accessibility to different groups of situations, and they develop and transform by the social encounters people experience. They are thus not actions repeated in a mechanical fashion. Bourdieu acknowledged his affinity with Dewey, suggesting that as with Dewey's concept of habits, habitus should be "understood as an active and creative relation to the world, [which] rejects all the conceptual dualisms upon which nearly all post-Cartesian philosophies are based: subject and object, internal and external, material and spiritual, individual and social, and so on" (Bourdieu and Wacquant 1992, p. 122).

In the Bourdieu-pragmatism framework, hence, the doings of people take place in a social context, and words, utterances and actions are understood as deriving their meaning through their use and their consequences in situated activities rather than being representations that are used and understood in a universal way.

Aesthetic judgements of feelings and emotions are therefore understood in relation to what is going on in the situation and through their consequences. Framed in this way, being interested is a process that unfolds during an activity with specific purposes, and, as will be discussed below, norms, values and facts are approached and explored as intertwined in this process.

3.2.1 Interest as Constituted in Action

In his writings on interest and learning, Dewey (1913/1996) describes interest as a directional process oriented towards near or distant goals in which actions and artefacts are valued and enjoyed based on what they bring to the fulfilment of this process. Dewey's action-oriented definition is thus in line with how interest usually is conceptualized within contemporary science education research. At the same time as being interested is a goal-oriented process, interest is not primarily the question of coming to closure but rather the totality of the process which is summed up when reaching consummation (Dewey 1913/1996).

Since actions and artefacts are distinguished and valued in relation to what they mean for in facilitating the process of interest forward, interest has a normative dimension. For example, a child who is experiencing interest when searching for tadpoles continuously makes decisions that will support the process, and, with or without the help of a more experienced friend, parent or teacher, such decisions successively become more and more informed and may ultimately develop into habits. What time of the year and day, for example, should one look for tadpoles? What kind of vegetation, pools and so forth may be signs of a favourable environment for tadpoles? What kind of stuff should be collected in order for the tadpoles to live and survive? To paraphrase Richard Rorty (1991), learning is an inevitable outcome of this process of acquiring habits of action for coping with the reality of tadpoles.

At the same time as interest is goal oriented, its directional nature reflects the rhythm of expectations and fulfilment rather than illustrating a predetermined movement to some fixed and final goal (Dewey 1913/1996). For example, the child searching for tadpoles chooses and revises tools, containers, paths, locations, companions and so on, in a way that serves her expectations of the search and of what is going on in the situation. What animals, vegetation, pebbles and small houses that finally end up in the container is in a similar way chosen for what they do for facilitating the process towards fulfilment (e.g. the tadpoles should have a cosy house, have vegetation to hide in, have algae to eat). As the rhythm of expectations and fulfilment is evident through the aesthetic judgements people make (Wickman 2006), the process of interest is possible to observe in classroom action (Anderhag et al. 2015a). Per-Olof Wickman (2006) has demonstrated that aesthetic judgements orient the participants in relation to purposes of the science activity, and in this way, they construe meanings regarding what is the case in terms of scientific facts, but also how actions, artefacts and utterances are conducive to purposes of the science activity. Aesthetic experience is thus closely connected to learning normatively and cognitively in science class. Norms may concern, for example, what is the preferred way to present data, which equipment is most accurate and so should be chosen for making measurements or, by using the example above, what characterizes a good habitat for tadpoles. The extent to which students can acknowledge and distinguish these aspects of doing science is thus important for their opportunity to participate successfully in the practice of science (Wickman 2006).

3.2.2 Interest as Taste

Making normative and aesthetic judgements that distinguish what is good, bad, interesting, beautiful and so forth in a specific social setting is usually described in terms of taste. People's taste is developed through upbringing and schooling, and distinctions of taste are central when participants of various social groups distinguish themselves in relation to other groups (Bourdieu 1984). Taste is thus important for an individuals' will and capability to participate in, and also given access to, different social practices. The habitual and social dimensions of taste have also been recognized by the animal pathologist William Beveridge (1957/2005), who in 1957 put forward the suggestion that a scientific taste may be an important disposition among scientists. As with Bourdieu (1984) and Dewey (e.g. 1929/1996), taste is to Beveridge (1957/2005, p.79):

> a sense of beauty or aesthetic sensibility and anyone who has it simply feels it in his [sic] mind that a particular line of work is of interest for its own sake and worth following, perhaps without knowing why. How reliable one's feelings are can be determined only by the results.

Even if neither Bourdieu nor Dewey actually studied the development of taste, they both, as did Beveridge, underscored its role for orienting action and therefore being central for how and what people learn as part of an activity. Its role in learning thus entails habits of distinguishing and valuing actions that are conducive to purposes, but also receiving the recognition of others as being a member of a certain social practice. In his studies on the selective structures of education, Bourdieu (1979) demonstrated how students' capability of recognizing and making certain distinctions of taste was central for how they were judged and rewarded academically. Students who distinguished the currently valued taste in a school – which in Western European schools typically signifies cultural and social capital, namely, a certain habitus – were described by teachers as brilliant and naturally talented, whereas those who did not were referred to as hard-working and learning by rote.

In the context of science education, students are judged in relation to implicit and explicit norms about what science *is* and whom it is *for* (Jobér 2012). Science is known to reproduce norms historically rooted in a middle-class white male context, and it has been demonstrated, for example, in the study of Carlone et al. (2014) that teachers may judge students according to gender, ethnicity and social class stereotypes, thereby failing to see their potential for science. It is also well known that some students have great difficulties in identifying themselves with the norms of the science classroom and can, regardless of whether they achieve well in science and even say that they like the subject, describe themselves as not being a science person (e.g., Archer et al. 2010).

As with taste in general, taste for science – or interest in science – is associated with home background (Anderhag et al. 2015a) which means that some students when entering school are already familiar with the norms and values projected in the science classroom and can thus make the distinctions that are functional and valued in science classroom practice. Most students, however, need influences from school in order to develop a taste for science that is recognized by others.

3.3 Analysing the Constitution of Taste in Classroom Action

The taste analysis builds on practical epistemological analysis (PEA) (Wickman and Östman 2002) and was originally developed to explore how teaching may support students in developing an interest in science (Anderhag et al. 2015a). PEA is grounded in Dewey's pragmatism and Ludwig Wittgenstein's studies of language and is used for examining situated meaning making (Kelly et al. 2012). PEA is based on four operational concepts: stand fast, encounter, relations and gaps. The unit of analysis is action as part of an activity. In an activity, what stands fast is what is immediately intelligible that which the interlocutors do not need to ask each other about in order to proceed. As the activity continues, knowledge gaps are noticed due to encounters with utterances, actions and artefacts. A gap can thus be anything that hinders the process in which the person is engaged, for example, encountering a new word ("Can you give me the *condenser*?"), an ambiguous outcome of an experiment ("Why did it not mix?") or a practical obstacle ("I cannot focus the microscope!"). An activity proceeds as gaps are continuously filled with relations to what already stands fast (e.g., Is this a condenser?). The taste analysis focuses specifically on situations of indeterminacy in which the interlocutors include certain relations while excluding others by making judgements of taste.

The aim of the first step of the taste analysis is to clarify the aims and purposes of the activity, namely, what are the students set to do and talk about? The proximate purposes or student-oriented purposes (Johansson and Wickman 2011) are evident as the science-related tasks the students are set to do by the teacher. If these purposes are intelligible, that is, the students know what to do and can act according to them, they are said to become ends in view of the students (Johansson and Wickman 2011).

The second step examines whether students' distinctions of taste are oriented towards or away from these purposes. Distinctions of taste are analyzed as the choices of action evident through:

(a) Language usage, namely, how the participants distinguish certain ways of representing content or actions as included or excluded
(b) Procedures, how ways of acting are distinguished as included in, or excluded from, the science classroom practice
(c) Ways of being, how the participants distinguish persons and manners as included in, or excluded from, the science classroom practice

The third step examines how participants experience and judge objects, actions and events aesthetically and whether such evaluations are oriented towards or away from the purposes of the activity. This step is motivated by the fact that students may achieve well in science without enjoying it. Even if the taste constituted in the classroom is consistent with classroom purposes and also enjoyed by students, it will be of little further use if it is functional and valued in this specific practice only. It is therefore necessary to analyse the extent to which the distinctions of taste are also continuous with other fields of science.

Steps four and five, therefore, examine how classroom taste, which usually originates from the teacher, also allows for personal contributions of the student as well as its continuity with a more current taste of science and so acknowledging other stakeholders (e.g., curriculum) and science practices (e.g., subsequent science educational levels, science outside school).

In the next section I will use data from previous studies in order to illustrate the role of norms and aesthetics for students' opportunity to pursue the scientific purposes of the activity. The examples also intend to demonstrate how it may look like when students' more personally oriented judgements of taste are acknowledged and made continuous with the current taste of the classroom.

3.4 Supporting a Taste for Science

The data come from previous studies where we examined how teaching can counterbalance home background by supporting students' interest in science (Anderhag et al. 2015b). The classroom practices studied were originally chosen because when considering their socioeconomic background, a relatively high proportion of students from these schools were shown to choose recurrently the post-compulsory natural science programme (NSP) in upper secondary school (Anderhag et al. 2013). That is, home background, which in Sweden has a strong impact on the choice of NSP, could not explain the high proportion of students from these schools who were choosing post-compulsory science.

The data presented come from two lower secondary science classrooms in two different schools. The data consist of classroom observations of students (age 13–14) measuring the volume of differently shaped objects by using a graduated cylinder filled with water and students (age 15–16) who are doing a practical on electrical circuits.

3.4.1 *Making the Norms of the Classroom a Shared Concern*

3.4.1.1 Implicit Norms of the Science Classroom

In the science classroom – as in any social practice – certain ways of talking, acting and being are usually considered as better than others. Students' awareness of these values is variable. The first excerpt presented below illustrates this. In Example 3.1, Karl and Eric were measuring the volume of a metal weight shaped as a cylinder. They did so by immersing the metal weight into a graduated cylinder filled with water, and in turn 1, Karl asked Eric to read the scale:

Example 3.1: You can't say it like that

1. Karl: Eric, since you got a good view there, check how high it is!
2. Eric: It has gone, like, two lines up.
3. Karl: [laughs] You can't say it like that.

In this short exchange, a gap was acknowledged, *How high is it?*, which Eric filled with the relation of *two lines*. Karl, who maybe had anticipated an answer expressed in millilitres, distinguished this relation as erroneous and so excluded from the activity: Two lines were not a correct way to refer to the change. This statement suggests that for Karl there are better and poorer ways of expressing oneself when reading a scale, which Karl in this particular case seemed to be aware of. Teaching, therefore, is to some extent also about clarifying to the students why some expressions and actions are valued the way they are in the science classroom. Example 3.2 illustrates how it may look like when a teacher is doing that.

3.4.1.2 Clarifying How Norms Are Conducive to Purposes

The second example below comes from the electric circuit lab where the students were working with an assignment where they draw and connect different numbers of bulbs and switches to a battery. Every electric circuit was supposed to generate a certain outcome; in Example 3.2 below, the students were supposed to draw and connect a circuit in which one bulb goes out while the remaining two continue to shine when the switch is pressed. Now Tova and Agnes wanted the teacher to check if their drawn circuit would function as intended:

Example 3.2: Really high standards on these drawings

4. Teacher: I'll just need to think a little. Here all three shine.
5. Tova/Agnes: Mm.
6. Teacher: Mm. And if you press that switch then...

7. Agnes: …that one goes out.
8. Teacher: that one. And if you press that switch, then…
9. Agnes/Tova/Teacher: …these two go out.
10. Teacher: Yes. Elegant! Then the thing is, it may sound a bit boring bringing it up, but, uhm, when you make circuit diagrams like these, then you would want, not like these, but rather as straight and as nice and as boring as possible. There is some sort of, uhm, if it were very advanced diagrams, then you would not want to have twisted wires. Can you be prepared for something?
11. Agnes: Draw on the whiteboard?
12. Teacher: Yes, and then it's clever if you have one of these straight; all wires are straight and only with 90° angles.
13. Agnes: But you can do like this, or?
14. Teacher: Yes. Uhm. If you only curve it. Like that.
15. […]
16. Teacher: So preferably not the wires, the bulbs in the corners either, it has to come out wires like these.
17. Agnes: Yes. Is that…
18. […]
19. Agnes: Really high standards on these drawings
20. Tova: Mm
21. Agnes: I think it'll be fine like this, and then you put one…
22. Tova: That's right
23. Agnes: Like that!

After they had jointly checked if the circuit would work as was expected (turns 4–9), the teacher made an aesthetic judgement of consummation in turn 10: "Elegant! The circuit is correct. Then, in a humorous way, he commented on how they had drawn the circuit. Instead of drawing *twisted* wires, it should be as *straight*, *nice* and *boring* as possible". The look of the circuit was then further distinguished in turn 12; it is *clever* to draw lines with straight angles and not placing the bulbs in the corners and so construing a conventional circuit diagram. The rationale for doing so was motivated by upcoming encounters with electrical circuits with a greater complexity (turn 10).

In summary, then, the teacher did, by distinguishing and aesthetically evaluating different procedures, support students in directing action towards purposes. Drawing circuits with nice and straight angles was included, whereas drawing twisted wires was excluded. This was the clever way to draw circuit diagrams. A notion that was also acknowledged by the students who comment on the *high standards* (turn 19) and redraw their circuit (turns 19–23). Learning to make conventional circuit diagrams was an aesthetic experience in which knowledge gaps became clarified through distinction of taste, namely, anticipated action was distinguished and aesthetically evaluated in regard to how it pertained to the purposes of the activity.

3.4.2 Making Everyday Distinctions of Taste Continuous with the Scientific Purposes of the Classroom

Even if norms and values are explicit in the classroom, it does not mean that the students can relate to them. Of importance for student capability and the will to participate in the activities of the science classroom is also the extent to which their personal judgements of taste are acknowledged and made continuous with the current taste of the classroom. A teacher could, for example, downgrade Eric's description of the change in the graduated cylinder (turn 2) as erroneous, but the teacher could also use it as a starting point for discussing the scale. Moreover, making conducive distinctions of taste is of little meaning if the students at the same time do not enjoy being part of the practice. For example, as implied by Example 3.3 below, students who recurrently feel that they are put down due to implicit or explicit norms of what characterize a science student are not likely to enjoy the science classroom practice.

3.4.2.1 Who and What Belong in the Science Classroom

In Example 3.3, the students were just about to start with their examination of the volume of one of their metal weights. In turn 24, Charles putted forward a suggestion on how they should distribute the work:

Example 3.3: I'm the brain

24. Charles: I'm the brain so I'll say what you should write and later I copy you.
25. Alfred: You? Brain!
26. Charles: No, but.
27. Alfred: Measuring volume.

Charles jokingly expressed a common stereotype of what characterizes the typical science student: being brainy. The joke was not well received as Alfred questioned the suggestion of Charles being the brain (turn 25). However, they both seemed content with the relation that was established, namely, that persons with brain know what to write in the science classroom.

A classroom that projects a narrow and elitist taste regarding who belongs and who does not belong, or what kind of language and procedures that is preferred, is likely to be excluding. That was however not the case in the classrooms that the examples come from: on the contrary the teachers were very careful in including the students' personal way of talking and being. In this way the teacher made the students' personal experiences continuous with what was going on in the activity. This is illustrated by Examples 3.4 and 3.5.

3.4.2.2 Acknowledging and Negotiating Personal Judgements of Taste

In Example 3.4, the teacher and a student were discussing an appropriate name for the metal weight the students were examining. Before this exchange, the teacher and the students had been talking about how they should do the measuring. During these talks the teacher stressed the importance of giving the different objects suitable names.

Example 3.4: Can you name it Jonny?

28. Teacher: Those yes, precisely. One at a time. And then you name it something. This is a…
29. Edit: We wrote small weight.
30. Teacher: But that is really good.
31. Anna: Can you name it Jonny?
32. Teacher: It is a small weight
33. Anna: Can you name it Jonny?
34. Teacher: You can name it Jonny as long as you tell that Jonny is the small weight.
35. Anna: All right.
36. Anna: But we have not done that.
37. Teacher: It works nicely, Jonny small weight.
38. Anna: Immerse Jonny small weight.
39. [Laughter]

In a playful way, Anna asked the teacher in turn 31 if it was possible to name the metal weight *Jonny*. The teacher did not downgrade her suggestion but instead used it for clarifying the scientific rationale for naming things: the name is not important as long as it captures the relevant qualities of the object (Jonny is the *small* metal weight). Anna's distinction was again acknowledged in turn 37 as the teacher made an aesthetic judgement summing up the situation: it works nicely, Jonny small weight. Anna found this funny, and they laughed when she jokingly included the not-so-scientific name in the scientific practice of measuring volume: Immerse Jonny small weight. The situation came to closure in turn 39, and the group continued with reading of the scale.

The final example, which comes from the electric circuit lab, illustrates how the teacher and a student negotiate distinctions. As with the previous example, also this teacher acknowledged students' distinctions of taste and carefully oriented these towards the scientific purposes of the tasks. In the excerpt, the teacher and the student were talking about the size of the circuit the student had drawn. Besides having straight angles and no bulbs in the corners, the teacher also wanted the circuits to be bigger. Before turn 40 Svante had questioned the teacher's evaluation of his drawing as being a little bit too small and that he could make it bigger. In turn 40 the teacher encouraged Svante to use more paper in order to make bigger drawings.

Example 3.5: So you're encouraging us to ruin the environment?

40. Teacher: ...you don't have to economize on your notebooks, they're...
41. Svante:...it is non-environmental friendly to throw away.
42. Teacher: That is so, yes.
43. Svante: That's no good [Teachers name].
44. Teacher: No.
45. Svante: So you're encouraging us to ruin the environment? [laughter]
46. Teacher: [Takes a deep breath and slaps his own hand]
47. Svante: Yes precisely!

Svante did not agree with the judgement the teacher had made on his drawing and questioned jokingly the teacher's appeal that they do not need to economize on their notebooks (turns 41, 43, 45). The joke distinguished the teacher's suggestion as inappropriate and so excluded from the science classroom: he should not encourage non-environmental conduct. The teacher supported this distinction, and the situation came to closure as he jokingly slapped his own hand.

When Anna and Svante were talking with their teachers about the distinctions made, their own personal contributions became part of the taste constituted in the classroom. As evident through the aesthetic judgements they made, they laughed and they were joking; they enjoyed being part of this practice. Participating in these activities was an aesthetic experience that influenced what the students learned about electrical circuits and measuring volume.

In summary, Examples 3.2, 3.4 and 3.5 illustrate situations where the teacher recognized the students' personal experiences as they were conveyed through the ways students distinguished ways of talking, acting and being. At the same time as students' more personally oriented distinctions of taste were acknowledged, the teacher also clarified how contributions may be aligned with the aims of the activity. This support helped the students in making informed judgements and so constitute a taste that was conducive to the scientific purposes of the activity.

3.5 Summary and Concluding Remarks

Being interested is generally unproblematic outside of school. A person experiencing interest acts according to her expectation of the object of interest and in regard to what is going on in the situation. That is, an interested person will simply act in a way that carries the process forward. In school, however, aims and goals, and sometimes also the means to reach them, are usually not for the students to decide. Not only do science classroom activities have implicit and explicit purposes but activities also encompass norms and values (e.g. nice, elegant, straight wires, a name should capture the scientifically relevant characteristics of the object) that the students are expected to recognize and relate to. Students, hence, rarely have full control over the directional rhythm of anticipation and consummation that characterize the process of interest. Indeed, previous research has shown that students may have great difficulties in relating to what is going on in the science classroom, both in terms of what norms and purposes mean for learning the science content (e.g. Säljö and Bergqvist 1997) and also in terms of what they mean for the individual as a participant in the social practice of school science (Carlone et al. 2014). I have argued in this chapter that taste may be a possible tool for exploring these dimensions of students' learning and interest in science.

Interest framed as taste is a social process in which the individual brings, by distinguishing and evaluating actions pertaining to situated purposes, some course of action to its accomplishment. In this view, taste is about capability of making informed decisions but also about social belonging and receiving recognition from others. The will and capability of pursuing activities in the science classroom may seem as a rather dull approach to interest. Nothing in it appears to suggest personal significance or positive emotions, which usually are associated with the interest construct. However, as argued throughout this chapter, making informed decisions is not only a question of rationality but it is also greatly governed by an aesthetic sense of the situation. Such a sense can be learnt and developed in school, and this process is possible to observe through the distinctions of taste the participants are making when doing science.

Finally, the teacher has a key role in providing situations in which the personal experiences of the students can be made continuous with purposes, norms and values of the science classroom. This does not mean that there is one and only taste for science, but rather on the contrary, every science classroom is unique, and there are a multitude of ways in which a functional taste for science may be supported and acquired.

References

Aikenhead, G. S. (1996). Science education: Border crossing into the subculture of science. *Studies in Science Education, 27*(1), 1–52.

Alsop, S. (2005). Bridging the Cartesian divide: Science education and affect. In S. Alsop (Ed.), *Beyond Cartesian dualism: Encountering affect in the teaching and learning of science* (pp. 3–16). Dordrecht: Springer.

Anderhag, P., Emanuelsson, P., Wickman, P.-O., & Hamza, K. M. (2013). Students' choice of post-compulsory science: In search of schools that compensate for the socio-economic background of their students. *International Journal of Science Education, 35*(18), 3141–3160.

Anderhag, P., Wickman, P.-O., & Hamza, K. M. (2015a). Signs of taste for science: A methodology for studying the constitution of interest in the science classroom. *Cultural Studies of Science Education, 10*(2), 339–368.

Anderhag, P., Hamza, K. M., & Wickman, P.-O. (2015b). What can a teacher do to support students' interest in science? A study of the constitution of taste in a science classroom. *Research in Science Education, 45*(5), 749–784.

Anderhag, P., Wickman, P.-O., Bergqvist, K., Jakobson, B., Hamza, K. M. & Säljö, R. (2016). Why do secondary school students lose their interest in science? Or does it never emerge? A possible and overlooked explanation. *Science Education*, 1–23.
Archer, L., DeWitt, J., Osborne, J., Dillon, J., Willis, B., & Wong, B. (2010). "Doing" science versus "being" a scientist: Examining 10/11-year-old schoolchildren's constructions of science through the lens of identity. *Science Education, 94*(4), 617–639.
Beveridge, W. I. B. (2005). *The art of scientific investigations*. Caldwell: Blackburn Press. (original work published 1957).
Bourdieu, P. (1979). *The inheritors: French students and their relation to culture*. Chicago: University of Chicago Press.
Bourdieu, P. (1984). *Distinction: A social critique of the judgment of taste*. Cambridge, MA: Harvard University Press.
Bourdieu, P. (1998). *Practical reason: On the theory of action*. Oxford: Polity Press.
Bourdieu, P. (2000). *Pascalian meditations*. Cambridge: Polity Press.
Bourdieu, P., & Wacquant, L. J. D. (1992). *An invitation to reflexive sociology*. Cambridge: Polity Press.
Bybee, R., & McCrae, B. (2011). Scientific literacy and student attitudes: Perspectives from PISA 2006 science. *International Journal of Science Education, 33*(1), 7–26.
Carlone, H. B., Scott, C. M., & Lowder, C. (2014). Becoming (less) scientific: A longitudinal study of students' identity work from elementary to middle school science. *Journal of Research in Science Teaching, 51*(7), 836–869.
Dewey, J. (1996a). Interest and effort in education: An introduction to social psychology. In L. Hickman (Ed.), *Collected works of John Dewey, 1882–1953: The electronic edition* (Middle works, Vol. 7, p.31). Charlottesville: InteLex Corporation. (original work published 1912).
Dewey, J. (1996b). Human nature and conduct: An introduction to social psychology. In L. Hickman (Ed.), *Collected works of John Dewey, 1882–1953: The electronic edition* (Middle works, Vol. 14). Charlottesville: InteLex Corporation. (original work published 1922).
Dewey, J. (1996c). Experience and Nature: In L. Hickman (Ed.), *Collected works of John Dewey, 1882–1953: The electronic edition* (Later works, Vol. 1). Charlottesville: InteLex Corporation. (original work published 1925).
Dewey, J. (1996d). The quest for certainty. In L. Hickman (Ed.), *Collected works of John Dewey, 1882–1953: The electronic edition* (Later works, Vol. 4). Charlottesville: InteLex Corporation. (Original work published 1929).
Fortus, D. (2014). Attending to affect. *Journal of Research in Science Teaching, 51*(7), 821–835. doi:10.1002/tea.21155
Hidi, S., Renninger, A. K., & Krapp, A. (2004). Interest, a motivational construct that combines affective and cognitive functioning. In D. Y. Dai & R. J. Sternberg (Eds.), *Motivation, emotion, and cognition* (pp. 99–115). Mahwah: Erlbaum.
Jobér, A. (2012). *Social class in science class*. Dissertation, University of Lund, Malmö.
Johansson, A.-M., & Wickman, P.-O. (2011). A pragmatist understanding of learning progressions. In B. Hudson & M. A. Meyer (Eds.), *Beyond fragmentation: Didactics, learning and teaching* (pp. 47–59). Leverkusen: Barbara Budrich Publishers.
Kelly, G. J., McDonald, S., & Wickman, P.-O. (2012). Science learning and epistemology. In K. Tobin, B. J. Fraser, & C. J. McRobbie (Eds.), *Second international handbook of science education* (pp. 281–291). Dordrecht: Springer Netherlands.
Koballa, T. R., & Glynn, S. M. (2007). Attitudinal and motivational constructs in science learning. In S. K. Abell & N. G. Lederman (Eds.), *Handbook of research on science education*. Mahwah: Lawrence Erlbaum Associates, Publishers.
Lemke, J. L. (1990). *Talking science: Language, learning and values*. Norwood: Ablex.

Potvin, P., & Hasni, A. (2014). Interest, motivation and attitude towards science and technology at K-12 levels: A systematic review of 12 years of educational research. *Studies in Science Education, 50*(1), 85–129.

Ramsden, J. M. (1998). Mission impossible?: Can anything be done about attitudes to science? *International Journal of Science Education, 20*(2), 125–137.

Rorty, R. (1991). *Objectivity, relativism, and truth*. Cambridge: Cambridge University Press.

Säljö, R., & Bergqvist, K. (1997). Seeing the light: Discourse and practice in the optics lab. In L. B. Resnick, R. Säljö, C. Pontecorvo, & B. Burge (Eds.), *Discourse, tools, and reasoning: Essays on situated cognition* (pp. 385–405). Berlin: Springer.

The Royal Society. (2008). *Exploring the relationship between socioeconomic status and participation and attainment in science education*. London: The Royal Society.

Tytler, R., Osborne, J., Williams, G., Tytler, K., & Cripps Clark, J. (2008). *Opening up pathways: Engagement in STEM across the Primary-Secondary school transition*. Australian Department of Education, Employment and Workplace Relations, Brisbane.

Wickman, P.-O. (2006). *Aesthetic experience in science education: Learning and meaning-making as situated talk and action*. Mahwah: Lawrence Erlbaum Associates.

Wickman, P.-O., & Östman, L. (2002). Learning as discourse change: A sociocultural mechanism. *Science Education, 86*, 601–623.

Per Anderhag has a PhD in science education and is working with school science development at Education Administration, City of Stockholm. His research interest is the difference teaching can make for students' interest and learning in science.

Chapter 4
The Heart of the Educator: Aesthetic Experience Shaping Knowledge, Identity, and Passion

Linda Hobbs and Leissa Kelly

4.1 Introduction

> *For much of my high school I wanted to be an English and Physical education teacher. Then my Year 11 and 12 Biology teacher introduced me to plant physiology and animal behavior. A passionate teacher, getting my hands dirty through dissections, experimenting with ants, and snorkeling off Portland harbor during classes totally changed my world, what I loved, and my future. I went on to study biology, returned to my first love of teaching but to teach science, and now teaching science education at university. (Linda Hobbs)*

Becoming a science educator involves both cognitive and affective/emotional dimensions as we come to know and develop passions for the subject matter and the artistry involved in educating others (Rubin 1985). For their students, a teacher's passion for the subject they are teaching is a key ingredient for capturing students' interest (Darby 2005). The enthusiasm that stems from this passion can be inspiring and lead to a memorable experience for the learner, as is demonstrated by Linda's experience above. But where does this passion come from, how does it relate to the teacher's conceptualization of the task of teaching science, and how they situate themselves in relation to the subject? In this chapter we explore how experiences that evoke strong cognitive and affective outcomes influence teachers' knowledge, professional identity, and passion for science education. Such experiences are called "aesthetic experiences" (derived from John Dewey's aesthetic experience, 1934/1980). Aesthetic experiences can occur at any time in the life of a person and can have various effects, most notably that they lay the foundation for future experiences. For some teachers, developing a connection with science is marked by such an aesthetic experience. For other science educators, a transformative experience

L. Hobbs (✉) • L. Kelly
Faculty of Arts and Education, Deakin University,
75 Pigdons Road, Waurn Ponds, Geelong, VIC, Australia
e-mail: l.hobbs@deakin.edu.au; l.kelly@deakin.edu.au

A. Bellocchi et al. (eds.), *Exploring Emotions, Aesthetics and Wellbeing in Science Education Research*, Cultural Studies of Science Education 13,
DOI 10.1007/978-3-319-43353-0_4

has the potential to inspire an interest in science that can lead to choosing science as a career.

In Linda's example, learning and doing biology involved both cognitive and affective dimensions, that is, there was an emotional response to coming to know and experience biology. According to Dewey (1934/1980), in such aesthetic experiences, the cognitive and affective are intrinsically linked through the process of coming to know, in the construction of identities, and leading to motivations for future learning and knowing. In this chapter we focus on two consequences of such experiences: that they result in coming to understand in such a way that it changes the way we see ourselves and the world around us, what Girod et al. (2003) call an "aesthetic understanding," and that they set the foundation for future experience. An aesthetic understanding involves not just the construction of knowledge, but also influences how we see ourselves in relation to that knowledge (our identity), as well as shaping our motivations and passions to engage further – coming to know can lead to wanting to know more. We propose that the aesthetic understanding lens can act as an effective framework for understanding and describing relationships between science educator knowledge, identity, and passion. We will show that the framework provides explanatory power in situations where the relationships between these elements may not be immediately apparent. Also, the framework acknowledges the recognition of all three – to speak of only one is to neglect its relation to the others. Furthermore, the framework is flexible in that the individual constructs can be explored in many contexts.

In the following sections, we describe the aesthetic lens as it emerges from Dewey and as it has been adapted by others and propose an analytical tool based on the knowledge-identity-passion (KIP) relationship. In order to illustrate the analytical power of this aesthetic understanding framework, we then apply a KIP analysis to describe the experiences of three types of science educators: a primary specialist science teacher, two preservice primary teachers, and a marine educator. The first two accounts illustrate how aesthetic understanding may arise as teachers or preservice teachers move into a subject area for which motivation or passion does not exist but can emerge due to a pedagogical commitment to their students. The third account illustrates how experiences that inspire an individual's passions ignite their need to know and seek further experience. These teachers were participants from three different research projects that examined, among other aspects, teacher identity; teachers from different contexts were chosen in order to draw out different experiences and the influence of those experiences.

The chapter will also discuss the practical applications of this framework in education policy, practice, and research. Working on the assumption that the key to quality education is passionate and knowledgeable educators (Darling-Hammond 2006), it makes sense to recognize in policy, practice, and research the significance of teacher knowledge, identity, and passion in relation to the roles in which they find themselves.

4.2 Dewey's Aesthetics

Dewey refuses to separate the cognitive and affective aspects of human nature. According to Dewey's philosophy, the individual acts as agent in their perception of an experience, and this agency involves both cognitive and affective dimensions. He termed this "aesthetic experience" signifying "experience as appreciative, perceiving and enjoying. It denotes the consumer's rather than the producer's standpoint" (Dewey 1934/1980, p. 47). This implies that "the individual acts as agent in their perception of the experience, and this agency involves both cognitive and affective dimensions" (Hobbs 2012, p. 2). It is through the combination of the cognitive and affective elements that the integrity of an experience is maintained (Dewey 1934/1980). Dewey (1938) described the quality of experience as having two aspects: "There is an immediate aspect of agreeableness or disagreeableness, and there is its influence on later experiences" (1938, p. 27). His notion of "aesthetic experience" can be associated with the first aspect. For Dewey, the individual actively perceives an experience: "not absence of desire and thought but their thorough incorporation into perceptual experience characterizes esthetic experience in its distinction from experiences that are especially 'intellectual' and 'practical'" (Dewey 1934/1980, p. 254). The second aspect he captures as "continuity of experience" where one experience lays the foundation for how we qualify and "live" future experiences (Dewey 1938). An aesthetic experience is marked by a unity that is "constituted by a single quality that pervades the entire experience in spite of the variation of its constituent parts" (Dewey 1934/1980, p. 206). Wickman (2006) further elaborated that in an aesthetic experience the cognitive and affective are intrinsically linked: to speak of knowing is to also speak of an emotional or affective response to that knowing. The effect of an experience is individually determined because the effect depends on past experiences and thus is a continuity of experiences.

An aesthetic experience changes the way we see the world and our place in it. An aesthetic experience combines conceptual knowledge with a deep awareness and appreciation for the inherent "beauty and power of ideas" (Pugh and Girod 2007, p. 12) (i.e., more aesthetic ways of knowing) that causes a transformation in the way we experience or perceive the world. As a result of this change in perception of both ourselves and the world, we invariably develop a renewed (or new) excitement and interest in the world around us (Pugh and Girod 2007). It is almost like a light switching on inside of us that adds clarity to our thoughts and understandings that were lacking prior to the transformation.

Not every experience provides the opportunity for transformation. Dewey states that "(t)he belief that all genuine education comes about through experience does not mean that all experiences are genuinely or equally educative" (Dewey 1938/1988, p. 13). Many experiences fail to develop beyond an inchoate stage, thereby constraining us to remain "unconscious" or deeply unaware of our surrounds. This is the nature of "ordinary experience" where, although we may be actively involved in the process, the value and significance remains external to ourselves. There is no coherence, flow, or development occurring, even though the activity(s) may happen over time (Girod and Wong 2002).

In contrast, in an aesthetic learning experience, the inner emotional world would be continuous with the outer world to the point where they could not be considered individually (Wickman 2006). According to Dewey:

> ...every experience affects for better or worse the attitudes which help decide the quality of further experiences, by setting up certain preferences and aversions, and making it easier or harder to act for this or that end. Moreover, every experience influences in some degree the objective conditions under which further experiences are had. (Dewey 1938, p. 33)

Dewey (1934) described criteria for an "educative experience" that included not only learning by doing; it also involved integrating the unique experience with mind and emotion and enabling it to continue until fulfillment or until the experience "is so rounded out that its close is a consummation and not a cessation" (Dewey 1934, p. 35). Such an experience is a whole and carries with it its own individualizing quality and self-sufficiency.

This concept is expounded upon in the principle of continuity of experience, or the experience continuum, which arises through the force of habit. The key characteristic of habit, according to Dewey, is that every experience that is encountered changes the individual that undergoes the experience. Because of this modification to the individual, the quality of any subsequent experiences is affected. From this point of view, the principle of continuity of experience means that every experience is shaped by previous experiences, but also modifies the quality of those that are to come (Dewey 1938/1988).

There is an expanding body of literature that draws parallels and linkages between the fields of arts and science by exploring the role that "aesthetics, creativity, passion, beauty and art play in the lives and learning of scientists" (Girod et al. 2003, p. 575). Some authors have extrapolated this further by using Dewey's philosophy on aesthetics to understand the impact aesthetic experiences have on science (or mathematical) education and learning (see, e.g., Girod et al. 2003; Jakobson and Wickman 2007). Girod and Wong (2002) theorized that, by borrowing "from aesthetic and artistic pedagogy" to teach science, we can "tap the power of aesthetic experience" (p. 200).

4.3 Aesthetic Understanding

Drawing on Dewey's philosophy, Girod (2001) referred to "teaching for aesthetic understanding" when describing students' experience of coming to know science content and the pedagogical principles that promote such understanding. Girod et al. (2003) illustrate how aesthetic understanding can be applied to an analysis of children learning science: "from the perspective of aesthetic understanding, science learning is something to be swept-up in, yielded to, and experienced. Learning in this way joins cognition, affect, and action in productive and powerful ways" (p. 575–576).

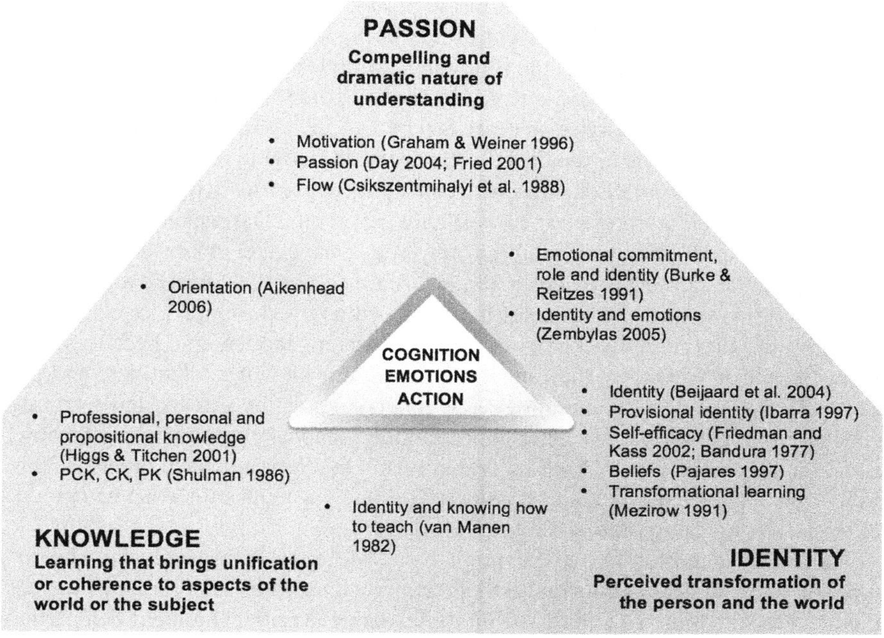

Fig. 4.1 KIP triangle and links with relevant constructs from the literature

While the work of Girod and others focused on children's science learning experiences, the framework has also been applied to the analysis of the experiences of science educators and in doing so shifts the emphasis away from the learning of science to learning to be a science educator. For example, the first author, Linda Hobbs (2012), applied this framework to explore how mathematics and science "teachers' experiences with the subject shaped their faculties and desires, and in turn, these faculties and desires shaped the way they engaged with, and saw themselves in relation to, the subject." (p. 726). The second author, Leissa Kelly, further applied the aesthetic understanding framework to the experiences, passions, and knowing of marine educators (Kelly 2013).

Girod et al. (2003) highlighted three interconnecting elements of aesthetic understanding: they are "compelling and dramatic," "unifying," and "transformative" (p. 578). These aspects of aesthetic understanding can be linked to the science education literature in the following way:

- Compelling and dramatic nature of understanding: passions
- Learning that brings unification or coherence to aspects of the world or the subject: knowledge
- Perceived transformation of the person and the world: identity

Figure 4.1 identifies a range of constructs that are related to knowledge, identity, and passions. Some constructs focus predominantly on either knowledge-, identity-, or passion-related ideas (situated at the vertices); others tend to explore relationships between these. The various constructs in Fig. 4.1 can be used to inform the KIP analysis; some are used in the following narratives.

While this reduction of aesthetic understanding potentially fragments the wholeness of an aesthetic experience as described by Dewey, by focusing on each aspect, it is possible to explore in detail the weaving of the cognitive and affective in an experience. It is the relationships between knowledge, identity, and passion and recognition that all are present in an aesthetic experience that makes it necessary to first identify each and then examine how they shape and relate to each other. Aesthetic understanding provides a framework for conceptualizing the relationships between knowing as a coherence of things to be known, developing one's sense of identities in relation to that knowing, and the passions that emerge because of that knowing, all of which lay the foundation for future knowing, identities, and passions. Therefore, as a methodology for understanding the effects and meanings attached to experiences, aesthetic understanding enables close analysis of knowledge, identity, and passion, both individually and in relation to each other; we have called this a knowledge-identity-passion (KIP) analysis. Questions that can guide a KIP analysis are suggested in Table 4.1.

Table 4.1 identifies two focal points, or "Moments," for analysis. The first moment is for understanding an aesthetic experience and its effect, that is, the aesthetic understanding that emerges from, for example, a critical moment or a particular learning experience. The second moment allows for examining an individual's journey or trajectory following an aesthetic experience, for example, how previous learning experiences or particular childhood experiences have shaped a learner's trajectory or career choice. Through these two moments, a KIP analysis can accompany narrative approaches to research if the intention is to examine both the immediate effect of a planned experience, as well as how previous experiences shape current and future cognition, emotions, and action.

In order to illustrate the power of a KIP analysis, we use the narratives of four science educators. The first is a generalist primary teacher who became a specialist primary science teacher for her school, an example of Moment 1. The second narrative juxtaposes the experiences of two preservice teachers teaching science during a primary science education unit, one for whom the experience changes the way she thinks about science and the other for whom the experience is building on a prior shift towards science, examples of Moment 1 and Moment 2, respectively. The third narrative is of a marine educator whose science education trajectory can be traced to a critical moment from his childhood, an example of Moment 2. At the end of each narrative is a summary of the relationships between knowledge, identity, and passion as demonstrated through the narrative.

Table 4.1 Research moments in a KIP analysis

Moment 1	*Identifying the aesthetic experience and its immediate influence*	1. What changes in knowledge are occurring?
		2. What shifts in identity are occurring?
		3. What passions and motivations are being ignited?
		4. How is the experience of coming to know causing transformation of identity and self, and ignited passions and motivations?
		5. How is the experience influencing action?
		6. How will this experience lay the foundation for future experience?
Moment 2	*Identifying continuity across experiences subsequent to an aesthetic experience*	1. How have prior experiences laid the foundation for current experiences in terms of:
		(a) what is becoming known;
		(b) actions undertaken;
		(c) notions of identity; and
		(d) motivation and passion to engage?
		2. What further transformations of identity are occurring?
		3. How are passions and motivations being ignited and morphed by further experiences?
		4. What new knowledge is becoming known?
		5. How is the experience of coming to know further shaping identity and passions and motivations?
		6. How have passions and motivation driven further efforts to come to know more?
		7. How have the developing identities directed the seeking out of certain knowledge?

4.4 Specialist Primary Science Teacher

The first research study involved investigating the outcomes of a 2-year primary science specialist program; this was a state-funded program[1] and involved periodic professional development with content focused on science pedagogy, content knowledge, school change processes, training in a curriculum package called "Primary Connections" (Australian Academy of Science 2014), and mentoring and coaching. Schools are elected to be part of the program and appointed two of their teachers (usually a generalist classroom teachers) as science specialists. Members of the team investigating the outcomes of the initiative each acted as critical friends to specialists at one school and undertook a series of interviews throughout the 2 years. The science specialist selected for this chapter was part of the second cohort undertaking this training. This cohort typically used the Primary Connections

[1] Professional development funded by the Victorian Department of Education and Early Childhood Development. Research funded by the Centre for Research in Educational Futures and Innovation, Deakin University.

resource to guide curriculum development and to drive pedagogical change of the other staff in the school. The specialists are tasked with creating real and sustained change in the amount and quality of science being taught in primary schools. Early research on this initiative is published in Campbell and Chittleborough (2014).

Hazel was one of two teachers who enrolled in the program from this primary school in regional Victoria, Australia. Near the end of the first year, the other specialist left the school, leaving Hazel to continue the program on her own. Hazel's narrative includes data from two interviews, one 4 months into the program and the other a year later. The change in her professional identity, her growing passions, and deepening knowledge are evident.

4.4.1 Hazel

After 4 months in her new role as science specialist, it became evident in her interview that there was some change occurring in her peers' perception of Hazel, and some early shifts in her views of science. When asked whether she had grown in her practice, Hazel initially struggled to recognize deep level change and was surprised to hear that others had noticed change. In particular, the Assistant Principal was affirming both in terms of Hazel's developing science knowledge, and the effect of her effectiveness in undertaking her role as science specialist, something which he believed was noted by the teachers with whom she was working:

> The breadth of knowledge [the specialists] have got now is so much greater and they are able to take on lots of different subjects. And with Science, their relationship with those staff is more than they would have if they were just going in there co-teaching.

Knowledge is positioned as key to leading change: the Assistant Principal stated, "People look at Hazel and Andrew as, you know, excellent resources for lots of curriculum. And also, well, just anything, anything to do with school," To this, Hazel responded with, "Oh, did they? (chuckles), I didn't see myself like that…" Here identity is being socially constructed, Hazel for the first time seeing a reflection of her newly developing professional self from her Assistant Principal.

At this time, Hazel referred to a shift in her awareness of science outside of school, particularly in her own garden where she would stop and look at things she might not have noticed otherwise: "going out finding about the other science things, which normally I would be just not worried about too much." "From there it obviously builds," Hazel stated. In the school playground, Hazel noticed a change in her interaction with children, as they identified her as someone with whom they can share their science discoveries. For example, she explained that "kids come up to you in the playground" and they relay stories of the mealworms in the classroom changing color. Similar to the Assistant Principal, Hazel's new awareness of her changing attitudes was being socially constructed through affirmation by the children. There is connection between her new role, developing knowledge and practice, and social confirmation and affirmation of the changes. There is a sense of experiences building over time, resulting in continuity of experience.

Twelve months later, Helen was much more certain about how her involvement in the program had changed her view of herself as a science specialist. In particular the science content had ignited a fire in her. When asked how she was going, she responded with:

> I really love it. It's been a huge shift for me. I really like looking more in-depth into the understandings and I go out of my way to find readings about that and investigate. It's probably moved into my home life, which is quite shocking for me. My daughter just loves science days. Is it science day at school today? And she comes home and talks about it. So she knows that's what I do in my work. That's been interesting. I'm growing tomatoes now, which is something I would never have done. ... Personally it's been a big shift in my discipline. I've never really thought I'd be looking at or doing science. I'm a lot more confident. I was always confident to make mistakes, but now I'm like well this is really interesting ... One grade produces different data and they're able to talk about why. And it's OK that I don't know it all and let's find out together.

Here, Hazel identified the shifts: greater content knowledge and a tendency to seek out deep understandings and develop resources, increased capacity and confidence to undertake and guide students through scientific investigations, and a repositioning of what she regarded as her discipline, from a classroom teacher to that of a science specialist. Evident in her responses is increased passion for learning about science and investigating scientific phenomena. Evident too is how her professional role and identity have become personal, for example, where she believes her daughter shares her own science experiences because "she knows that's what I do in my work."

Hazel reflected on factors that were leading to these shifts: "PD [professional development] throughout the program. Having conversations with the other staff. I really love going to the local secondary college to speak to the staff and find out about Science...The leadership stuff."

The PD in particular became increasingly influential as the connections were made between the PD and her own practice:

> Initially when we had the PD I felt like it was over my head. Now I'm making links back to it. The particle stuff, we did the role-play, then we used it in the 5/6 unit ... That real content stuff that at the time I didn't think it linked to anything but now I see the links.

Initially the PD provided the foundation for future experience, where connections could be made as the context demanded. The knowledge of what and how to teach gained through the PD, as it is played out in practice, appears to have contributed to the shift in her self-efficacy (Bandura 1977) and identity and is an agent in her developing passions and motivations to seek further experiences.

In addition, her experiences in her new role as science specialist were consolidating and extending her identity as a leader of people to include leader of content knowledge. Again, affirmation by others served to catalyze this change:

> Being seen as a leader in the school and being valued as a leader. I've always been on the leadership team, and I've liked those roles, but its not just being a leader of people but now it's a leader of content knowledge, that's been a change in me. And leadership will actually come to me about the curriculum, about the science, and that's been a shift in maybe how they view me as well.

Another factor leading to change was taking on sole leadership of the program after the other specialist left the school, giving her full control over the direction of the program:

> … when I was sharing I didn't feel like I had true ownership over it. So probably for me taking it all and doing it myself has been really nice because I've been able to take it in the direction that I wanted to go, which is really bad because we like to work in teams…. I was going to have an area in the juniors and would have been able to do that my way. But now being able to have consistency across the whole school is really important. Probably the way I like to work too is that I'm happy with people to have input but I like to do it my way as well.

Autonomy and control "to do it my way" gave creative freedoms, latitude, and permission to steer the initiative so that, in her mind, the program was more coordinated. In order to ensure sustainability of the curriculum changes, the leadership decided to have the Assistant Principal undertake the next round of PD. For Hazel this introduced a sounding board without compromising her control: "… I was really happy when the AP came on board in that role. Because I can still drive the content and units the way I wanted to, sounds really selfish doesn't it, but that sounding board was really important."

4.4.2 KIP Summary

Hazel's growth in knowledge through firstly the professional development sessions and as she moved into her new role resulted in shifts in how she perceived her professional self, and new passions and motivations that transformed her work but also enabled new ways of operating in her personal life. Moving from generalist to specialist could be seen as a boundary crossing (Akkerman and Bakker 2011) between two spaces marked by sameness and uniqueness. Unique, to each space, was the dominating teacher knowledge – generalist knowledge of the classroom teacher versus specialized science knowledge of the specialist. The boundary crossing created a need for learning (Hobbs 2013), which has led to an understanding that is aesthetic in nature – an aesthetic understanding of what it means to teach science. It is aesthetic in nature because of the effect it has had on her perception of herself and her view on her world: seeing science everywhere, links to family life, reshaping the way she connected with those around her, and others affirming her newly developing identity.

4.5 Preservice Science Teachers

The second narrative draws on data generated from a research study[2] involving science teacher educators from five universities analyzing their use of school-based delivery of science education units in Bachelor and Masters level teacher education

---

[2] Research funded by the Australian Office for Learning and Teaching. Project reference ID12-4212.

programs (Hobbs et al. 2013). The school-based practice involved preservice science teacher (PST) planning, implementing, and reflecting on a science teaching and learning sequence in local primary schools, over a period of 6–8 weeks. This model of practice provides authentic engagement between the preservice teachers and children, with a high degree of responsibility placed on the PSTs to link theory to practice. The analysis included a number of processes that both evaluated current practice and shaped frameworks and models to inform future practice. Interviews with a range of key stakeholders were used to inform project outputs, including interviews with the preservice teachers (PSTs) at the completion of the program. Interview data from two PSTs were selected to be part of this narrative as they offer differing entry points in terms of interest in science: Erin who came into the unit loving science and Bryony who was apprehensive about teaching science to begin with.

4.5.1 Erin

Erin completed a unit on sustainability with her two partners. In her reflections on a critical moment in her teaching when she discovered something about herself, Erin explained how at school she hated science, but that, on completing her Victorian Certificate of Education (VCE) at a Technical and Further Education (TAFE) institution, she was inspired by Biology – "I actually loved it and I got really into it" – and chose to continue studying it at university when completing her education degree. From there emerged her passion for wanting children to enjoy science:

> So I got a real passion for wanting kids to understand how good science can be because I don't think it's taught enough in primary schools and it's not taught in a way that's interesting for them. There can be a lot of writing or a lot of just learning definitions and things like that so I guess I have a lot of passion for kids to want to learn why science is so interesting and why we should learn it instead of just that it's another thing that we have to learn.

This transformative learning experience at TAFE was a turning point that set a foundation for future experiences. Erin's desire to know the value of science takes on a pedagogical dimension as she draws on her own learning experience to shape the learning of her students. The experience in a sense continued into the current experience of teaching children about biology-related topics:

> Oh I loved science before but I hadn't been so involved in the environmental side. I did Biology but we focused more on Human Biology so I knew more on that side than the environment. But I love that kind of getting involved in what's involved in the world and kids are involved in that every day so it's good for them to learn about it.

This desire to know more was also reflected in how she engaged with the children. Erin referred to an eager student who constantly asked questions and provided answers. Erin recognized him as a fellow lover of science: "it was really good to have another kid who was really engaged in it and wanted to know more."

Erin contrasted her passion for science with that of her peers, who initially "didn't get involved in it." She noticed that the pedagogical motivation that comes from seeing children getting excited through hands-on activities made it "more exciting": "it was fun when we were doing models and things like that where they would come out and they'd all 'ooh aah' about what we were doing and ask questions." The response of the children was fundamental in shaping her positive response to the experience. The children were demonstrably stimulated during the hands-on experiences and gave them thank you cards with comments about "what they'd enjoyed about the experience with us and how they'd learned more and that they thought that it was really cool and hands on and they loved it." Such feedback from the children affirmed Erin's (and her peers') ability to plan engaging and effective learning experiences in science. Erin was proud of their achievements with the unit.

So, while Erin's critical turn towards science preceded this unit, the school-based teaching experience provided an opportunity for shifting her passion for learning science to a passion for teaching science, with affirmation that she can do it effectively. There was continuity across the experiences over time, with prior experiences shaping how she approached and existed in the current experiences.

4.5.2 Bryony

In comparison to Erin, Bryony was less enamored with science to begin with and had never taught science before on her placements. Bryony's critical moment came at the end of the unit when evaluating the quality of the children's learning, and her own learning from the unit. Whereas Erin worked well with her partners to plan a hands-on and well-targeted unit, Bryony felt she and her partner could have been more purposeful in their planning. However despite this realization at the end of their unit when conducting an evaluation of children's learning, she was also surprised at her personal response to teaching science: "I also realized how much I enjoyed teaching science which I was a bit shocked about as well which was exciting because I would really like to do it again."

Similar to Erin, Bryony explained that the children had an influence on her attitudes towards student learning. The teacher had warned Bryony and her partner about one particular child with behavior problems; this child had been kept out of some of the earlier lessons. As a result of this prior warning, the child's actual response to the unit surprised her:

> But he was the one who retained the most information and interest in the topic and I guess that really was a pretty shocking moment because we shouldn't have listened to the teacher who told us about this kid in the first place; but also how a subject can really transform someone's interest and for that kid I guess it must have been science and I'm guessing they don't do a hell of a lot of science at the school. So that could be a real breakthrough for him if he actually gets sent in that direction. So I think that was a pretty good turning point in my mind about student learning.

Bryony's recognition of the transformative possibilities of science for children is highlighted ("how a subject can really transform someone's interest"), but also for Bryony. Preservice teachers bring their prior science learning experiences into these primary science education units, as well as perceptions of the value and complexity of learning science. Research has shown that preservice teachers often see science as something that is difficult to learn (e.g., Lloyd et al. 2000) and even more difficult to teach (e.g., Harlen et al. 1995). Having to "know" the science content can lead to apprehension of teaching it (Howitt 2007). However, for Bryony, this experience of teaching science shifted her view of science and enabled new possibilities to be entertained:

> I have learned how fun science can be and how it doesn't have to be that complicated. I guess I'd always imagined Bunsen Burners and so on from science at high school and I was again quite surprised at how basic it can be, but how deep the learning can be at the same time.

Evident in these reflections by Bryony is a shifting of her view of what constitutes learning and teaching science, marked by a disconnection from prior experiences of science as complex and connection to a form of science learning that had pedagogical meaning; seeing science through the children's eyes, particularly a supposedly difficult student, has had the effect of raising the possibility of engaging children through the context of science. There is therefore connection between this new version of science and her pedagogical commitment to her students.

There was also evidence that the experience has laid new foundations for her future, instilling a desire to "do it again." The question remains, however: will she seek out chances to teach science again? Has this experience been enough – will the transformative possibilities of science learning inspire her to teach science on her placements and when she is a teacher?

4.5.3 KIP Summary

Evidence that an aesthetic experience has taken place for Erin and Bryony includes a shift in what constituted valuable knowledge and making connections between past, current, or future experiences and what is perceived as worth noticing; knowledge and experience; how they see themselves, their role, and capacity to undertake the role; and motivations and passions to continue to pursue future experiences. For example, knowledge that was seen by the preservice teachers as once difficult and unattainable became, for Bryony, accessible and had purpose for meeting a pedagogical imperative and, for Erin, fascinating and worthy of pursuing further both personally and in terms of how it shaped her pedagogical approach.

There were differences in what caused the change. Erin's passion for learning science stemming from her TAFE experience laid the foundation for how she approached the school-based teaching experience and was extended and transformed further through the process of learning more, planning, and implementing learning experiences that aligned with her own experiences of being turned on by hands-on and inquiry learning. This contrasted with Bryony's experience where seeing science through her students' eyes highlighted the potential of science to enable her to excite, inspire, and connect with her students. In accordance with Dewey's perspective, the aesthetic experience is subject to how the individual engages with the experience. But common to both Erin and Bryony was a transformation through this school-based experience. Both exhibited a commitment to science and teaching science in the future, although perhaps this commitment was more personal for Erin.

4.6 The Marine Educator

The discussion in this chapter so far has centered on framing aesthetic understanding within the context of the formal teaching profession, specifically in relation to in-service and preservice teachers. In this section, the focus moves to how experiences that inspire an individual's passions (in this case, a marine educator) ignite their need to know and seek further experience. The third research study was doctoral research investigating marine educators' narratives in order to explore their role in regard to their interactions and relationships between and with policy, community, and education within the marine education context (Kelly 2013). Narrative methodology was used and involved interviews with the marine educators over a period of 14 years to capture the changing stories and directions of the educators. The marine educator selected for the narrative below was formerly a science teacher and then worked as a marine educator in Australia and Canada.

4.6.1 Patrick

During interviews, Patrick related a learning experience that provided an example of an aesthetic experience resulting in an aesthetic understanding. Patrick explained during interviews that he was about seven years of age when he and his father were shoveling sand out of a riverbed and discovered a Christmas beetle. Not knowing what the species was, they sent it off to the museum to have it classified. Patrick reflected on the effect of this experience:

> Looking back on it, it was like the perfect interpretive technique. He did the whole excitement thing and didn't know what it was and ended up putting it in a matchbox and we sent it off to the Museum in Victoria because he was sure this was a new species and they'd name it after me. Very, very cool! and I've still got this image of him, beetle in the redhead matchbox and we never heard back from the Museum, of course. But it didn't matter. The idea that it's possible still that, at some level there's a matchbox deep in the bowels of the Museum that's yet to have my name put on it, is pretty great!

On the surface, finding this beetle may seem simple and uneventful, except that Patrick's father made it eventful by showing excitement and curiosity over the find and building a significant amount of anticipation into the experience.

Patrick sought further experiences that built on this new understanding of the world, such as "hanging out in nature, turning over logs, and throwing sticks at snakes... totally immersed as a child in places that were full of natural stuff, and it happened with a lot of people who got excited about it." These experiences inspired an interest in biology that resulted in Patrick taking the proactive step of seeking out a biology night class at another school when it was not offered at his high school. This led Patrick to complete a biological degree. He then went on to a career in teaching, although Patrick saw his decision to become a teacher as "a bit of an accident...I always wanted to be a ranger."

Working as a teacher, he quickly developed an appreciation of what he wanted – and didn't want – for his students. Patrick described a negative experience during a class visit to an informal education institute, which led Patrick to question the benefits of such learning experiences. He believed that his students had not had a value-added experience and that he could have done the same thing back in his own classroom; Patrick explained that this caused him to directly challenge the institution. In doing so, he began to develop a personal attitude about the role of such informal conservation organizations and the unique position they have in value adding to a visitor's experience:

> ...it reminds me that there is this need in non-school settings that whatever you're providing for people, whether it be students or public, it needs to be something that they can't do for themselves. The organizations, particularly with conservation, they've got to take that extra step to really value-add to the experience and whether it's in a fish tank or back in the real work or in a marine environment, there's got to be this significant value adding, and that's got to be purposeful.

This understanding would later underpin Patrick's personal and professional philosophy and his approach to his work. After completing a Park Management course (and realizing that that was not what he wanted to do after all), Patrick explained that he continued to seek out other opportunities to build on his previous experiences. He found his niche when he undertook a Masters in Ecosystem Management that enabled him to marry his passion for environmental management with his education practice. It was here, he claims, that his interest in utilizing computers and systems to inform environmental management was born and where his personal philosophy of how education sits within that framework initiated:

> (T)hat whole computer end of things and really looking at systems to inform environmental management... And marrying that with my education practice, I've come to see education as one of the really important set of tools for environmental management. That's a small glimpse into my philosophy there.

However, it wasn't until part way through his career as a teacher that Patrick began his professional interest in the marine world. Patrick related a story of how, after taking part in a sand dune revegetation workshop with the Queenscliff Marine Discovery Centre, he jokingly told the leader that he wanted his job. He spoke of how this assisted in his applying for a position at the Centre but that it was his personal enthusiasm and passion for the marine environment that helped him to overcome his lack of any personal in-water experience.

Patrick's childhood experience with the Christmas beetle (described above) and the sense of discovery and curiosity that was inspired as a consequence gave him an appreciation of small things and delight in discovering the unique. This same passion underpins his approach to his work and informs the way he encouraged his staff to interact with visitors to the Aquarium in which he worked:

> I really encourage staff to look for those opportunities and to create those special moments… almost using obscure animals to set up what can be very unique moments of discovery for the participants. Everyone has seen a bloody dolphin. It's a dolphin. Big deal! If you've got something that is possibly brand new or no one knows anything about it, then there's that sense of excitement. Personalizing the experience - I think that's really good!

It is this understanding and the values that are generated from this view of the world that informed Patrick's pedagogical approach and that shapes his sense of identity in the workplace. He took the principles and values he developed while teaching, particularly in regard to the necessity to value-add to a visitor's experience, into his position of manager of an internationally acclaimed aquatic institution. In doing so, he saw his role as creating change by challenging his staff (and others) to reflect on and question the way they approach curriculum development and their pedagogical practice.

> Functionally my role is to stir people up and catapult them into the next step forward. That's my lot in life, my task! Challenge people in terms of their approach to programs. The way they develop curriculum. The way that they develop products and the slant they put on them, and the factors they consider within the design.

4.6.2 KIP Summary

Patrick's experience as a child provided an aesthetic experience that laid the foundations for a way of being in the world that was informed by a deep-seated sense of discovery and curiosity. In doing so it altered the way that Patrick experienced and interacted with the world.

Patrick's pursuit of knowledge enabled him to gain insights into the world and his place within it. His work as a teacher enabled him to develop both a personal pedagogical understanding and a sense of identity as an educator. In following his childhood dream (whether it was, as he claims, a result of "watching too much 'Skippy' as a child" or not) to become a Park Ranger and study Park Management provided him with the opportunity to explore his deep passion for the environment and his need to care for it. In undertaking the Masters in Ecosystem Management, Patrick found that he was able to wed the knowledge that he gained over time through his experiences and studies to create a situation where he was able to combine his identity as an educator and an environmental advocate.

Patrick became passionate about the necessity of value-adding or giving meaning to the experiences that students have when visiting nonschool education facilities – a philosophical understanding that is reflective of his own childhood experience in a nonschool environment. It was this passion that underpinned both his personal and professional life that shaped his sense of identity as a change agent.

4.7 Aesthetic Experiences That Lead to an Aesthetic Understanding: What Is Common?

What were the aesthetic experiences for these science educators above? The aesthetic experience is key to reaching an aesthetic understanding. We contend that an analysis focusing on aesthetic understanding recognizes the relationships between knowledge, identity, and passion – these are very complex constructs in themselves and are useful places to start when analyzing the effect of experiences on participants. What is the nature of an aesthetic experience in educational contexts? It is an experience that triggers the need to know more, an experience that presents as an opportunity for learning and lays the foundation for future learning and experiences (Dewey 1938). Our data has shown that fulfilling one's role requires some professional commitment, but when the professional becomes personal, identities shift, and there is coherence between the parts and a desire to know and experience more, it can be confidently assumed that what catalyzed this change was an aesthetic experience.

For the preservice teacher, Bryony, the aesthetic experience appeared to be seeing science through the children's eyes, borne from her commitment to her students. Teaching science, somewhat successfully, was the way she saw science by shifting her from her own (somewhat fearful) experience of learning science to the positive experience of observing her students learn science. Out of a pedagogical commitment, she began to see science as a rich context for engaging and inspiring children. This shift from science learner to teacher of science was also evident in Erin's narrative, where there was a relatively high degree of self-efficacy arising out of what Bandura (1977) calls "mastery experience," that is, firsthand experience of success. Both were at the stage of still defining their role, with science as one part of what they will need to teach. As a result they were both developing what might be called provisional identities (Ibarra 1999) – trying out the label of "teacher of science" to see if it fits until they have further experience that confirms that it does. So on the one hand, for these preservice teachers, the aesthetic experience is the experience of having to teach science and seeing science through their children's eyes.

On the other hand, learning science content can take on new meaning and value when it becomes personal for the teacher. Beginning to notice science around them illustrates a shift in what is valued and acknowledged as important, not just for their children but also for themselves. This is particular relevant to science education where preservice teachers often do not have the opportunity to teach science or see science being taught before they enter teaching (Howitt 2007). This was also evident in Hazel's narrative. For Hazel, the experience is part of her professional role as science specialist – having to take on the role of teaching science was an expectation. The experience became personal and aesthetic in nature when there was a shift in her perception of the nature of her role, from teacher who has to teach science to teacher who loves to learn science and help others to teach it. The aesthetic experience is a process, similar to the preservice teachers, where, over time, there is a shift in the way Hazel notices and perceives of science and the way she sees herself in relation to science has changed. She begins to see connections between science and her everyday and professional life. Key to this change has been the professional development introducing knowledge, followed by planning, enacting, and leading changes to science teaching in her school. Affirmation from peers, outcomes such as student learning and students' excitement towards science are salient in fostering a strong sense of herself as someone who can lead change in the school. More significant though has been the new ways she sees science, the effect being a transformed understanding of what it means to be a teacher.

Another key outcome of an aesthetic experience is the continuity across experiences and how they lay a foundation for future experience. This was evident in Erin's TAFE experience of learning science, and how it switched her onto science in a way that shaped her approach to the school-based experience. This continuity was even more pronounced in the marine educator's narrative. Patrick's early aesthetic experiences as a child resulted in a desire to seek out further knowledge and experiences that built on his fledgling understanding of the natural world. He was able to move along the continuity of experience by seeking out further knowledge through academic courses he chose to undertake and by establishing connections with the environment through his recreational pursuits, for instance, revegetating sand dunes. These aesthetic understandings led to a world view that informed both his personal and professional attitudes, as well as Patrick's approach to his work.

The aesthetic dimension of teaching is fundamental to the way we think of the science educator. Teaching varies across subjects because the subject matter differs. But the teacher's aesthetic understanding of what it means to be a science educator is based on more than content knowledge. The aesthetic dimension of teaching is fundamental to how teachers develop an appreciation for the subject they teach as in the case of the preservice teachers and Hazel, respond to the pressures of change as in the case of Hazel, and make decisions about how to maximize impact through education as was the case of Patrick.

4.7.1 Passion for the Work of Educating

We have showcased here the experiences of a primary teacher, science preservice primary teachers, and a secondary science teacher who became a marine educator. Represented across the narratives are trajectories involving shifts between science learner, preservice teacher, in-service teacher, and passionate educator for the environment. There are challenges for science education at each of these points. For the science learner, a general drift from science as a subject of choice (Chubb 2014) is being positioned as threatening society's ability to be innovative and competitive in the twenty-first century. Both the preservice teachers indicated an early lack of interest in science, and this is not uncommon of those entering primary education programs in Australia (Marginson et al. 2013). Disconnection of the school versions of science from children's everyday lives has been shown to disengage students (Tytler 2007). The analysis here has shown that a learner's interests and passions provide motivation in learning, thereby reflecting the compelling and dramatic nature of understanding (Girod et al. 2003). It is possible to draw on students' interest and acknowledge what motivates them in life and within the learning experience. This may come in the form of appreciating the beauty of disciplinary ideas and modes of inquiry or capturing the elegance and personal satisfaction involved in solving problems (Hobbs and Davis 2013). For preservice teachers who have limited interest in science, the possibility of "seeing science through their children's eyes" raises the pedagogical value of science. For in-service teachers who take on an expanding role such as a science specialist, a passion for learning science is driven by and drives both pedagogical change and a personal interest in the subject. Darby (2009) highlighted two imperatives driving practice: "pedagogical imperative" where a teacher's passion for their students drives learning, such as was demonstrated by Bryony, and a "personal imperative" where passion for the subject drives practice (see Fig. 4.2), which became the driving force for Hazel and which drove Patrick to move into educational roles where he could raise awareness to

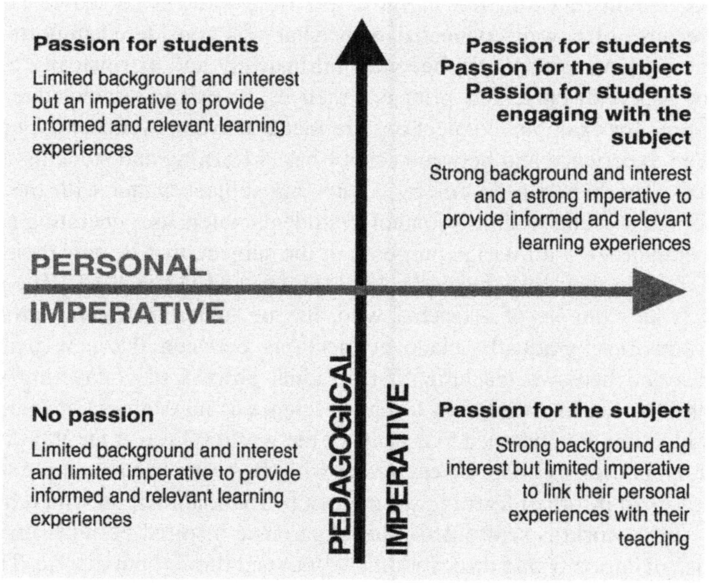

Fig. 4.2 Passion model (from Darby 2009)

environmental concerns. "When we view teachers as passionate beings we unleash the possibility for them to embrace innovation, and to be desirous in their dealings with students so that they seduce students into caring about the subject" (Hobbs 2012, p. 727).

Missing from Fig. 4.2, however, is a passion for the art of teaching, a "teaching imperative." Rubin (1985) depicts the artistry of teaching as being more than motivation and dramatization: "It is an extraordinary level of performance, bred out of personal commitment which elevates the state of the art" (p. 159). According to Eisner (1979), the experience of teaching is essentially aesthetic due to this artistic quality: "teaching is an art in the sense that teaching can be performed with such skill and grace that for the student as well as the teacher, the experience can be justifiably characterized as aesthetic" (p. 153). This affective part of teaching is critical for dealing with the aforementioned issues in science education.

4.7.2 A Connection with the Knowledge and Pedagogy of Science

If passion is the underpinning of a good science educator, then a connection with the knowledge of pedagogy and science is the substance. Such connection can be lacking for primary teachers; research has shown, in fact, that a lack of science pedagogical content knowledge can result in primary teachers avoiding teaching science or adapting inappropriate teaching strategies from other disciplines (Appleton 2003). As demonstrated in the narratives, a little knowledge led to substantial change because of how it connected with what was considered important. In an aesthetic experience knowledge becomes intrinsically and extrinsically connected with some aspect of ourselves, prior experiences, or existing knowledge. For science learners, for example, connections are made between events and ideas within the learning experience and between school-based learning and students' lifeworld experiences. Teachers can do this by connecting subject matter with the personal experiences of students, relating content to students' interests, generating new interests, and emphasizing utilitarian purposes of the subject, that is, how their learning will help them in their lives beyond school (Hobbs and Davis 2013). The narrative of Hazel is an example of a teacher who, having undertaken some professional learning activities, gradually made connections between the new pedagogical approaches and her own teaching. This gradual process of connecting with and expanding her own knowledge of teaching science is an example of learning that brings unification or coherence to aspects of the world (Girod et al. 2003).

Similarly, Patrick's childhood encounter with the beetle is an example of how an experience in an informal learning setting can be foundational on which his understanding of the world is scaffolded. This experience inspired both his imagination and a sense of curiosity and discovery that continued throughout his life. This experience formed a foundling connection with small and unique animals and laid the foundations for a desire to seek out future opportunities for similar experiences. The resultant connections that were formed with the natural environment initiated a deep understanding that inspired a strong emotional response. Through these experiences, Patrick's view of the world was altered, and, as a result, so were his subsequent experiences, until new habits or ways of seeing and dealing with the world became firmly established as part of his psyche and value system. As Patrick's understanding of the world around him altered through his experiences, so did his personal sense of identity.

4.7.3 An Identity That Is Developing

So where passion is the underpinning, connection with the knowledge is the substance, and then a perceived transformation of oneself and the world is what holds a teacher together. The process of becoming a subject teacher is aesthetic in nature in two ways. The notion of "becoming" refers to developing an identity based on a confidence that the qualities of one's teaching and the nature of one's assumptions about teaching are appropriate and suited to being a teacher of the subject.

The notion of "becoming" also refers to a sense of attraction. The teacher is attracted to what the subject has to offer them and their students. But also teachers can become attracted to the pedagogical benefits associated with teaching as the children display passion in their learning, as was seen with Bryony. Identity is constructed and morphed through experience, and this becomes aesthetic when there is transformation of the person and how they perceive of the world. Through such experiences and beyond, there is a re-storying of who they are and the type of person/learner they are and want to be. There is recognition that disciplinary knowledge has a place in their lives, allowing human experience to enter the learning process, thereby situating the story within the lifeworld of the student.

Preparing teachers to teach a subject becomes a process of not only building their knowledge of content and pedagogy and assisting them in developing pedagogical content knowledge. Nor is it simply enculturating them into the ways, traditions, beliefs, and practices associated with the subject. But it is also a process of "becoming" where teachers increasingly see themselves in relation to subject matter and it's teaching. The school-based teaching experience and the science specialist program allowed the teachers to "try out" what it feels like to be a science teacher – to know the content and pedagogy, the role as it is defined by the task and as affirmed by others, and being compelled to engage deeply and seek further experiences. This trying out also involves developing "provisional selves," which, according to Ibarra, are "temporary solutions people use to bridge the gap between their current capacities and self-conceptions and the representations they hold about what attitudes and behaviors are expected in the new role" (Ibarra 1999, p. 765). The teachers are in the process of constructing a worldview within which they can situate themselves. This was particularly evident for Hazel who was initially experimenting with a new provisional "science specialist" self and then, gradually, overtime was able to situate herself into the role with increasing responsibility, affirmation by peers, and greater confidence in her knowledge and abilities.

4.8 Practical Applications of the Aesthetic Understanding Framework

Aesthetic understanding is a powerful lens for linking knowledge, identity, and passion, both to support reflection on practice and to inform the structure of learning experiences. For the four educators of science represented here, learning was underscored by an experience that had jolted them from one way of looking at the world and themselves: for Erin and Bryony, a new appreciation for science and the potential it has for students having seen science through the eyes of her students; for Hazel, opening up a scientific worldview that has potential in shaping her ongoing teaching career as a change agent and advocate for science; and for Patrick, an experience that set the course for his entire life.

Aesthetic understanding can inform teaching and policy in a number of ways. Student learning experience designed for deep meaning and aesthetic response is perhaps the most typical application (see, e.g., Girod et al. 2003). Aesthetic understanding can be seen to underpin transformative learning. As promoted by Mezirow (1991), transformative learning can be psychological (changes in understanding of the self), convictional (revision of belief systems), and behavioral (changes in lifestyle). Darby-Hobbs (2013) describes transformative experience as leading to "greater potential for students to say 'I am the type of person that can appreciate this', 'This is part of my future', and ultimately see that 'I am part of the solution'" (Darby-Hobbs 2013, p. 94). Certainly for Patrick and Hazel, there was transformation of self and perspectives on and within their world. Essentially, aesthetic understanding has the potential to examine what lies behind the actions of passionate educators. In addition, as illustrated through Hazel's narrative, aesthetic understanding enables interrogation of teachers' personal response to innovation and change.

For science educators, creating conditions for aesthetic experiences involves firstly identifying not only students' prior knowledge but also students' attitudes and identity in relation to science. Then knowing how to ignite students' passions and motivation is the next challenge; tapping into students' interests, and generating new interests through challenge, such as open investigations, can disrupt standard perceptions of science and their relationship to it (Tytler 2007). This notion of disruption could characterize the preservice teachers' context, where the school-based teaching experience disrupted previously held views of science (for Bryony) and provided the conditions for new interests to be generated (for both). Teaching involves creating conditions for an aesthetic experience that leads to transformation and then moving students forward in their journey by exploring things further and making them aware of the opportunities for future experiences. A continuum of experience becomes the focus of instruction rather than discrete and disconnected learning experiences.

For teacher learners, such as preservice teachers, aesthetic understanding provides the basis of a tool to support reflection on changes in knowledge, identity, and motivations and the relationships between these. Questions that can prompt reflection include:

- What do you know (focusing on the developing connections between knowledge of science, teaching, and learning)?
- What drives you in your teaching (focusing on developing interests in the subject, engaging with children, and the practiced art of teaching)?
- How do you see yourself as a teacher of science (focusing on developing a professional identity that is social historically constructed through experience and embracing what is possible as a provisional identity)?

Teachers' stories of science are located to a place and time, and such stories are constituents of their professional identity (Goodson 1997). It can be valuable to assist preservice teachers to connect science to their previous experiences with science in informal and social contexts, rather than simply focusing on the formal education experiences, which can often be quite negative. Patrick's narrative shows the power of such informal experiences. Perhaps a barrier for preservice teachers and students is being able to recognize the science in their everyday experiences.

Further, it is possible to see the potential for the aesthetic understanding framework to inform the structure, content, and approach of teacher continuing professional development. For example, a primary science specialist program that involves the presentation of new knowledge, which is then integrated into practice of a new role, can lead to identity shifts and motivation to continue in this role. Bredeson (2002) uses a framework from architecture to describe professional development that attends to the function, structure, and beauty of professional development programs. He states that "beauty comes from the artistic arrangement and use of materials and systems to create learning spaces that engage teachers and administrators in growth opportunities that meet their needs and change them as people and professionals" (p. 667). He refers to the "artful designs for learning" (p. 667), the "hoped for result" in terms of the interactions between teachers, provision of spaces and processes to reflect on practice, and capacity building that leads to substantial change in practice. Ultimately, he states that "beauty in professional development may be expressed in enhanced motivation, positive emotions, and renewed feelings of empowerment" (p. 667). Hazel's experience through the primary science specialist program appears to be achieving these outcomes.

The framework can also be useful in describing teacher change processes; for example, a focus on the compelling nature of experiences, and the unity that arises, can help to identify which elements of an intervention result in real transformation of identity and practice, but also the challenges that teacher change can involve. Simmons et al. (2008) suggest that the "self lies at the center of teachers' interpretations, explanations, and understanding expressed through their beliefs and classroom actions" (p. 948). As teachers adapt to different educational environments, they construct their knowledge and beliefs "from the perspectives of self-in-relation-to-social context" (p. 948). Expectations placed on teachers by school change processes or professional development programs require teachers to adapt to different educational environments. Simmons et al. explain that "how the environment in which one functions, especially with regard to the expectations of others, contributes to teachers modifying their actions and eventually their beliefs" (p. 932). This was particularly evident for Hazel who, when becoming the sole specialist in her school, flourished because of an expectation of increased autonomy in directing the program. Facing imposed change, a once confident teacher, who felt that they had an aesthetic understanding of what it means to teach the subject, is forced to reconsider whether their passions and commitments are relevant and useful, whether their understandings of what and how to teach are still coherent, and, therefore, whether they can perceive of themselves in the same way in relation to the subject.

Most importantly, the framework is valuable in raising attention to the value of aesthetic experiences in shifting science learners back onto science. For example, Erin's experience of science at TAFE changed her attitude to science. It is a reminder for policy makers and curriculum developers to focus not only on the content of science, but that true learning comes when students can identify with the content and can see a place for themselves in the ongoing narrative of the subject and its possible applications to other parts of their lives, for example, into Erin's teaching.

4.9 Methodological Possibilities and Limitations

As a methodology for studying teachers and teaching, the aesthetic understanding framework has a number of benefits. This framework suits research focusing on the aesthetic dimensions of being a teacher: "finding out who teachers are, what they do, and how they know is crucial to understanding the action world of contemporary teaching" (Black 2002).

Using the aesthetic understanding framework, Avraamidou (2015) identifies the implications that aesthetic understanding from Hobbs (2012) has for educational theory, calling for "an emphasis on the affective domains of learning and on the central role of emotions when conceptualizing science teacher identity" (p. 17). In her own research into science teachers, she noted that "participants' desires, passions and emotions in relation to science shifted their identity work" (p. 17). Similarly, by recognizing the cognitive and aesthetic dimensions of teacher identity development, an analysis by McIntyre and Hobson (2015) highlighted the important role that mentors can play in inspiring and fostering beginning teachers' passion for the subject "who initially lack confidence in that subject, and that this helps them to identify both with the subject and with themselves as teachers of that subject" (p. 13). In doing so, McIntyre and Hobson reinstates content knowledge back into our thinking about teacher identity development and our conceptualization of teaching and teachers.

A unique characteristic of aesthetic understating is the positioning of passion in relation to identity and knowledge. When passion is the focus of analysis, the framework enables close attention to teachers' desires and passions for their students. Burgess (2014), for example, drew on aesthetic understanding when describing history teachers who are passionate about improving educational outcomes for aboriginal students.

Another characteristic of the framework that has had traction within the literature is the notion of connection; Anderson and Risner (2013), for example, used the aesthetic dimensions of teaching to interpret the work of arts educator, referring to the transformative and connective nature of artistic work.

One of the limitations of the aesthetic understanding framework that has been noted in the literature is that the analysis focuses on the individual in response to experiences but potentially takes little account of the broader context of the teacher (Hsieh 2015). While passion and knowledge are related to identity, the framework does not immediately situate the teacher in context, in particular into their social context (Bennison 2015). A teacher's orientation to their practice is potentially missing from the framework (Hsieh 2015). Hsieh represented the aesthetic understanding framework as offering discrete areas that influence teacher identity development. On the contrary, a KIP analysis highlights an intersection of many factors, especially when the aesthetic experience, the nature of that experience, and the onward influence of that experience on identity development are part of the analysis.

It may also be important to take account of the range of identities (Wenger 1998) that a teacher carries with them and to recognize how these identities are shaped by their many experiences, not just those that might be labeled as aesthetic in nature (Bennison 2015). We hope that the second "moment" for research presented in this chapter provides scope to incorporate relationships between changing context and the development of competing identities (e.g., in the case of Patrick's teacher versus environmental manager identities) and how the developing knowledge, identities, and passions shape action.

4.10 Conclusions

According to Girod and Wong (2002), educative experiences are much more than simply events that occur. Instead they possess a forward movement that has a degree of unity of its elements that freely flow "without seam and without unfilled blanks, into what ensues" (Dewey 1934, p. 36). In doing so, these experiences take on a sense of possibility or of anticipation of what could be and of how things could develop and unify, thereby becoming "an act of thinking and meaning" (Girod and Wong 2002, p. 203). Learning, therefore, becomes something to be experienced. It is at moments such as this that emotion, cognition, and action are fused creating a deep and memorable experience that is meaningful for the participating individual – a dramatic experience. "This renewed seeing, if you will, provides drama to ordinary events and interactions and compels us into further engagement with the world" (Girod et al. 2003, p. 579).

Working on the assumption that the key to quality education is passionate and knowledgeable educators, it makes sense to recognize in policy, practice, and research the significance of teacher knowledge, identity, and passion in relation to the roles they find themselves in. In keeping with a Deweyan perspective on aesthetics, the affective elements are entwined with the cognitive experience. The result of such an experience is not just empowerment, but also transformation of one's identity. The transformative nature of gaining knowledge is a consequence of an affective response. In his analysis of McWilliam's pedagogy of desire, Zembylas (2007) uses a Deleuzo-Guattarian perspective on productive desire. Here, desire is seen as:

> an "immanent principle" of creativity and movement [that] enables a new view on affect that does not assume simple feelings but immanent becomings (Deleuze and Guatteri, 1994). In this manner affect in education may be redefined as a landscape of becoming in which forces, surfaces and flows of teachers/students are caught up in a desiring ontology and consequently, a pedagogy of desire is explored as a transformative practice. (p. 332)

When we view teachers as passionate beings, we unleash the possibility for them to embrace innovation and to be desirous in their dealings with students so that they seduce students into caring about the subject. Teacher preparation, professional development, and teacher development are dealing with not only issues of content and pedagogy but also issues of identity, passion, and seduction. On this basis, it is fair to conclude that the process of becoming a teacher of a subject is essentially aesthetic in nature because it is fundamentally about transformative experience. It stands to reason then that an aesthetic framework is needed to capture the many nuances in teacher growth, shifting commitments, and the catalytic influence of aesthetic experience.

References

Akkerman, S. F., & Bakker, A. (2011). Boundary crossing and boundary objects. *Review of Educational Research, 81*(2), 132–169.

Anderson, M. E., & Risner, D. (2013). "Art life in action": Case study of a teaching artist in theatre. *Teaching Artist Journal, 11*(3), 139–146.

Appleton, K. (2003). How do beginning primary school teachers cope with science? Toward an understanding of science teaching practice. *Research in Science Education, 33*(1), 1–25.

Australian Academy of Science (AAS). (2014). *Primary connections*. Canberra: AAS.

Avraamidou, L. (2015). *Stories of self and science: Preservice elementary teachers' identity wok through time and across contexts*. Pedagogies: An International Journal. doi:10.1080/15544 80X.2015.1047837.

Bandura, A. (1977). Self-efficacy: Toward a unifying theory of behavioural change. *Psychological Review, 84*(2), 191–215.

Bennison, A. (2015). Developing an analytic lens for investigating identity as an embedder-of-numeracy. *Mathematics Education Research Journal, 27*, 1–19.

Black, A. (2002). Making sense of what it means to teach: Artful representations as meaning-making tools. *Teacher Development, 6*(1), 75–88.

Bredeson, P. V. (2002). The architecture of professional development: Materials, messages and meaning. *International Journal of Educational Research, 37*, 661–675.

Burgess, C. M. (2014). *Shifting sands: The narrative construction of early career aboriginal teachers' professional identities at the cultural interface*. Unpublished doctoral dissertation thesis, University of Sydney, Sydney, Australia.

Campbell, C., & Chittleborough, G. (2014). The "new" science specialists: Promoting and improving the teaching of science in primary schools. *Teaching Science, 60*(1), 19–29.
Chubb, I. (2014). *Science, technology, engineering and mathematics: Australia's future*. Canberra: Australian Government.
Darby, L. (2005). Science students' perceptions of engaging pedagogy. *Research in Science Education, 35*, 425–445.
Darby, L. (2009). Translating a "relevance imperative" into junior secondary mathematics and science. *Eurasia Journal of Mathematics Science and Technology Education, 5*(3), 277–288.
Darby-Hobbs, L. (2013). Responding to a relevance imperative in school science and mathematics: Humanising the curriculum through story. *Research in Science Education, 43*(1), 77–97.
Darling-Hammond, L. (2006). Constructing 21st century teacher education. *Journal of Teacher Education, 57*(3), 300–314.
Deleuze, G., & Guattari, F. (1994). *What is philosophy?* New York: Columbia University Press.
Dewey, J. (1934/1980). *Art as experience*. New York: Berkley.
Dewey, J. (1938). *Experience and education*. New York: Touchstone.
Eisner, E. W. (1979). *The educational imagination: On the design and evaluation of school programs*. New York: Macmillan Publishing Company.
Girod, M. (2001). Nobody likes soap in their eyes: Portraying a more inviting science by teaching for aesthetic understanding. *CESI Science, 34*(2), 20–24.
Girod, M., & Wong, D. (2002). An aesthetic (Deweyan) perspective on science learning: Case studies of three fourth graders. *The Elementary School Journal, 102*(3), 199–224.
Girod, M., Rau, C., & Schepige, A. (2003). Appreciating the beauty of science ideas: Teaching for aesthetic understanding. *Science Education, 87*, 574–587.
Goodson, I. F. (1997). Representing teachers. *Teaching and Teacher Education, 13*(1), 111–117.
Harlen, W., Holroyd, C., & Byrne, M. (1995). *Confidence and understanding in teaching science and technology in primary schools* (SCRE research report 65). Edinburgh: Scottish Council for Research in Education.
Hobbs, L. (2012). Examining the aesthetic dimensions of teaching: Relationships between teacher knowledge, identity and passion. *Teaching and Teacher Education, 28*, 718–727.
Hobbs, L. (2013). Boundary crossings of out-of-field teachers: Locating learning possibilities amid disruption. In J. Langan-Fox & C. L. Cooper (Eds.), *Boundary-spanning in organizations: Network, influence, and conflict* (pp. 7–28). New York: Routledge.
Hobbs, L., & Davis, R. (2013). Establishing connection through narrative in Mathematics, Science & Technology. *Research in Science Education, 43*(3), 1289–1305.
Hobbs, L., Jones, M., King, J., Chittleborough, G., Redman, C., Campbell, C., Kenny, J., & Herbert, S. (2013). *Science Teacher Educator Partnerships with Schools (STEPS): Developing an Interpretive Framework for Primary Science Teacher Education.* Paper presented to the Contemporary Approaches to Research in Mathematics, Science, Health and Environmental Education, Deakin University Melbourne, 28–29 November, 2013. Accessed at http://www.deakin.edu.au/research/src/crefi/events/car-2013.php#papers
Howitt, C. (2007). Pre-service elementary teachers' perceptions of factors in an holistic methods course influencing their confidence in teaching science. *Research in Science Education, 37*(1), 41–58.
Hsieh, B. (2015). The importance of orientation: Implications of professional identity on classroom practice and for professional learning. *Teachers and Teaching: Theory and Practice, 21*(2), 178–190.
Ibarra, H. (1999). Provisional selves: Experimenting with image and identity in professional adaptation. *Administrative Science Quarterly, 44*, 764–791.

Jakobson, B., & Wickman, P. (2007). The roles of aesthetic experience in elementary school science. *Research in Science Education, 38*, 45–65.
Kelly, L. (2013). *Marine educators: Linking personal commitment, education and public policy*. Unpublished doctoral dissertation thesis, Deakin University, Waurn Ponds, Australia.
Lloyd, K. L., Braund, M., Crebbi, C., & Phipps, R. (2000). Primary teachers' confidence about and understanding of process skills. *Teacher Development, 4*(3), 353–370.
Marginson, S., Tytler, R., Freeman, B., & Roberts, K. (2013). *STEM: Country comparisons: International comparisons of science, technology, engineering and mathematics (STEM) education*. Final report. Australian Council of Learned Academies, Melbourne, Vic.
McIntyre, J., & Hobson, A. J. (2015). Supporting beginner teacher identity development: External mentors and the third space. *Research Papers in Education*. doi:10.1080/02671522.2015.1015438.
Mezirow, J. (1991). *Transformative dimensions of adult learning*. San Francisco: Jossey-Bass.
Pugh, K. J., & Girod, M. (2007). Science, art, and experience: Constructing a science pedagogy from Dewey's aesthetics. *Journal of Science Teacher Education, 18*, 9–27.
Rubin, L. J. (1985). *Artistry in teaching*. New York: Random House.
Simmons, P., Emory, A., Carter, T., Coker, T., Finnega, B., Crockett, D., et al. (2008). Beginning teachers: Beliefs and classroom actions. *Journal of Research in Science Teaching, 36*(8), 930–954.
Tytler, R. (2007). *Re-imagining science education: Engaging students in science for Australia's future*. Camberwell: Australian Council for Educational Research.
Wenger, E. (1998). *Communities of practice: Learning, meaning and identity*. Cambridge: Cambridge University Press.
Wickman, P. O. (2006). *Aesthetic experience in science education: Learning and meaning making as situated talk and action*. Mahwah: Lawrence Erlbaum Associates, Inc.
Zembylas, M. (2007). Risks and pleasures: A Deleuzo-Guattarian pedagogy of desire in education. *British Educational Research Journal, 33*(3), 331–347.

Linda Hobbs is a Senior Lecturer in Science Education at Deakin University, where she teaches primary science education. She has a range of research projects, which mainly focus on the teacher. Her latest research is in exploring the issue of teachers teaching across subject boundaries, and the implications for teacher learning, identity, and support needs. Another research agenda of hers is exploring the efficacy of school based approaches to primary science education.

Leissa Kelly has recently completed a PhD in marine education working with teachers and educators in the marine environment. Leissa has worked as a science educator in a number of formal and non-formal science education settings and facilities, including schools, universities, zoological gardens and parks, and hospitals. Leissa is currently project officer on a number of STEM-related projects forging links between education researchers, scientists and school teachers and students.

Part II
Emotion and Affect in Science Education

Chapter 5
Interaction Ritual Approaches to Emotion and Cognition in Science Learning Experiences

Alberto Bellocchi

There is ongoing concern internationally about student disaffection with science, technology, engineering, and mathematics (STEM) disciplines. Research focusing on student engagement with STEM disciplines has offered one approach for addressing this concern (e.g., Olitsky and Milne 2012). In this context, science educators such as Stacy Olitsky and Catherine Milne (2012) have sought to study commensurate sociological and psychological constructs of emotional, behavioral, and cognitive engagement in an attempt to understand and ameliorate student disaffection with school science. Paralleling this focus on engagement has been the perceived disconnection between emotion and cognition in studies of science education that have tended to focus predominantly on the latter in the past (Alsop and Watts 2003; Fortus 2014).

Approaches to science education research that draw on perspectives from the sociology of emotion and interaction ritual theory (IRT) may hold one of the keys to understanding and addressing student disaffection. This is due to the emphasis on the interrelationships between social practices, emotion, and knowledge construction made possible by these theoretical traditions. These approaches also offer scope for addressing the study of emotion and cognition in integrated ways.

The remainder of the chapter is divided into three major parts. In Part 1, I present a discussion of studies offering a foundation for understanding science learning through the sociology of emotion and interaction rituals. I will discuss Randall Collins' (2004) theory of interaction ritual chains, which brings together emotion with social practices at a microsociological level. This theory has been adopted in an increasing number of science education studies that have yielded fruitful understandings about what it means to learn science in school settings (e.g., Milne and

A. Bellocchi (✉)
Faculty of Education, Queensland University of Technology,
Brisbane, QLD, Australia
e-mail: alberto.bellocchi@qut.edu.au

A. Bellocchi et al. (eds.), *Exploring Emotions, Aesthetics and Wellbeing in Science Education Research*, Cultural Studies of Science Education 13,
DOI 10.1007/978-3-319-43353-0_5

Otieno 2007) and what it means to become a science teacher (e.g., Rinchen et al. 2016). In Part 2, I trace the roots of interaction ritual theory to its origins in Durkheim's (1915/1964) social epistemology of knowledge and sociology of religion. My elaboration of Durkheim's social epistemology of knowledge paves the way for my final argument for a microsociology of learning in Part 3.

5.1 Part 1: Sociology of Emotion and Interaction Rituals for Science Education Research

Studies in science education adopting theoretical perspectives from interaction ritual theory and sociology of emotion presented in Part 1 of this chapter reveal common views about learning science or learning to become a science teacher as socially situated practices (e.g., Olitsky 2007). Within these perspectives, embodied actions, emotions, and cognitions are treated as interrelated phenomena. Learning science thereby involves a process of growing participation in group activities focused on common science practices, science/science teaching objects, symbols, and ideas (Olitsky 2007; Bellocchi et al. 2014). These elements are important to the wider science community (i.e., scientists, curriculum bodies, educators) because they represent canonical knowledge and practices within scientific disciplines. From an interaction ritual perspective, however, the same elements must gain salience within small groups such as school or university classes, if learning experiences are to be successful. In addition to adopting accepted practices and symbols, the creation of science symbols, ideas, practices, and objects is another important aspect of learning science from this perspective. Evidence of learning in the studies I will discuss in Sect. 5.1.2 comes from observations of growing or ongoing participation by individuals and groups in interactions focused on science practices, objects, ideas, or symbols. As a socially situated activity, science learning takes place in specific contexts (e.g., school classrooms, outdoors, teacher education classes), at particular times, with the possible inclusion of specific artifacts, and within particular physical environments (school buildings, universities, museums, outdoors, laboratories). All of these elements are important in shaping learning experiences. Before expanding this initial synopsis through detailed elaboration of selected studies, I present an explanation of IRT next.

5.1.1 Interaction Ritual Theory

Interaction rituals are focused encounters during which people are initially involved in a collective action or event (Collins 2004). During this involvement, some transient emotional stimulus forms an initial ingredient for directing attention to the actions or event. As individuals find themselves under these conditions, they are primed for four ritual ingredients including the formation of a group through bodily copresence, some sort of delineation between who is a member of the group and who is not, a common focus for visual and cognitive attention, and a common mood among the group. Through a feedback system, the common mood and mutual focus of attention intensify one another. As such, an individual becomes more focused on the action or event as he or she develops a common mood with others. Greater focus of attention in the group parallels higher levels of attunement to the sentiments of others, and further intensification of a common mood. Eventually a heightened level of emotional arousal within the group may be achieved, which is called *collective effervescence*. This state of elevated group emotion spills over into four possible ritual outcomes, which include a sense of membership to the group (i.e., solidarity), heightened emotional energy within individuals, some representations (e.g., symbols) of the social group, and feelings of morality in adhering to the group, it's practices, symbols, and ideas.

Throughout this chapter, I will use the terms *ritual* or *interaction ritual* interchangeably with the phrases *focused encounter* and *focused interaction*. In the sociological sense adopted here, a ritual involves a communion between participants of an encounter resulting in a set of common thoughts and feelings that bind people together in a group (Collins 2004). Because of this theoretical attention on the common emotional and cognitive awareness within a group of people during shared practices, and due to the need for people to come together to achieve a shared focus on an action or event, the unit of analysis proposed originally by Erving Goffman (1967) was the *encounter*. Collins also takes this up as the unit of analysis in his theory. Instead of considering the individual as a *being* with fixed properties that is able to move across social situations without changing, for Collins, the individual is reconstituted into something potentially different with each encounter he or she traverses. This is an important analytical focus particularly in education research when we consider, for example, that students can be seen by some researchers or teachers to be a particular *type* of student based on fixed traits (e.g., a *student with misconceptions*). An alternative view, through IRT, is to understand the individual as someone who is constructed as one who possesses a misconception *during* moment-to-moment exchanges unfolding in classroom encounters. In this case, a *misconception* is an attribute of the social practices that evolve over time during classroom interactions. The misconception exists between the student and the teacher or other students as a property of the interaction, not as a property of the individual student's mind. This opening discussion of IRT is elaborated further through a review of empirical studies in science education.

5.1.2 Empirical Studies of IRT in Science Education

In one study focusing on student interest and engagement in science, Stacy Olitsky (2007) explored the ways in which 8th grade students in one US classroom developed feelings of membership to the *science class group*. Drawing on IRT, Olitsky

(2007) conceptualized science learning as a process of growing participation in the practices and sentiments of a science class (i.e., the *group*). Membership to the group was characterized by the development of a common mood among interaction participants who shared a visual and cognitive focus directed at science ideas and materials (e.g., laboratory equipment, science demonstrations, science concepts).

In her study, elements of successful interaction rituals were observed when iconic sports personalities, popular movies, or television shows were incorporated into physics instruction. All of these elements were familiar to the students as symbols imbued with emotions in contexts separate to the science class, such as television shows or during football games. When these elements were introduced into a physics lesson, they became familiar symbols that were used to access unfamiliar physics practices and ideas. In doing so, students gained access to physics and formed common feelings about physics ideas and practices that generated sentiments of membership to the physics class. This was made possible by the transfer of emotion from one symbol (i.e., sports icon) to another (i.e., physics concepts), a process that Durkheim (1915/1964) calls *contagion* (see also Collins 2004).

In contrast, when canonical science concepts were used solely for instruction in Olitsky's study, there was less mutual noisemaking in the class in the form of laughter, there were lower levels of synchrony in body movements, there was less attention focused on the physics ideas, and fewer students contributed to class discussions. All of these interactional phenomena provided evidence that the rituals had failed to produce a common emotional experience as predicted by IRT (Collins 2004). On a different occasion, however, when the instruction was focused on students solving chemical equation problems, entrainment through shared noisemaking—that is, collective emotion—was achieved. As students attempted the problems on the board, others shouted suggestions to assist them in solving the problem. The class became united in their efforts to assist the student working at the board. Although the chemical equations and the balancing equation problems consisted of symbols and practices that were not experienced typically by these students outside of school, they became resources for establishing a common cognitive, visual, and emotional focus. An important outcome was that the feelings of entrainment and emotional energy during the balancing equations practices were repeated over time indicating that membership in the science class group and engagement with the topic was sustained. This episode of classroom interactions illustrated that successful interaction rituals can develop without the inclusion of other symbols (e.g., sports stars) with which students are familiar in nonscience contexts. What we can understand from this balancing equations episode is that learning science involves the attachment of group emotion to scientific symbols (i.e., formulae of chemical elements/compounds). The attachment of emotion to science symbols is mediated through coordinated group practices such as answering a science problem.

An important distinction was evident in the balancing equations episode when compared to other classroom instruction involving the initiate-respond-evaluate structure of interactions. In an initiate-respond-evaluate episode during the physics topic, one student who possessed the right kind of cultural capital (e.g., used scientific terminology correctly) gained higher status in the group by answering teacher-initiated questions correctly. Olitsky observed this during a physics lesson in which the teacher had greater confidence with the concepts than she did when compared to the earlier balancing equations example. Due to her confidence with physics, the teacher tended to ask known-answer questions relegating the students' role during interactions to one of providing correct or incorrect responses. This practice maintained the typical classroom power structure where teachers have the higher status, and higher emotional energy, than students due to possession of physics concepts (i.e., valued symbols in science class). Students who can answer questions correctly gain higher status and higher emotional energy for future interactions over other students when the teacher confirms their correct answers to questions. In contrast, the classroom interactions were dialogic during the balancing equations episode as students and the teacher contributed ideas collectively. This meant that students developed enhanced feelings of participation in the equation lesson as no single student held privileged knowledge, and neither did the teacher.

What the equation lesson illustrates is that it is possible for students to form feelings of group membership during science practices without the need for other discourses (e.g., football) with which they may be more familiar and with which they had previously formed emotional connections in other contexts. As shown through these examples from Olitsky's study, IRT provides a theoretical perspective for investigating the connections between collective emotions, shared symbols, and practices in the forms of physics and chemistry concepts and practices such as balancing equations. In doing so, the theory can inform research that seeks to explore the interrelationships between emotion, cognition, and embodied practices.

More direct exploration of the interplay between emotions and learning science concepts has also been achieved in science education research. Catherine Milne and Tracy Otieno (2007) adopted IRT in a study of high school chemistry involving science demonstrations. In the context of IRT, science demonstrations and the predict-observe-explain technique are social practices that pertain to the broader scientific practice of inquiry. Science demonstrations became the center of visual and cognitive focus for a group of high school chemistry students in Milne and Otieno's (2007) study. As the teacher repeated demonstrations overtime, there was an observed increase in student engagement through willing participation in responding to questions. When interaction rituals like these are repeated and they call up similar sentiments in a group over different occasions, an *interaction ritual chain* is formed (Collins 2004). Students who, in previous lessons, did not engage with chemistry symbols (i.e., language and concepts) began to adopt and apply them in an effort to explain observations during the demonstrations in Otieno's class. As predicted by IRT, individuals who did not typically participate in class discussions or did not respond to teacher questions became entrained during the demonstrations and gained the confidence to offer responses to teacher questions. Examples like this illustrate that individuals gain emotional energy overtime by participating in science demonstration rituals. In this case, emotional energy was evident from the student's confidence—a form of emotional energy (Collins 2004)—to take social action by proposing a prediction for the potential outcomes of the demonstration.

Students in Milne and Otieno's study transitioned from descriptions of observed phenomena during the demonstration to using chemical symbolic representations of submicroscopic processes in the demonstration lessons. Evidence of learning came from the students' willingness to adopt symbolic representations for explaining chemical processes after the demonstrations. From an IRT perspective, chemical symbols are valued objects that come to represent the *chemistry class group* and the group's practices. Adopting these symbols is a sign that students are willing to be identified as part of this group and that they have attached emotional energy to the symbols in the shared practices of science demonstrations earlier in the ritual chain. If the chemical symbols and language were not valued by students as representations of their science practices and membership to the chemistry class group, they would not be inclined to use them. The sense of solidarity derived from being part of the chemistry class group was established initially during the first science demonstration ritual. Willing use of these symbols by students is further evidence that emotional energy had built up during the science demonstration practices and carried over to subsequent interactions as part of the ritual chains that had formed in this class.

In a study of preservice science teachers, Alberto Bellocchi, Stephen Ritchie, Kenneth Tobin, Donna King, Maryam Sandhu, and Senka Henderson (2014) found that science demonstrations also fostered emotional engagement and group membership in a preservice science teacher education class. When the teacher, Donna, invited preservice teachers to predict the outcomes of a demonstration or to suggest possible explanations for observed phenomena, the students responded in similar ways as the high school science students in Milne and Otieno's study (2007). That is, there was evidence of emotional entrainment through synchronous vocalizations and body movements, and students contributed collectively to developing explanations for observed phenomena. During this time, students were learning about the use of science demonstrations as engaging approaches for teaching science in high school subjects. An interesting outcome of that study was that lower-intensity teaching episodes were considered by preservice teachers to be high-quality learning experiences as well as the higher-energy demonstration episodes. Lower-energy classroom episodes consisted of the teacher reflecting on her own teaching practice during the course of a lesson. When preservice teachers witnessed these reflections, they found them useful for informing their own practices. They also reported feelings of empowerment, which is a form of confidence in taking social action and indicative of growing stocks of emotional energy.

5.1.3 Summary of Part 1

In Part 1 of the chapter, I have discussed three empirical studies to illustrate applications of IRT in science education research. My intent was to show how learning science and learning to be a science teacher can be understood through IRT and to demonstrate the scope of this theory for exploring emotion and cognition as interrelated phenomena. Applications of IRT in science education research have sometimes required the introduction of learning theories to commensurate this sociological theory of interaction with studies of science learning. For example, Olitsky focused on learning as legitimate peripheral participation in a community of practice and then accounted for the micro-processes in classroom interactions through IRT. Milne and Otieno focused on constructs of engagement that include behavioral engagement, cognitive engagement, and emotional engagement to understand chemistry learning. IRT was used to explore the emotional dimension of this engagement at the micro-interactional level and provided an overarching theoretical anchor for data analysis. My own work (Bellocchi et al. 2014) has also traced these lines without explicating theories of engagement or situated and embodied learning. For example, we accepted that *learning to be science teachers* was taking place in our preservice science class and focused our analytic efforts on understanding learning experiences from the perspectives of IRT. In many respects, these studies have provided foundational work related to applications of sociology of emotions and IRT in science education that can pave the way for something else.

That *something else* is the possibility of a more direct exploration of emotion and cognition during situated classroom practices. I present next a discussion of the origins of IRT in Emilé Durkheim's study of religious ideas to illustrate how a closer connection with his work can bring the study of emotion and cognition in more direct grasp of science education researchers and help to bridge the perceived disconnection between knowing and feeling highlighted by some education researchers (cf. Alsop and Watts 2003).

5.2 Part 2: What Else Can We Learn About Science Learning Through IRT?

In the studies discussed in Part 1, learning was evident during the use of science practices and symbols by students and preservice teachers or by preservice teachers adopting ideas about science teaching. It is important to note that Collins' development of IRT did not focus on *learning*, but instead on understanding social processes and macrosocial structures from the perspective of localized microsocial interactions. A key aspect of his work was to consider emotional energy as the initial ingredient for sustaining the success of social interactions, practices, and valued symbols in social groups. I propose that one way of extending the theory further into studies of learning is through greater integration of Durkheim's social epistemology of knowledge.

Collins (2004) developed IRT by drawing heavily on Durkheim's (1915/1964) germinal work *The Elementary Forms of the Religious Life*. Durkheim's study initiated interest in ritual theory, the sociology of knowledge, and the sociology of religion. Goffman (1967) subsequently provided a means for applying Durkheim's analysis of religious rituals to develop understandings of everyday encounters.

By treating any encounter as a ritual encounter, Goffman investigated all aspects of social interaction, from the mundane greeting in the street to formal occasions such as political speeches and teaching, as ritual practices. Like Collins, however, Goffman was not interested in understanding learning or the construction of knowledge in the scholastic sense. His focus was to understand what micro-interactional processes form the basis for social life and practices. For this reason, I suggest that science educators and other educational scholars interested in developing holistic understandings of learning and knowledge construction will find value in Durkheim's original study when considered in conjunction with subsequent developments by Goffman and Collins.

5.2.1 A Social Epistemology of Religious Ideas

Durkheim's study was focused predominantly on totemic rituals and beliefs among Australian Indigenous people, although he also drew on examples from a wide range of other indigenous groups to frame his arguments. The choice to focus on indigenous practices and beliefs was premised on the assumption that ritual practices associated with totemism represented the basic features of all other religious practices present in his time. That is, by studying the totemic beliefs and practices of Australian Indigenous people, he believed he could study the origins of practices and beliefs in the most basic religion that still existed. His analyses were focused on uncovering the social processes that lead a group of people to believe in the existence of a sacred force. Focusing on the idea of sacred force was important because it is essential to any religious system of beliefs or practices. More generally, Durkheim wanted to establish how any idea develops, not in the minds of individuals or experientially through the senses, but through social practices. He used religious practice and ideas as the basis for achieving these aims. Due to this focus, Durkheim's social epistemology of knowledge provided the link between practices, thoughts, and feelings that can offer a strong epistemological foundation for studies seeking to connect social practices, emotion, and learning/cognition in education research.

In the original text, *The Elementary Forms of the Religious Life*, Durkheim (1915/1964) devotes initial attention to elaborating the social basis of various aspects of human intellect (e.g., time, class, causality) through detailed analysis of ritual practices. Once he completes this analysis, Durkheim then presents an argument for the origin of the idea of sacredness that is derived from social practices. This was an unfortunate sequence in his work because the outcome of his study was that collective emotion experienced during social practices generates within the individual a feeling of moral force. This moral force then becomes the idea of sacred force that is attributed to sacred things (e.g., totems, symbols) existing outside of the individual. Categories within human intellect are formed subsequently due to the differentiation between sacred and profane things. For example, as I will discuss further in Sect. 5.2.3, the idea of time is borne out of the temporal separation between sacred rituals and profane activities (i.e., nonreligious aspects of life). The sequence of Durkheim's text can make his overall argument about the social epistemology of knowledge seem obscure as Anne Rawls (2009) has noted previously. To avoid this challenge, I have sequenced my discussion of his work by beginning with the origin of the idea of sacredness before presenting examples of Durkheim's social explanation for the origin of human intellect.

5.2.2 Establishing a Social Epistemology of Knowledge

Durkheim outlined his intention of establishing a social epistemology of knowledge as follows:

> But our study is not of interest merely for the science of religion...For a long time it has been known that *the system of representations with which men [sic] have pictured to themselves the world and themselves* were of religious origins...But it has less frequently been noticed that religion has not confined itself to enriching the human intellect...*it has contributed to the forming of the intellect itself.* Men [*sic*] owe to it not only a good part of the substance of their knowledge, *but also the form in which this knowledge has been elaborated.* (Durkheim 1915/1964, p. 9, emphasis added)

In his analysis, Durkheim distilled all religions into a system of practices that sustain the idea or belief of *sacredness*. Sacred things are set apart from *profane* things and from those things that are forbidden. Those who adhere to these practices and beliefs constitute a *church*, that is, a union of people who represent a single moral community. In the first italicized section of the above quote, Durkheim refers to our system of representation, or cosmology, which is to say the ideas we form about natural phenomena and ourselves. His dual goal of establishing sociology of religion and a social epistemology of knowledge were interrelated. Part of Durkheim's argument for this dual focus was premised on the view that religions provided societies with their first cosmologies. Because religious practices and beliefs are social affairs, he then sought to demonstrate that human intellect itself (i.e., forms of knowledge and representation) also has a social origin. By studying religious practices, he wanted to understand the foundations of all cosmologies, and through this, as the second italicized section indicates, he believed his study would lead to understandings about the social development of human intellect.

In the last italicized statement presented in the quote, Durkheim refers to the way in which knowledge or the intellect is formed. In particular, he was interested in uncovering the roots of what Aristotle called *categories of the understanding*. These included, totality, cause, time, space, personality, and class, which were thought to form the foundations for all aspects of human thought. Durkheim's justification for studying these categories was presented as follows:

> Logical thought is made up of concepts. Seeking how society can have played a role in the genesis of logical thought thus reduces itself to seeking how it can have taken part in the formation of concepts. (p. 432)

He reasoned that the same processes leading to religious thought must have also shaped human intellect, and therefore the categories of understanding. The rationale for this was that religious practices and beliefs provided societies with their first cosmologies. During and before his time, the two dominant ways of dealing with questions of the categories of understanding were the *rationalist* (or *a priori*) method of assuming that knowledge is the product of an individual's rationality. In this view, we are born with fundamental categories as part of our natural mental constitution. We then impose the categories on our sensory experiences of natural phenomena. The other dominant view of the categories was that the individual was the craftsman who assembled his or her knowledge from different pieces of experiential information (an empiricist view).

The problem that Durkheim identified with the rationalist arguments was that they did not offer an explanation about *how* the individual developed the categories in the first place. Simply stating that they were part of our mental constitution did not address the question of how this came to be. The rationalists did not explain where the categories came from, accepting instead that they existed in the mind and were imposed upon objects and experiences by the individual's mind. In contrast, his issue with empiricist arguments was that sensory experiences of objects were not enough to provide an individual with *ideas* about abstract concepts such as causality. For example, observing that one thing preceded another would only add more information about each object. To develop the fundamental category of causality was not possible through sense experience. The idea of causality had to come from somewhere else other than the objects that were being experienced.

In contrast to the rationalist and empiricist perspectives and in order to identify the social origin of the categories of the understanding, Durkheim sought first to explain how social (religious) practices gave rise to the categories. To establish that the categories had a social origin based on practices was not enough. He also needed to demonstrate how it was that *religious* ideas came to *be* in the first place. Another way of stating this is that he needed to establish how the ideas of sacredness and profanity were developed and sustained through social practices, as these were the two fundamental categories of religious thought.

Departing from the arguments of both the rationalists and empiricists, Durkheim proposed that the fundamental categories of understanding originated neither from the individual mind nor from sensory experience alone but were foremost the collective representations of society and the mental state of the group that arose as a result of social practices. By placing the origin of categories in the social realm, he opened the possibility for explaining how an individual could come to accept an idea for which there was no direct empirical basis. For example, the idea of *classes* of objects or animals could not be based on any inherent property of the things themselves. Observing that one animal had similar features to another one did not imply the idea of class, and Durkheim did not accept that the idea of class was innate to the mind. So his focus was to establish how human societies came to the very idea of *class* through social practices given that *class* is neither an inherent property of the objects being classified nor one that is innate. Based on Durkheim's argument, the idea of class had to have a social origin. To explain the categories of understanding such as class and causality, Durkheim first had to identify the social origins of the most fundamental religious idea of sacred force.

5.2.3 Feelings and Ideas of Sacred Forces

Durkheim found that all things considered sacred were endowed with a sacred force. The type of question he asked was, "Given that the objects and symbols representing sacred things do not possess an inherent sacred force, from where did societies develop such an idea?" His answer was that the only forces that could be experienced directly by members of a society were the group sentiments during collective practices such as religious rites or rituals (cf. the focused encounters in Goffman and Collins). In his work, *society* is understood to mean a group of people who engage in shared practices and beliefs (cf. Rawls 2009). During rituals, feelings were intensified because the group was gathered and focused on a single object and idea (a communion or sharing of intimate thoughts and feelings), and their actions became attuned to those of others through the rhythmic coordination of bodily motions and vocalizations. Individuals eventually lost their sense of self and experienced the collective effervescence (i.e., emotion) of the group as discussed previously in relation to Collins' conceptualization of IRT. Individuals became aware of the group/society as this feeling of heightened emotional energy, which originated from collective emotional experience during group practices.

It was through heightened emotional experiences that collective sentiments of the group generated within an individual a sensation of being transported outside of himself or herself. At this time, group sentiments directed at an object (e.g., a totem) were attributed *to* the object itself in a process described as *contagion* by Durkheim. Symbols that came to represent the groups' practices became imbued with group sentiments. In future situations, the symbols and objects themselves invoked the same feelings of emotion and respect within individuals that were produced in the group's first ritual encounter. It is through these collective sentiments that Durkheim found the sacred force of society that imbued with sacredness all manner of things including practices, objects, ideas, symbols, animals, plants, and group members. The *idea* that there was a force in those objects came from an individual's experience of the collective emotion during effervescent rituals that he or she then projected onto practices, objects, and ideas. The efficacy of the sacred force for causing all manner of observable phenomena, both natural and psychological, was not something mystical. It was none other than the collective sentiment of moral respect for group practices, symbols, and beliefs that was experienced by individuals as a moral force initially during ritual practices. Individuals later experienced this force as the drive to sustain the moral order of the group even when sacred rituals were not enacted. For the force to remain salient both to individuals and the social group, rituals had to be repeated. If this was not the case, the sentiments would dissipate over time during profane life and lose their efficacy as scared forces.

Once the experience of collective emotion was no longer considered by an individual to belong to his or her consciousness, the path was laid for the mind to accept abstraction. As Durkheim argued, in order to arrive at the idea that everyday phenomena have underlying patterns, the mind first had to become free from the limits of the senses. Sense experience, in and of itself, could not provide explanatory concepts like *forces* as causal agents of experiential phenomena, nor the *relationships* that were thought to exist between two or more observable phenomena. Individual minds first required the intellectual faculty for establishing causality and relationships before these processes could be attributed to objects. The fundamental categories of understanding were present in social life and it was social (or ritual) practices that provided the substrate for individuals to lay these ideas on top of sensory experience. In other words, it was not that the rationalists were wrong in thinking the fundamental categories preceded sense experience according to Durkheim. The issue was that rationality could not be explained in scientific terms (i.e., empirically), and thus the question of how it came to be that individual minds were capable of rationality could not be accounted for by the rationalist argument that this was an innate ability. By identifying the origin of fundamental categories first in social life and then by identifying the energizing mechanism by which these categories gained their salience for groups, Durkheim developed a social origin of the categories and an explanation for the way in which they are borne out of social practices before they become ideas in the minds of individuals. Because social practices are real phenomena, they are available for empirical study. In this way, Durkheim established a social epistemology of knowledge that could serve as the foundation for empirical studies in sociology.

5.2.4 Social Practices as the Origins of Categories of the Understanding

Having outlined Durkheim's social epistemology for the fundamental religious idea of sacred force, it is now possible to consider how he used this foundational explanation to account for the categories of understanding. Durkheim exemplified how the fundamental categories were first present in social life before becoming a part of individual thought. For example, *time* emerged from the temporal distribution of different types of sacred ritual practices to honor sacred objects (e.g., totems). The concept of *space* was derived from the geographical location at which different rituals were performed as distinct from locations where profane life took place. Classification was made possible by the existing social separation between individuals within tribes, clans, and subgroups within clans each represented by a different totem (i.e., a sacred object). Furthermore, the representation of each of these social divisions (e.g., clans) with sacred symbols (e.g., totems) created another system of classification that divided all things that were the same as one another (i.e., belonging to the same totem) from those belonging to a different group (i.e., a different totem). This also paved the way for notions of *similarity* and *difference.*

5.3 Part 3: Science Learning as Durkheimian Ritual Practices

I have presented a discussion of Durkheim's social epistemology in the preceding sections to illustrate foundational ideas that hold the most value for extending Collins' IRT into a sociology of learning. I now take up these ideas in Part 3 to suggest how they might be applied in science education or general educational research.

There is a limitation to Durkheim's study that he could not eliminate in order to achieve his ultimate goal of understanding the origins of religious ideas (and categories of understanding). If we take one step back in his reasoning, the choice to focus on totemic practices of indigenous cultures was based on the assumption that these practices were fundamental to all modern religions and that the groups he studied had the simplest social structures known and still practiced in Durkheim's day. By studying indigenous religious practices, he reasoned that he was studying the same thing as modern religion but in a less complicated form. In doing so, he thought he could access the essence by which religious ideas are formed and thereby establish the social origins of the categories of understanding.

What he could not do, however, was to study the beginnings of a *new* religion, which would have been the definitive way of answering his questions about the origins of religious ideas and sacred forces in ritual practices. This did not go unacknowledged as Durkheim noted that there was no such thing as the *first* religion. It is also unlikely that historically older rituals were devoid of the ingredients that were identified in the societies that he studied. An issue implicit to the approach he took to his study, however, was that once ideas became represented as collective practices or symbols, they were then subject to being modified from their original forms through social construction. This is partly the reason why Durkheim argued for the study of practices rather than symbols (i.e., social constructs), as symbols can lose their connections to empirical reality as they are socially manipulated over time. Such was his rationale for not studying modern religions that Durkheim saw as involving too many practices and beliefs that were not foundational to establishing or sustaining the idea of sacredness. Yet the indigenous groups that were his focus had well-established practices and thereby numerous socially constructed ideas to go with them. This did lead to a focus in his analyses on less complicated practices than those of modern religions, but certainly did not provide the foundational perspectives he was seeking to achieve.

The implication of this limitation for conducting research in educational contexts, following Durkheim's epistemology, is the need to focus on the one element that preceded all others with regard to fundamental categories and ideas of sacredness, that is, *society*. But society, in the macro sense of the term, is likely to have been a construct too large for achieving the focus of attention required for practices and ideas to be shared quickly and for those ideas to gain salience for a group or for individuals. It is likely that before practices could become established into what may be recognized formally as a *religion* or as *totemism*, smaller groups began to share practices and ideas. I posit that these smaller group practices were the likely starting points for what later became more widely shared religious (or any other) practices. Studying the formation of ideas though localized social practices in small groups may move us one step closer to Durkheim's goal of understanding the *foundation* of human intellect. Although his argument focused strongly on empirical evidence from social practice, in order for a practice to be considered *social*, it had to belong to a society, that is, to a social group. That which keeps smaller societal groups together is the social bond between individuals (Scheff 1997; Turner 2007) and solidarity within larger groups (Collins 2004). To study the origins of human intellect, it follows that one should begin with the study of social bonds that make, break, and sustain a social group during social practices.

5.3.1 Social Bonds and Social Solidarity

Social and emotional ties constitute social bonds and they are important because larger societal groups cannot exist without them. Social bonds also form the foundation for the formation of social solidarity within larger groups (cf. Scheff 1997). In Durkheim's (1915/1964) formulation, the sacred force experienced by ritual participants is none other than the moral pressure they feel within themselves due to their social bonds to others and due to the solidarity they feel toward the larger group during situated social practices. Without this social and emotional connection, a moral sentiment cannot arise because it has a social origin not an individual one (Durkheim 1915/1964). Most importantly, social bonds and social life are mutually constitutive so that one cannot precede the other. There is no root cause as it were, in this context, other than the coherent social group (i.e., the society) that has formed by establishing emotional and social connections overtime during common practices. In feeling a social bond to others and in feeling solidarity with larger societal groups, individuals feel the moral force that Durkheim described in relation to religious groups. Moral force is derived from social emotions and it is this aspect of social bonds that is central to the formation of a *cohesive* group. Although any collective of individuals can gather together and engage in similar practices, such a gathering does not necessarily imply that a moral order has been established as a result of the formation of social bonds. It is important then to understand the extent to which ideas depend upon the cohesion in groups engaged in shared social practices.

I propose that one way of studying the origins of an idea or concept is by attending to the formation, maintenance, and disruption of social bonds and solidarity within groups of different sizes. Bonds involve dynamic social processes that fluctuate over time; they involve ritual ingredients including common emotions and common visual and cognitive foci that are salient during social encounters (Scheff

1997; Turner 2007). I predict that changes in patterns of social bonds during a learning experience/practice will coincide with changes to group morphology, practices, and changes in emotional energy being directed at different ideas. As such, the salience of different concepts for different groups and individuals within those groups is expected to arise from changes in social bonds, morphology, and emotional energy. When individual students, for example, begin to form social and emotional connections to the teacher, to other students, or groups of students (including those with whom they do not associate typically), it is expected that these transient individuals will begin to form new ideas as they engage in different or new social practices.

There is an important difference to note here between the rituals we are likely to observe in everyday educational contexts and those at the center of Durkheim's study. The religious rituals performed by indigenous people were described as highly effervescent. Undoubtedly these highly emotionally charged times gave rise to collective ideas and influenced the minds of individuals. But much of social life unfolds in less dramatic ways, and it was this mundane reality that Goffman recognized when he applied ritual analysis to everyday undramatic encounters. By focusing on the rituals that sustain everyday life, it became possible to extend Durkheim's analysis beyond the investigation of religion or fundamental categories at highly effervescent times. However, neither Goffman nor Collins took up the study of mundane rituals to understand the formation of concepts in the scholastic sense. It is precisely here that I see scope for education researchers to make a contribution. As education researchers we are in a position to capture moments of practice when students are forming initial ideas through formation, maintenance, and disruption of social bonds. Only then are we in a position to address Durkheim's original goal of exploring the *origins* of ideas and knowledge. This does not imply just those ideas that constitute fundamental categories; *any* idea in any context could be understood by the methods that Collins, Goffman, and Durkheim used to understand their subject matter through inquiries founded on Durkheim's social epistemology of knowledge. One way to achieve this goal is through a microsociological focus on learning experiences/practices.

5.4 Toward a Microsociology of *Learning* (Science)

We are now in a position to consider what a *microsociology of learning* could look like through integration of Durkheim's social epistemology of knowledge with the modern theory of interaction rituals developed by Collins and its contemporary applications in science education research (e.g., Milne and Otieno 2007). I propose first that a shift in terminology from use of *knowledge* to use of *learning* is necessary. This is an important shift for a dynamic microsociological focus to take hold because *knowledge* can suggest some fixed object, which is complete and enduring. What I am interested in is something more fluid that enables ideas to change as interactions, practices, and social bonds change on a moment-by-moment basis.

The idea of process implied by *learning* captures this more effectively than focusing on knowledge. An essential elaboration to IRT that we can make from Durkheim's epistemology is that ideas or concepts originate from social practices and morphology and that social emotions are central to the way ideas are formed first in the group and how they subsequently take hold in the minds of individuals. As groups and their practices change, common ideas will change accordingly. Focusing on fluctuations in social bonds and social solidarity provides an empirical basis for establishing whether or not a cohesive social group has formed and whether or not a moral order has developed within that collective. Depending on the development or lack of development of a moral order, we can then proceed by studying the significance of social practices in the development of group and individual ideas.

If we accept that learning is a process involving formation or changes in practices, ideas, social bonds, or sentiments, then we can also add that it is constituted through growing participation in group activities focused on any or all of these elements (i.e., feelings, ideas, and practices) of an interaction ritual *when a moral order is being or has been established.* Moral order is evident when a high level of respect is afforded to practices, ideas, objects, and symbols. Formation of an idea then is shaped by these elements and becomes represented in objects, symbols, symbolic actions, and bodily movements or sounds. It is only when students are engaged deeply in science practices that the ideas they are forming will become apparent. I use the term *science* here to refer to scientific practices and concepts collectively. It is essential then to start with the investigation of those learning experiences that most closely resemble the practice of scientists if we are interested to learn about the development of scientific concepts in students. Most commonly in school settings, this occurs when students are engaged in extended scientific research projects (i.e., inquiry projects). What is important in analyzing interactions in such practices is to identify when groups are gathered; whether or not social bonds are formed, maintained, or disrupted; what the group is doing; and around what initial concepts or proto-ideas the collective attention is focused. These ideas may or may not be the concepts valued in the formal/intended curriculum because ritual analysis and microsociology of learning is about active and evolving social processes.

5.4.1 Lived/Enacted Curriculum

Microsociology of learning may be best understood in traditional educational terms as the study of enacted or lived curricula rather than being conceived as studies of the intended curriculum, although the two are not mutually exclusive. It may in fact be best to bracket initially the formal or intended curriculum so that it does not get in the way of seeking to understand the processes by which student groups form social bonds, practices, and ideas in the context of science (or any other school subject). Moreover, the lived or enacted curriculum may not align with the intended one at all, but we can investigate the interplay between practices and ideas that develop during classroom interactions in relation to the intended curriculum post hoc if this is considered to be a desirable line of inquiry. The most salient aspect to determine through a microsociology of learning is what binds individuals or groups together in a science class and how experiences of collective practices and emotion are achieved. Once this has been achieved, we can then isolate and trace ideas that are charged up by group sentiments over time. In this way, we begin by studying the practices and ideas (i.e., science concepts or concepts about science) of student groups rather than imposing existing ones from a curriculum onto the group.

A dynamic microsociology of learning becomes possible when a group is studied for an extensive period of time (e.g., through prolonged video ethnography) such that the fluctuations of collective emotional energy and individual emotional energies can be observed and so too can changes in social bonds and solidarity be tracked over time across a range of social practices. These fluctuations may be representative of emotional change (Bellocchi and Ritchie 2015; Bellocchi et al. 2014), which in turn is indicative of the ideas that are coming to form a central focus (i.e., being learnt) of the group's attention.

I have argued elsewhere in the context of an intended curriculum that emotional change was indicative of the transformation of conceptual understanding about scientific energy concepts (Bellocchi and Ritchie 2015). A class of 8th grade science students was learning about energy in a Physics course through laboratory exercises involving different devices whose functioning could be understood in terms of energy changes. After these experiences, students were asked to share their responses to prearranged questions focused on the energy changes operating in the devices. One student offered an answer to a teacher question about the first device. His response was not consistent with the canonical science explanation (i.e., the intended curriculum) expected by his teacher. This led to a range of emotions for the student such as frustration and embarrassment when he was told his answer was incorrect. While he was focused on holding his ground that his answer was correct, the student did not show changes in his canonical understanding of the concept. As classroom discussion ensued with other students responding to new questions, he then realized the incongruity between his first response and what was expected. His emotions changed first to surprise and later in the lesson to pride and triumph when he answered correctly other questions involving the same energy change as the one related to the first question. That study provided evidence that emotional change (i.e., a shift from anger/irritation, to surprise, and then to pride and triumph) was associated with conceptual change in the students' understanding of energy concepts. Although the focus of the study was on the intended curriculum, the same methods could be applied to study enacted and lived curriculum. In the latter type of study, we could ask, "During what social practices did the student form his initial (i.e., non-canonical) ideas about energy?" and "Were social bonds formed, maintained or disrupted during those interactions?"

It is possible to adopt the same methods as those used in a range of other studies discussed in Part 1 (see also Ritchie et al. 2016), to focus on ideas that are emergent and contingent upon group interactions, social bonds, solidarity, and practices. At the center of such investigations lies the importance of the formation and disruption of social bonds that are dependent on emotional and social ties formed among groups of students who gather to focus their visual, emotional, and cognitive attention on some common object (a science idea or practice). Observable fluctuations in social bonds are also a dynamic way of monitoring how ideas are formed within groups and sustained or not sustained by individuals in subsequent encounters.

5.4.2 Future Directions for Microsociology of Learning in Science Education Research

The microsociological view of learning presented in the preceding section, with the central focus on practices, social bonds, and emotional changes, offers a holistic view of learning where practices, emotion, social bonds, and cognition have equal footing in understanding learning. Although the ideas that are formed by groups are still directed at experiential phenomena, or an external reality (including ideas themselves as social realities), an understanding of this reality is constructed through an emotionally charged set of social practices. This emotional energy can be both of the more intense effervescent variety as seen in studies of dramatic emotions like pride and triumph (see Bellocchi and Ritchie 2015), or it can be focused on the more subdued but ever-present emotional energy that ebbs and flows in interactions (see Davis Chap. 7 this volume).

There are now a growing number of studies focusing on specific (high-intensity) emotions in science education (e.g., Bellocchi and Ritchie 2015). A different direction for this kind of work is possible with a subtle shift in focus on the way in which we conceptualize emotion and school science practices. The term emotion and specific emotion labels such as joy, happiness, anger, fear, and love can create the impression that there is a singular objective reality available for investigation. That is, we can assume to study *anger* or *love* as some real object that has fixity both spatially, temporally, socially, and culturally and as a concept. In a similar way, investigating science learning as a *social practice* or studying *the practice of science* may suggest a singularity in the idea of scientific practice. What I have come to understand from my own work on learning science focusing on emotion and social practices is that we can learn more if we shift our thinking about the aforementioned constructs by focusing on their adjectival or verb forms. Such investigations would focus on, for example, in the case of emotions, *loving* or *love* in the verb form or *anger* (in the verb form) in the present situation and how this leads or does not lead to a situated or enduring idea of *love* or *anger* (in the noun form) for the interaction participants. This idea can be extended more generally to the study of *emot-ing*[1]

[1] I am not using the term *emoting* based on the English dictionary derivation of *emote*. In contemporary usage, emote means to act out an emotion, for example, in the context of theater. At the same time, I am not excluding this performative perspective from my overall argument. I have simply chosen to avoid the performative aspect in the present discussion. I acknowledge that emotional/affective performativity is yet another important aspect in the study of emoting and affecting in education research that should not be dismissed or excluded from empirical investigation.

rather than emotions or *affect-ing* rather than affect. In the context of scientific social practices, the focus becomes the study of *practicing science* rather than a singular scientific practice. In this way, *scientific practices* and *ideas* are studied as emergent and contingent realities within educational contexts.

A microsociology of learning asks questions such as "How are social and emotional bonds established, maintained, and broken during interactions focused on science practices?" "What (science) concepts/practices arise from *this* situation?" "Where is emotional intensity directed during the practice/interaction?" "Who is assembled when these social practices are generated or circulated?" "Who is excluded?" "In what way, if any, do the practices/ideas formed by those excluded from the group differ from those who form part of the group?" "How is a sense of time, place, or space developed around the practices and ideas of the group?" "Are the same practices and ideas circulated again in some other place or time?" "Who (re)assembles in these later interactions?" I think the first two questions are the most important ones initially with regard to pathways that connect emoting and learning/cognition in science education research. The moral or political authority of the scientific establishment and that of the curriculum authorities seeks to sustain, knowingly or otherwise, the primacy of formal curricula. But if students and teachers do not accept this formal curriculum, it's efficacy for sustaining the place of scientific practices and ideas at a high social status is lost. Similarly, efforts to engage disaffected students with school science through a preordained curriculum are not likely to have widespread success if the same practices and ideas championed through formal curriculum are not borne out in situated local practices that lead to lasting social bonds, for example, in school classrooms. Scientific practices need to be efficacious in addressing the needs of the smaller groups before they are likely to gain salience for larger social groups. Science needs to bind small groups through social bonds and large groups through solidarity if it is to hold a high status within student groups or society in the wider sense of the term. A microsociology of learning founded on social epistemology can deepen our understandings about these issues.

5.4.3 Concluding Remarks

My goal in this chapter was to articulate some of the basic assumptions and applications of IRT for science education research (Part 1) and to suggest new directions in which this research may head by connecting more closely with Durkheim's social epistemology of knowledge (Part 2). A microsociology of learning was proposed in Part 3 that is founded on Durkheim's social epistemology and IRT and foregrounds the role of social bonds and emotional change in learning. This chapter is an attempt to take contemporary IRT and supplement it further with aspects of Durkheim's work as well as more recent developments about social bonds (e.g., Scheff 1997), solidarity (Collins 2004), and empirical studies in science education that can move our thinking forward and perhaps engage a wider range of researchers with these perspectives. The choice to discuss Durkheim's work and IRT in the context of developing holistic understandings of emotion and cognition can be received as establishing a dualism between mind and body. This choice in terminology was a difficult one to make because I did not want to convey such a view of learning (or being). My choice in the use of these terms has been purely to make the perspectives presented in this chapter accessible to a wide audience, including those who may come from lines of inquiry in science education different from my own. From a personal perspective, I accept that we are inherently emotional so that we may experience higher or lower levels of emotional energy, but not a complete absence of it that would make it absent during cognitive processes. I have commenced some of the microsociological work discussed in Part 3 in my own empirical studies of science classrooms and preservice teacher education. Beginning with a focus on the social epistemology of knowledge and extending into the study of social bonds, I seek to develop further a microsociology of learning as outlined initially in this chapter. This has so far been an exciting and fruitful journey.

Acknowledgments I thank Anne Rawls, Jonathan H. Turner, James Davis, and my coeditors, Cassie Quigley and Kathrin Otrel-Cass, for their suggestions and feedback during the preparation of this chapter.

The Australian Research Council Discovery and Linkage Grants, contract grant numbers DE160101053 and LP110200368, supported this research. Any opinions, findings, and conclusions or recommendations expressed in this chapter are those of the authors and do not necessarily reflect the views of the Australian Research Council.

References

Alsop, S., & Watts, M. (2003). Science education and affect. *International Journal of Science Education, 25*, 1043–1047.

Bellocchi, A., & Ritchie, S. M. (2015). "I Was Proud of Myself That I Didn't Give Up and I Did It": Experiences of Pride and Triumph in Learning Science. *Science Education, 99*, 638–668. doi:10.1002/sce.21159.

Bellocchi, A., Ritchie, S. M., Tobin, K. G., King, D. T., Sandhu, M., & Henderson, S. (2014). Emotional climate and high quality learning experiences in science teacher education. *Journal of Research in Science Teaching, 51*, 1301–1325. doi:10.1002/tea.21170.

Collins, R. (2004). *Interaction ritual chains*. Princeton: Princeton University Press.

Durkheim, E. (1915/1964). *The elementary forms of the religious life*. (J. W. Swain, Trans.). London: George Allen & Unwin Ltd.

Fortus, D. (2014). Attending to affect. *Journal of Research in Science Teaching, 51*, 821–835. doi:10.1002/tea.21155.

Goffman, E. (1967). *Interaction ritual*. New York: Anchor Books.

Milne, C., & Otieno, T. (2007). Understanding engagement: Science demonstrations and emotional energy. *Science Education, 91*, 523–553. doi:10.1002/sce.20203.

Olitsky, S. (2007). Promoting student engagement in science: Interaction rituals and the pursuit of a community of practice. *Journal of Research in Science Teaching, 44*, 33–56. doi:10.1002/tea.20128.

Olitsky, S., & Milne, C. (2012). Understanding engagement in science education: The psychological and the social. In B. J. Fraser, K. Tobin, & C. McRobbie (Eds.), *The second international handbook of science education* (pp. 19–34). Dordrecht: Springer.
Rawls, A. (2009). Epistemology and Practice: Durkheim's. In *The elementary forms of religious life*. Cambridge: Cambridge University Press.
Rinchen, S., Ritchie, S. M., & Bellocchi, A. (2016). Emotional climate of a pre-service science teacher education class in Bhutan. [Special Issue] *Cultural Studies of Science Education, 11*(3), 603–628. doi:10.1007/s11422-014-9658-0.
Ritchie, S. M., Hudson, P., Bellocchi, A., Henderson, S., King, D., & Tobin, K. (2016). Evolution of self-reporting methods for identifying discrete emotions in science classrooms. *Cultural Studies of Science Education, 11*(3), 577–593. doi:10.1007/s11422-014-9607-y.
Scheff, T. J. (1997). *Emotions, the social bond and human reality*. Melbourne: Cambridge University Press.
Turner, J. H. (2007). *Human emotions: A sociological theory*. London: Routledge.

Alberto Bellocchi is a researcher and academic at the Queensland University of Technology, Brisbane, Australia. He is currently the recipient of a 3-year funded research fellowship focusing on the interplay between social bonds and learning science. His broader research program addresses teaching and learning within university preservice teacher education classes and high school science classrooms.

Chapter 6
Emotional Events in Learning Science

Stephen M. Ritchie and Jennifer Beers Newlands

6.1 Contextualizing Emotional Events

As any classroom teacher can attest, classroom transactions are imbued with emotion. Yet, until recently, little research has explored the emotional arousal of teachers and students in science classrooms (Rosiek and Beghetto 2009). As some have asserted, "outside of attitudes and interest, there is a lack of research investigating the broad array of emotions that may impact science learning" (Sinatra et al. 2014, p. 420). These authors reviewed research mostly informed by conservative psychological methods that relied heavily on high-inference measures such as self-reporting surveys and interviews. In contrast, Michalinos Zembylas (2004) took a very different path in his 3-year ethnographic study of an elementary teacher's classroom that reaffirmed the pivotal role of emotions in teaching and learning science.

Stephen Ritchie and his colleagues (e.g., Bellocchi and Ritchie 2015; Ritchie et al. 2013) have taken up the challenge of adding to the limited stock of classroom research on emotional transactions because recent empirical discoveries from neuroscience had shown that emotions and cognition are mutually dependent in learning (e.g., Davidson and Begley 2012). Furthermore, we found the sociological theories of interaction ritual chains (Collins 2004) and human emotions (Turner 2007) influential in our 7-year research program because these compatible theories recognized the importance of emotion in interactions at both the individual and collective levels. This research program has deviated from that of Zembylas' program (2004) and even farther from others involved in emotion research (e.g., Sinatra et al. 2014).

Fundamental to the research program was our focus on analyzing interactions from identified (emotional) events. This focus involved the study of "individuals'

S.M. Ritchie (✉) • J.B. Newlands
School of Education, Murdoch University, South Street, Murdoch, WA, Australia
e-mail: s.ritchie@murdoch.edu.au; jsbnew@gmail.com

A. Bellocchi et al. (eds.), *Exploring Emotions, Aesthetics and Wellbeing in Science Education Research*, Cultural Studies of Science Education 13,
DOI 10.1007/978-3-319-43353-0_6

engagement in specific activities, rather than averaged across multiple occasions" (Patrick and Middleton 2002, p. 28). For us, an *emotional event* was much more than any classroom episode identified for discourse analysis. It was a unit of analysis associated with cultural transformation in the classroom that originated from ruptures in the everyday practices of teachers and/or students to produce salient emotional outcomes. While the emotional events used in our classroom research varied in duration and complexity, their salience was identified in consultation with stakeholders through methods such as cogenerative dialogue and emotional diaries (see Ritchie et al. 2016). In this chapter we would like to explore our use of event as the unit of analysis by tracing its origins beyond William Sewell's (2005) pivotal work on historical events.

6.2 History of Events

Sewell (2005) argued: events are "sequences of occurrences that result in transformation of structures" (p. 226). This was our starting position for our program of research. For example, in the study of beginning physics teachers who were implementing student-led scientific investigations (Ritchie et al. 2013), one of the teachers observed a group of students who had deviated from the required task of connecting light bulbs in series. Instead, the group had begun to explore parallel combinations of light bulbs in the circuit. The group attracted the teacher's attention because they were animated in their actions and their voices were audibly louder than their previous work. Initially, the teacher was somewhat angered that his students had gone off task. Yet, he soon recognized the powerful learning opportunity afforded to his students through their exploration of the apparatus rather than simply following the instructions passively. Upon this emotionally positive realization, the teacher advised all students in the class to abandon the "recipe" he had provided in favor of a free exploration of circuit components. This was a *turning point* in his teaching practices because it touched off "a chain of occurrences that durably transform[ed] previous structures and practices" (Sewell 2005, p. 226). An event, then, "transforms structures largely by constituting and empowering new groups of actors or by reempowering existing groups in new ways" (Sewell 2005, p. 110). In this example, the teacher transformed the structure of the classroom activities to encourage students to explore circuit configurations. He learned the value of such student experiences within one lesson, but it had an enduring effect across lessons. His pedagogical approach to lab work changed thereafter.

The conceptualization of events we adopted in our research was developed from the macro-level analyses of historical events, particularly the French Revolution, by both sociologists (e.g., Sewell 2005) and philosophers (e.g., Žižek 2014). Interrogation of what happened before, during, and subsequent to the French Revolution helped scholars to formulate and refine a theoretical conception of an event. There is general agreement that an "Event is a radical turning point" (Žižek 2014, p. 159) or "rupture" (Sewell 2005, p. 227) or "breach[…] in everyday social practices" (Moore 2011, p. 304) and not just any kind of happening or episode. As Slavoj Žižek (2014) argued:

> In an Event, things not only change, what changes is the very parameter by which we measure the facts of change, i.e., a turning point changes the entire field within which facts appear. (p. 159)

This aligns with Adam Moore's (2011) position that, "While an event is an incident that results in the transformation of a cultural system, it is in turn shaped by the terms of the structure it transforms" (p. 298). Marshall Sahlins (1985) first identified this dialectical relationship between structure and event, and Sewell (2005) later adopted this when he recognized that the concepts of structure and event are mutually dependent. From Sewell's (2005) extensive macro-level analysis of the French Revolution, the following additional features characterize events and differentiate them from ordinary happenings:

- Events rearticulate structures and produce more events.
- Events are cultural transformations that are characterized by heightened emotions and shaped by particular conditions.
- Events are acts of collective creativity that are punctuated by ritual.
- Events are spatial as well as temporal processes.

All of the characteristics described by Sewell also can be identified in the event depicted earlier about the transformed practice of the physics teacher (Ritchie et al. 2013). Briefly, the structure for doing lab work in the physics classroom was rearticulated by the teacher following his observations of a group of students who ruptured the existing structure of recipe-like work, and this rearticulation had a cascading effect on the practices of other groups of students; successful interactions between students in these groups led to an elevation of positive emotional energy at the collective level and positively valenced discrete emotions at the individual level (e.g., happiness); the first innovative group created new ways for connecting circuit components as they were enacting usual rituals for connecting these components; and this group's creativity preceded a change in practice of other groups in different places around the laboratory, after being directed by their teacher to observe the innovative group's actions.

The analysis of classroom events depicted in this example, however, differs from Sewell's macro-level exploration of the events in that the transformation of teaching practice was examined at the meso- and microlevels. Additionally, most change in society is not on the same scale or as dramatic as the French Revolution; rather, societal change usually occurs in gradual or incremental shifts. The gradual or evolutionary and temporal terms typical of social change are similar to what one would expect to be associated with teaching and learning events in a classroom. Thus, in order to justify our continuing use of events in research, we pause momentarily to reframe the concept of event as a suitable unit of analysis for classroom studies.

6.3 Reframing Event

Both Sahlins and Sewell accept that events are different from uneventful happenings in that they transform structures rather than reproduce social order. Typically, such visible transformation occurs in intensive bursts. Yet, institutional (e.g., school, classroom) structures can change gradually in incremental shifts as they are enacted. Accordingly, Moore (2011) argued that gradual change in structures is as likely to occur as the punctuated ones and these gradual changes can accumulate over time, resulting in more significant structural transformations. In other words, events are just as likely to be constitutive of the reproduction of structures as their rupture, and this may occur in increments of smaller events that impact locally rather than more widely. More specifically, "the eventfulness of life can be found in instances of social reproduction as well as change" (Moore 2011, p. 311). Thus, what is important in identifying events is the salience of the happenings for stakeholders, as revealed through their retellings or narratives of the events during particular activities, including conversations and diary entries of participants in the case of classroom experience. Moore (2011) explained this through four interrelated points:

> First, it is through narratives that heterogenous experiences are configured into an intelligible plot. Second, events are made comprehensible through their emplotment in narratives, which brings them together from the manifold of life into an understandable story. Third, events differ from mere occurrences in that they contribute to the progress of a narrative. They become, in effect, points in the plot that either carry the story along a preexisting narrative arc or signal a change. In other words, they are not only experientially but also semantically significant happenings. Finally, it is through the social acts or narration and emplotment that this status accrues to experiences and events. In sum, socially narrated events are an integral part of processes of both order and change, actively making and remaking social relations and identities. (pp. 306–307)

Moore's (2011) expansion of the meaning of event is important in classroom studies because we are less likely to observe dramatic turning points in children's understanding of science concepts, for example, although a recent study (Bellocchi and Ritchie 2015) has indeed captured a visible epiphany for one student who finally "gets it" after struggling with the meaning of energy transformation across several lessons. A narrative composed by classroom researchers across lessons is usually required to illustrate the temporal nature of coming to know for individuals, and this can transpire suddenly or in fits and spurts.

In addition to temporality, the contingent aspects of events need to be highlighted in the narrative that establishes the salience of the events being examined. In this regard, classroom studies that investigate single episodes, or even a selection of episodes, run the risk of "studying incomplete sequences [and] greatly restrict the number of cases" (Sewell 2005, p. 83). For this reason, it is important to recognize that "what has happened in an earlier point in time will affect the possible outcomes of a sequence of events occurring at a later point in time" (Sewell 2005, p. 100). Thus, in order to capture the temporally contingent nature of events in classroom studies, salient episodes or fragments of events need to be linked together so that researchers' reports look "more like well-made stories or narratives than like laws of physics" (Sewell 2005, p. 111).

We would now like to illustrate the contingent and temporal nature of *eventful* classroom learning for a group of preservice science teachers (PSTs) in which we capture the transformative dimensions of their activity.

6.4 Investigating Classroom Events: An illustrative Case

Cogenerative dialogue and emotional diaries were two methods we found most helpful in identifying salient events. Cogenerative dialogue, or cogen, is a reflective conversation about what happens in class, where members from the research team join the teacher and several selected students to discuss what works well and what improvements can be made (Tobin and Roth 2005). For each study, we typically employed this method between three and six times. The composition of the participants in cogens could change based on their availability or willingness to participate. This created an opportunity to expand the perspectives voiced and capture different views on the same issues. Students also completed emotional (emo) diaries at the end of each lesson. The emo diaries required students to identify the circumstances that led to their experience of emotions. Students typically recorded between three and six emotional experiences for each lesson, and by searching for patterns in their declared experiences, we could identify both coherence and contradiction.

The use of cogens together with emo diaries helped us select events for analysis as they provided evidence of temporality and contingency in what was occurring at particular points in time. These methods also allowed the stakeholders (students and teachers), rather than the researchers, to identify aspects of classroom happenings that were significant. This identification then guided the researchers to look at what was occurring at certain points in time in the audio and video data. Analysis of data involved viewing video-recorded lessons and replaying salient clips of these recordings several times at normal and slow speed to glean microlevel interactions as well as gestural and facial expressions. Transcripts of selected events were made for the purposes of conversation analysis, supplemented by visual (e.g., gestures and facial expressions) and audio (e.g., speech parameters such as pitch, volume, and speech rate) observations.

We choose one event from a study of two preservice teachers (PSTs) conducting an extended experimental investigation (EEI) on the variables that affect electroplating in a science methods' course in an Australian university in this chapter to illustrate what characterizes it as an event. Both PSTs were undergraduates enrolled in a Bachelor of Education degree that prepared them to teach high school science classes. Alex was a chemistry major but Calvin (student names are pseudonyms) had very little chemistry background. EEIs are relatively open-ended investigations where the students rather than the teacher design the experiment (see Ritchie et al. 2013). Even though the event was recorded in week 7 of the 10-week course, it had its origins in the previous year. At that time, the lecturer (Alberto—this is a real name because Alberto Bellocchi was a CI on the awarded research grant from which the event was recorded) had expressed frustration with the PSTs' confusion with the task requirements. As one of the PSTs commented during the previous year's cogen with Alberto (see appendix for transcription conventions used):

> He was getting a bit frustrated because they weren't listening to him…. They just didn't want to accept his answer ((two other students nod in agreement))…. They didn't want to accept his answer so they continually (.) they just wanted to argue.

The widespread expression of confusion and frustration by Alberto's students provided impetus for him to redesign the task for the current year in such a way that the PSTs would gain firsthand experience with EEIs before considering the implications of this experience for teaching inquiry in schools. During the second scheduled cogen of the current year, Alberto commented:

> It brings memory back from high school time. I enjoyed doing experiments with kids and it was transformative for my teaching. It just brought it all back and I was thinking oh it's so cool. This is the way it should be. I'm just walking around and I'm so happy , listening to different groups. It's been always hard for me to back off . Now I have to remind myself to back off and let them do the intellectual work. But it's fun as it brings back that pedagogy. I'm really enjoying what we are doing. It's a good change.

We see evidence in the video clips of Alberto's expression of happiness during his interactions with Calvin and Alex as well as his strategy for backing off.

Figure 6.1 shows all three smiling upon hearing Calvin's phrase "we're trying to be positive." This in itself is suggestive that there was some initial frustration or even annoyance experienced by Calvin in the laboratory preparations. Calvin's emotional response is verified by both students' emo diary entries where they identified negative emotions in the first hour of their work, as represented in Table 6.1.

Despite the previous frustration demonstrated in the students' emo diaries, there is also a lot of overlapping speech in video clips from the event, which is consistent with mutual focus, solidarity, and a shared emotional state of happiness (Collins 2004). The three are also smiling, laughing, and joking, as shown in Fragment 6.1. In particular, Alberto smiles (see Fig. 6.1) as he acknowledges the students' smiles in Turn 04.

Fragment 6.1 Smiling

Turn	Speaker	Transcript
1	Alberto	What's going on?
2	Calvin	Good [not much]
3	Senka	[He's smiling that means]
4	Alberto	[Its all good] Yeah you're smiling what's that what's that all about?
5	Calvin	[Well]
6	Alex	[He's] always smiling ((laughs))

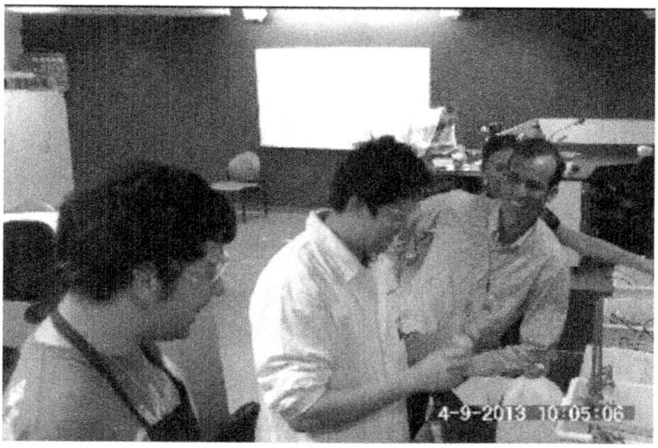

Fig. 6.1 Expressions of happiness

Table 6.1 Summary of emo diary entries after the first hour

Student	Emotion	Explanation
Alex	Frustration	Forgot the proper setup of the equipment so I lost valuable time
Calvin	Annoyed	Experiment went wrong

7	Calvin	((laughs))
8	Alberto	Yeah I know eh ((laughs))
9	Calvin	We're trying to be positive but na we've been able we've been able to fff: we've now that we've really got into it we've been able to observe a few more differences as well
10	Alberto	Okay

One minute later, Alberto avoids interrupting Calvin's observations upon hearing a possible misconception (see italicized text in Turn 21 of Fragment 6.2).

Fragment 6.2 Traveling ions

Turn	**Speaker**	**Transcript**
21	Calvin	Not differences we've been able to make a few more observations obviously the temperatures there there is a change in temperatures when you're putting it on like that's generic but like we're able to see a travel of uhm like a current that's traveling from one uhm one plate to the other (0.1) and you know you can see the oxygen gas bubble up but yeah you can you can see the like the ions that are traveling

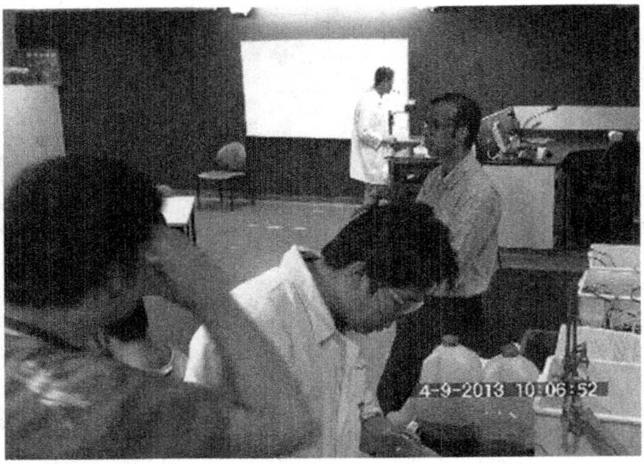

Fig. 6.2 Alberto backs off

22	Alberto	But any mass difference in the
23	Alex	((Nods))
24	Calvin	Yea there is a definite mass difference yeah [so far]
25	Alex	[Its too sss] like its too small to be like completely true so we'll do two more test then I'll do the first one again
26	Alberto	Uh huh right
27	Alex	like four volts

Upon completion of the discussion between Calvin and Alex beyond Turn 27, Alberto then says: "Ah, I just want to go back to the comment you made earlier. You said you could see the ions move (see Turn 21). Tell me more about that." Now, Alberto knew that it was more likely that Calvin had seen bubbles of hydrogen and oxygen gas at the electrodes, and any visible movement in solution other than gas would have more to do with the associated movement of the suspended bulk matter because ions would have been too small to see (personal correspondence). Calvin then became flustered in response to the question and Alberto commented: "You do what you have got to do and when you think you are seeing it, just tell me and I'll watch. So, don't rush it, do your thing and I'll hang around until you're ready." Alberto then physically backed off by repositioning his body adjacent to Calvin and Alex, as shown in Fig. 6.2, where he waited in silence without close scrutiny of Calvin's work. Alberto soon moves to another group, while Alex and Calvin complete their observations.

Alberto approached this group again 27 min later. Alex has his head down and is focused on a task with a neutral facial expression. Calvin's face is not in the frame, but audio data shows that he begins to express happiness, wonderment, and elation verbally, as follows: "look at the physical properties!" From prosodic analysis the pitch of his voice is also noticeably higher than usual. Slightly off camera, Alberto

Table 6.2 Summary of emo diary entries after the second hour

Student	Emotion	Explanation
Alex	Happiness	Got really good results
Calvin	Excitement	Experiment showing positive results
Calvin	Happiness	Results showed a definite change in physical properties

nods his head as he says "wow" and smiles. Referring to his experimental results, Calvin says, "it's wonderful" and shows the results of the experiment to the camera before verbally expressing the emotion of happiness: "I was so happy. I was really really happy about that." Calvin and Alex both identified happiness in their emotion diaries for the second hour of their laboratory session, following their obvious success, represented in Table 6.2.

Clearly, achieving success with the EEI produced happiness. The intensity of this emotion is likely to be linked to the resolution of their previously experienced negative emotions (Turner 2007). Just as importantly, the transformation of their emotional states was aligned with Calvin's improved fluency with the chemistry. In particular, during the interactions that followed the students' expression of happiness, a possible alternative conception for what they were observing arises as Alberto quizzes the PSTs in Fragment 6.3.

Fragment 6.3 Activity

Turn	**Speaker**	**Transcript**
31	Alex	`What I'm thinking that's causing it is just because the bubbles are all moving up in one direction and it's just being pushed out`
32	Alberto	`What's being pushed out?`
33	Alex	`'cos all the bubbles will just keep rising up and its pushing everything from the base that's why it's causing this stream of gas to come up [to the surface]`
34	Alberto	`[Ahh:] so you're saying it's a stream of [gas]`
35	Calvin	`[I did] yeah okay`
36	Alex	`But (0.2) but there's no doubt that there's a current going through from one side to the other because that's how the electrons would travel that's what I think so`
37	Alberto	`Do you think you would see the ions actually moving between the two?`
38	Alex	`Hmm I don't think you will actually be able to see the ions`

In Fragment 6.3, despite seeing something visibly moving (Turn 31), Alex, rather than Calvin, rules out the possibility of seeing the ions (Turn 38). Alex attributes the movement to the "stream of gas" and how this would "push out" the bulk matter of the saturated $CuSO_4$ solution. In addition, Alex remarks that there is no doubt that an electric current flows between the electrodes in solution because that is how "electrons" would travel (Turn 36). This statement was not unpacked further through probing questions; however, Calvin appropriated this explanation later in his description of the outcome of the experiment to another student who passed their workspace. In this exchange, Calvin said: "You know we've been able to see the inquiry process in action 'cos … *we've been able to see like what appears to be a current flowing through from one plate to the other*" (emphasis added). Here Calvin has abandoned his previous view that ions are visible, instead utilizing Alex's explanation that a current appears to flow between the electrodes. In other words, by changing his explanation for a different audience without prompting, it could be argued that conceptual change had occurred. At the minimum, in this moment, positive emotions were associated with cognitive activity that was considered salient by Calvin. An equally important outcome for Calvin expressed in this comment was that this experiment gave him firsthand experience with the inquiry process. There was additional evidence of the impact that this cognitive and emotional experience had on Calvin in a subsequent conversation with yet another student as shown below in Fragment 6.4.

Fragment 6.4 Inquiry

Turn	**Speaker**	**Transcript**
41	Sonam	Get the f+*k out you guys look like you're doing proper science there, there like figures with decimal places and everything
42	Calvin	((laughs))
43	Sonam	You guys are getting loose
44	Calvin	((laughs)) Ah look it it's really interesting what we've come up with so far and we're actually like doing this, we're actually seeing inquiry in progress in this, in this particular experiment 'cos we've you know have questions that we're asking as well
45	Sonam	Yeah
46	Calvin	That you know we can answer ourselves by just adjusting the experiment just a little bit more
47	Sonam	Yeah
48	Calvin	Which is just you know is just amazing
49	Sonam	Well I look forward to your poster presentation

Fragment 6.4 not only reinforces Calvin's previously asserted affirmation about learning about inquiry firsthand (Turn 44), but also it indicates that by engaging in this inquiry, they had the agency to transform the structures in place for doing the activity themselves through changing aspects of the design (Turn 46).

Our narrative has linked salient fragments temporally and contingently, which collectively provides clear evidence of a transformative activity that matches Sewell's notion of an event. Taken individually, the fragments or episodes presented may be ignored or deemed "uneventful" because they appear as mundane classroom happenings. Yet, the transformative activity unfolded incrementally over time, and its significance is garnered within the context of the entire narrative rather than presenting as a visible disruption to the usual social practices within the classroom (cf. Moore 2011). In this regard, the interpretive narrative presented has established strong temporal links between video-recorded observations, previous declarations, and observations. The student and teacher experiences across these observations were transformative in that they afforded these participants the agency to transform structures, which, in turn, impacted on subsequent practices. Successful transactions were accompanied by the expression of the discrete emotion of happiness and greater fluency in chemistry and appreciation of the pedagogical impact of such activity.

6.5 Implications for Classroom Research

In keeping with Moore's (2011) ideas on events, it is important to remember that event analysis in classroom research requires careful attention to constructing a coherent narrative for readers. Events may change schema and practices, alter how people interact, or modify how structures afford agency. The analysis of these events often begins with a contradiction to what is expected, making the narrative more straightforward and the salience of the event easier to observe. On the other hand, when events happen in a more protracted manner, it is essential to consider the temporal framework and frames of reference being used to determine the effect of events (Moore 2011). Even the most mundane aspects of an interaction may be important for the narrative and should not be simply ignored or glossed over to create a more coherent story. Events have both temporal and contingent qualities that do not always allow for a seamless narrative. Thus, when viewing video and audio data and selecting events for further analysis, it is important to keep track of the narrative in its totality.

We are cognizant that conversational analysis of extended transactions that constitute an event creates a challenge for researchers who need to comply with editorial constraints imposed by publishers to limit page lengths of manuscripts. In this regard, very long transcripts may only partially reveal the collective compelling evidence of transformed understanding of a particular phenomenon or practice. Moreover, this evidence may be present in a fragmented rather than fluent unfolding of interactions; such is the nature of naturalistic rather than contrived research.

Participants respond in knowing ways that are never fully captured by an onlooker without considerable unpacking of how the participants responded to each other in the moment. Again, we contend that this would be most likely addressed through careful attention to the emplotment of narrative accounts.

Some colleagues unfamiliar with and unwilling to engage in the theoretical and philosophical underpinnings of what constitutes an event, upon reading reports involving analysis of events, may remark naively, "we all know what an event is, simply tell us what you did and why you did it." This is an unsophisticated comment because it fails to recognize the temporal, contingent, and transformational properties of an event as illustrated in this chapter. Similarly, methodological concerns about researcher selection of classroom episodes for analysis are not applicable to events because they are not just any happenings, but rather identified as salient by classroom participants, which also satisfy specific conditions for the purposes of classification as events.

As a result of the temporal ties between the fragments that constitute an event, it is only through narrative that their meaningful association can be told in a way that makes sense to colleagues. We reject the assertion that it is possible for "the data to speak to you" independent of theoretical lenses and temporally linked fragments of events. We do not rely on a single data source and attempt to balance data derived from high-inference methods (e.g., questionnaires, diaries, interviews) with less inferential measures such as conversation analysis of classroom interactions, supplemented by prosodic analysis, facial and gestural analyses, to support our assertions (Ritchie et al. 2016).

Acknowledgment The data analyzed in this chapter were recorded in a study under an Australian Research Council grant (DP120100369) awarded to Stephen Ritchie, Kenneth Tobin, Alberto Bellocchi, and Donna King.

Appendix: Transcript Conventions

Symbol	Meaning
::	Colons are used to indicate prolonged sound with each representing 0.1 seconds
(0.2)	The number in parentheses, e.g., 0.2, is the time of the pause, in seconds, in this case, 0.2 seconds
[text]	The square brackets indicate the starting and ending points of the overlapping speech with another speaker
((text))	The text enclosed by the double parentheses is author commentary, typically about context or observations

References

Bellocchi, A., & Ritchie, S. M. (2015). "I was proud of myself that I didn't give up and I did it": Experiences of pride and triumph in learning science. *Science Education, 99*, 638–668. doi:10.1002/sce.21159.

Collins, R. (2004). *Interaction ritual chains*. Princeton: Princeton University Press.

Davidson, R. J., & Begley, S. (2012). *The emotional life of your brain*. New York: Hudson Street Press.

Moore, A. (2011). The eventfulness of social reproduction. *Sociological Theory, 29*(4), 294–314. doi:10.1111/j.1467-9558.2011.01399.x.

Patrick, H., & Middleton, M. J. (2002). Turning the kaleidoscope: What we see when self-regulated learning is viewed with a qualitative lens. *Educational Psychologist, 37*, 27–39. doi:10.1207/S15326985EP3701_4.

Ritchie, S. M., Tobin, K., Sandhu, M., Sandhu, M., Henderson, S., & Roth, W.-M. (2013). Emotional arousal of beginning physics teachers during extended experimental investigations. *Journal of Research in Science Teaching, 50*, 137–161. doi:10.1002/tea.21060.

Ritchie, S. M., Hudson, P., Bellocchi, A., Henderson, S., King, D., & Tobin, K. (2016). Evolution of self-reporting methods for identifying discrete emotions in science classrooms. *Cultural Studies in Science Education, 11*, 577–593. doi:10.1007/s11422-014-9607-y.

Rosiek, J., & Beghetto, R. A. (2009). Emotional scaffolding: The emotional and imaginative dimensions of teaching and learning. In P. A. Schutz & M. Zemylas (Eds.), *Advances in teacher emotion research. The impact on teachers' lives* (pp. 175–194). New York: Springer.

Sahlins, M. (1985). *Islands of history*. Chicago: University of Chicago Press.

Sewell, W. H., Jr. (2005). *Logics of history. Social theory and social transformation*. Chicago: University of Chicago Press.

Sinatra, G. M., Broughton, S. H., & Lombardi, D. (2014). Emotions in science education. In R. Pekrun & L. Linnenbrink-Garcia (Eds.), *International handbook of emotions in education* (pp. 415–436). New York: Taylor & Francis.

Tobin, K., & Roth, W.-M. (2005). Coteaching/cogenerative dialoguing in an urban science teacher preparation program. In W.-M. Roth & K. Tobin (Eds.), *Teaching together, learning together* (pp. 59–77). New York: Peter Lang.

Turner, J. H. (2007). *Human emotions. A sociological theory*. London: Routledge.

Zembylas, M. (2004). Young children's emotional practices while engaged in long-term science investigation. *Journal of Research in Science Teaching, 41*, 693–719. doi:10.1002/tea.20023.

Žižek, S. (2014). *Event. A philosophical journey through a concept*. Brooklyn: Melville House.

Stephen M. Ritchie has been interested in the complexities of classroom teaching and learning throughout his research career. His recent research has focused on the emotional engagement of students in science classes and the emotional experiences of beginning science teachers. He has returned to his substantive position of Dean of Education after acting as Provost at Murdoch University during 2015.

Jennifer Beers Newlands is a foundation Principal of a secondary school and a part-time doctoral student at Murdoch University in Western Australia. Her research focuses on the teaching and learning of STEM for students who are at-risk of disengaging with school.

Chapter 7
Emotions, Social Beings, and Ethnomethods: Understanding Analogical Reasoning in Everyday Science Classrooms

James P. Davis

7.1 Introduction

My experiences of teaching secondary school science grounded me in the fact that science classrooms are full of emotion. But when we read pedagogical theory and the interdisciplinary theory that supports pedagogy, the emotional dimension is often absent. Our pedagogical strategies are firmly grounded in objective, rational scientific theory, through disciplines such as the cognitive sciences. This is certainly the case with our dominant theories about analogical reasoning that objectify cognition while excluding emotion (cf. Gentner 1983).

7.2 Analogical Reasoning and School Science Context

These dominant theories of analogical reasoning in science education focus on inter-domain analogy by drawing from the cognitive sciences *theory of structure mapping* (Gentner 1983). This theory is important in defining our notions of analogical reasoning in science pedagogy as evident by the foundation it provides to the teaching with analogies (TWA) model (Glynn 1994) and the focus-action-reflection (FAR) guide (Treagust et al. 1998). This dominance of the cognitive sciences in science pedagogy addresses the cognitive dimension of inter-domain analogical reasoning with undisputed rigor.

In contrast, my study was not concerned with the cognitive dimension as a unidimensional approach to analogical reasoning. From a science education perspective, I wanted to know *something more* about analogical reasoning. For this reason

J.P. Davis (✉)
Faculty of Education, Queensland University of Technology, Brisbane, QLD, Australia
e-mail: jp.davis@qut.edu.au

A. Bellocchi et al. (eds.), *Exploring Emotions, Aesthetics and Wellbeing in Science Education Research*, Cultural Studies of Science Education 13,
DOI 10.1007/978-3-319-43353-0_7

I focused on the boundaries of analogical reasoning, looking for what was being excluded. Initially what I found was the exclusion of school science context. The lack of school science context was evident in the underlying research context where the theory of structure mapping was developed (Gentner 1980). I am referring here to Dedre Gentner's original work as part of a commissioned research consultancy for the US Department of Navy (Gentner 1980). That study sourced evidence from contexts unrelated to school science such as laboratory situations. Similar studies from nonschool contexts were also used in support of the theory of structure mapping. For example, Shawn Glynn (1994) cites research on scientific representation, conducted in the context of doctoral students doing physics research (Chi et al. 1981). The clear genealogical links between science education and the nonschool contexts of cognitive sciences theory development are problematic in terms of the theory's ecological validity for application to school science.

This problem of ecological validity became more evident when I explored some of the boundaries of this theory as proposed from within the cognitive sciences field (Gentner and Markman 1997). For example, Gentner and Markman (1997) highlighted limitations of structure mapping when explaining analogy as being similar to similarity. They acknowledged analogy to be too restrictive to fully explain analogical situations involving embodied imitation, everyday conversation, and situations involving empathy. In addition to these limitations, the science pedagogy literature cites many examples of everyday analogies because they are considered to be emotionally appealing in terms of excitement and student interest (Harrison and Coll 2008). In the context of school science where emotion seems prevalent, the theory of structure mapping seems to be a poor fit, given its lack of explanatory power with respect to emotion.

This chapter explains my approach to understanding emotion as a contextual factor in analogical reasoning within everyday science classrooms. Since emotion is fundamental to being human (Turner 2007), I start this chapter by exploring ontological presumptions. In the section *ontology, human dualism, and analogical reasoning*, I explain the notion of being human, built into the cognitive sciences view of analogical reasoning. I also show how different notions of being human make different theories of analogical reasoning possible. In the section *the sense of social being*, I expand upon an empirical, social approach to understanding being human. The section on *emotional energy as a collective, lived experience* outlines a social approach to understanding emotion, predating psychology. I explain *social practices as ethnomethods* as these were the indigenous resources of everyday interaction enacted by me and my students (cf. Garfinkel 1967). Ethnomethods enabled my students to co-construct occasions of analogical reasoning, and I analyzed these ethnomethods to understand interactional situations within the classroom. Finally, in *ethnomethodology and undramatic emotional energy*, I illustrate the locality of emotion within practices of analogical reasoning. In this section I identify a relationship between the emotion of respect and social practices of objectivity in analogical situations. I illustrate how practices of objectivity may be thought of as a form of undramatic emotional energy that suppresses other emotions when enacted through scientific practices.

Finally, where possible, I will illustrate analytical concepts using examples drawn from my data. These examples are presented using transcript conventions commonly associated with conversation analysis (CA) where speech gestures are shown in a chronological sequence of turns. These CA conventions originate from ethnomethodology (cf. Silverman 1998).

7.3 Ontology, Human Dualism, and Analogical Reasoning

Ontology is the study of *being* as the basis of fundamental existence as humans (cf. Hume 1777/2007). It is a notion people tend to take for granted, but it is important for two essential reasons. Firstly, different people define their being in different ways, and secondly definitions of being influence the way people think about the universe around them.

Researchers can identify particular ontological perspectives by analyzing the way people think, as ontology is often embedded in the language people use. Since theory is defined through language, ontology is also embedded in theoretical understandings often without people realizing. Becoming aware of ontology is therefore important when researchers import or inherit theories from different disciplinary or historical contexts. By adopting particular theories, researchers are also adopting implicit ontological perspectives. The purpose of this section is to explore the ontological foundations of the dominant view of analogical reasoning in the science education literature and to demonstrate how a change in ontology may change the way science education researchers understand analogical reasoning.

7.3.1 A Priori Ontology

An example of an embedded ontology is the a priori ontology I identify within the theory of structure mapping (Gentner 1983). This theory proposes a mapping of the analogical mind in terms of the *mental schemata* enabling analogical reasoning to exist as a way of thinking. This proposal suggests the mind exists as an independent object prior to the experience of reasoning, as evident by the language of science education theory (cf. Glynn 1994). In addition, the theory of structure mapping presumes concepts, and concepts about those concepts, exist prior to analogical reasoning taking place (Gentner 1983). An example of how language embeds this ontology is the proposal that "a theory of analogy must describe how the meaning of an analogy is derived from the meanings of its parts" (Gentner 1983, p. 155). This statement proposes an analogy, and its component concepts exist prior to the experience of reasoning, making it possible for the analogical mind to explore the universe through preexisting analogical concepts.

The implications of this a priori ontology as a foundation to the theory of structure mapping are that concepts are treated as though they exist prior to reasoning. Through the language of both Gentner (1983) and Glynn (1994), concepts are treated as fixed knowable objects. By understanding analogical reasoning through structure mapping of preexisting concepts, Gentner (1983) and Glynn (1994) propose to map our analogical minds. This treats peoples' minds as a priori objects with an innate capacity for analogical reasoning.

This a priori ontology is often associated with the cognitive sciences and is consistent with the ontology proposed by Immanuel Kant (1787/1996). This form of ontology presumes being arises from the innate capacity to think. This means at birth, a person is defined as a mind with the capacity to reason in different ways. The a priori existence of the psychological mind gives rise to the idea of individuality through a human dualism consistent with Kant's description of the body and a soul. This ontological perspective appears in everyday language when people say things such as "that boy is a natural sportsman" or "she is a born leader." Through such statements people presume innate capacities via a presumed existence as a unique soul, born, or maybe reincarnated, at the moment of biological emergence.

7.3.2 Empiricist Ontology

In contrast, in my study, I adopted an empiricist ontology consistent with the perspective of David Hume (1777/2007). Hume's ontological account described being as a dualism between the body and mind, where the mind was quite different from the Kantian soul. As an empiricist Hume (1777/2007) was concerned with knowing reality from the accumulation of particular experiences of perceptions. From this perspective Hume conceived of the mind as being *composed* of many particular perceptions of the reality individual people experience. If science education researchers were to develop a theory of analogical reasoning using Hume's empiricist ontology, they would need to use a form of inductive reasoning. This could be achieved by studying particular subject's experiences of phenomena that become knowable as objects and concepts used to construct analogies. This empiricist ontology opens the possibility for concepts and the pathways of reasoning to be constructed through actual lived experiences. To illustrate the practical implications, this perspective would make it possible for sportsman and leaders to be trained through experience, rather than being born with purely natural or innate capabilities.

Moving beyond these two examples (Kant and Hume), Emile Durkheim (1912/1915) identified a long and diverse history of ontologies he described as a human sense of dualism. Durkheim explained how different senses of dualism could arise in terms of what humans have in common across time and across diverse cultures. The common thread holding this sense of dualism together was society itself. That is, all humans with a sense of dualism belonged to a society capable of generating ideas, understandings of which could be passed from generation to generation. On this basis Durkheim (2005) conceived of a generalizable phenomenon of human dualism he described in terms of a social ontology. This social ontology may be expressed as the duality between a biological being and a social being (Rawls 2004).

7.4 The Sense of Social Being

The fundamental social dualism identified by Durkheim (1912/1915) is a dualism between the individual biological being and the collective social being. This fundamental notion of dualism defines a person's biological being in terms of their primordial, embodied existence. The biological being is the source of individuality as through the body; specific and fundamental perceptions of the world are experienced and interpreted as the physiological self (Rawls 2004). Thus people possess basic senses and basic capacities to interpret these senses to ensure a probable chance of survival beyond the reproductive maturity of our biological self.

In contrast, a person's social being develops from the moment in their biological life where they engage in social interaction with other humans. This social interaction opens a whole new world beyond a mere biological existence (Durkheim 1912/1915). Social interaction involves all of the motions and sounds of a person's body, made for interpretation by others (Rawls 2004). Interpretation of embodied motions and sounds gives rise to embodied responses of the other as further motions and sounds. These responses are then interpreted by the self, and the cycle of action and reaction continues throughout the life cycle.

The responsive, or reflexive, motions and sounds of social interaction appear as everyday talk and gesture between members of a society (Garfinkel 1967). It is through this talk and gesture that people make sense of the world. This is not simply by one person talking and the other listening; it appears as an ongoing sequence of interaction where sense making appears through shared understanding. To illustrate this, I have drawn an example from my study where year 10 students were observing tea leaves in a beaker of water being heated. This was part of an inquiry leading toward the concept of thermohaline ocean currents in a unit on global systems. At turn 1, Max makes an emphatic sound of "awh::" and begins to move his hand in a circular motion as shown below in the beaker example:

Transcript 7.1 Beaker example

1	Max	awh:: we got some uhm:: whats the word:: circ: ((Max stands, Max's face orientates to Mr D, and Max moves his right hand in a vertical circular gesture))
2	Simon	circulation
3	Mr D	circulation how do you know that

In this example, making sense or knowledge of what is happening in the beaker is co-constructed as shared understanding. That is, Max makes motions and sounds at turn 1 thereby sharing a sense of what he feels about the phenomenon in the beaker. In terms of language, Max's sounds and motions are mostly incoherent. Regardless of this, Simon makes sense from what he observes in the beaker and from Max's utterances and gestures. Simon makes the sound "circulation" at turn 2. In this way, Simon has shared his synthesis of prior shared meanings in a new shared meaning. Finally, at turn 3 I agreed with Simon's synthesis of the situation. This agreement is evident in the way I mimic Simon's sound of "circulation" sharing with both students, my agreement of their co-constructed meaning. In this way, all three members of this small society (Max, Simon, and me) have demonstrated how social interaction establishes meaning, shared understanding, and knowledge through back-and-forth, reflexive, social interaction.

This example illustrates how an idea such as *circulation* appears as a generalization from a perception of particular observed phenomena. The idea enables different biological beings to share their perceptions of phenomena they experienced through observation as individual physiological beings. The idea that this particular observed reality could be generalized as *circulation* became possible through a back-and-forth exchange of talk and gesture. Each moment of the exchange was observed and interpreted by the other as confirming or disconfirming previous shared meanings. In this way, each member of this small society is, metaphorically, holding a mirror up to the other, as a way of confirming or disconfirming ideas. Contributing to the reflexivity of social interaction enables each society member to co-construct shared understanding. As shared understandings, ideas are the product of society, and for this reason, they come to define society (Durkheim 1912/1915).

An important feature of the way in which the shared idea of *circulation* was established in Transcript 7.1, above, was the incoherence of Max's motions and sounds. Max's utterances of "awh:: we got some uhm:: whats the word:: circ:" together with the energetic circulating hand motions were observed as a moment of excitement. While Max's language may have been incoherent, he was enacting a feeling about his idea. This enactment involved an emotional experience drawing the attentiveness of Simon and me toward Max. I will now explore this emotional dimension as the experience of shared or collective emotion.

7.5 Emotional Energy as a Collective, Lived Experience

An important part of a person's social being is their ever-present experience of emotion. Everyday understandings of emotion tend to form as people reflect on their experiences and on what they observe as other people's bodily expressions of emotion. Through these reflexive experiences, people come to know emotion as ideas, evident in the rich language they use to describe emotion (Turner 2007). This rich language tends to describe how one person perceives another person's expression of emotion or how people label emotion. Such language forms conceptualizations of emotion from specific instances to enable a level of generalization in thinking about emotion. In essence labels help us think about emotion as an idea.

For researchers, one of the challenges in studying emotion is identifying the construct as an observable phenomenon, distinct from secondary ideas of the phenomenon. For example, John Dewey (1894) suggests Charles Darwin's (1872) study of *The Expression of Emotion in Man and Animals* was an example of the *psychologists' fallacy*. The *psychologists' fallacy* was described by Dewey as the confusion between labels of emotion as secondary reflections and the phenomenon of the experience of emotion. My study elaborated this approach to the experience of emotion by studying situated, lived emotion, experienced through people's sense of biological and social being.

Dewey (1894) suggested emotion experienced at an individual level comprises of the elements of internalized feeling, embodied movements, and external object or vague idea. While rudimentary, these elements are consistent with contemporary component models of emotion (Scherer 2005) and are used in my study because of their coherence with my sociological perspective. In my study I used these elements to operationalize the experience of emotion as a holistic phenomenon where elements are non-serial and interdependent. That is, all three elements are considered to erupt spontaneously and evolve concurrently as a whole pulse of emotion.

The internalized feelings arise through and with the physiological actions of the body (Dewey 1894). These may include increased heart rate, perspiration, heavy breathing, facial gestures, or other bodily movements. Embodied movements in relation to feelings occur concurrently. For a feeling to be perceived by the self, internal bodily movements are essential in terms of physiology. For example, feeling embarrassed may involve being flushed and sweaty, which requires physiological motion. Such physiological motion produces the subjective feeling. More prominent gestures or motions such as lowering the head or a nervous smile may also be observable by other people. But to the self these are all part of the emotional experience.

In a temporal sense, the locality of embodied motions in relation to the feeling may vary. They may appear to be preparatory, intermediate, or as an accomplishment of the emotion (Dewey 1894). For example, if a person is walking down the street and hears an unanticipated loud noise immediately behind him/her, the first response may be to jump or run. This embodied reaction is quickly experienced as a feeling possibly known as *fear*. The jump or run is a preparatory bodily motion as it precedes the feeling of fear. The internal bodily motions of increased heart rate, vasodilation, and gastrointestinal constriction are intermediate as they are the physiological source of the feeling of fear, immediately following the initial jump or run. The bodily motions of heavy breathing, crying, or laughter following the feeling of fear are the final accomplishment of the emotion. Hence bodily motions are located at different temporal points in relation to the feeling that together form the experience of emotion. In this sense, these embodied motions are not mere expressions. Together with the feeling, they are the emotional experience.

The final element of emotion is what Dewey (1894) described as the intellectual element appearing as the focus of the emotion. Emotion is always about, or focused toward, an idea. The idea is most often vague, but it is situated within the pulse of emotion as an experience of movement, feeling, and concept (Dewey 1895).

A notable feature of the intellectual element as an idea is its tendency to be vague, at least at the beginning of an emotional experience (Dewey 1894). The experience of emotion may be observed as a response to these vague ideas, which researchers may think of as embryonic concepts. That is, the idea may start out vague and may become clearer as the lived experience progresses. This vagueness of the idea suggests the spontaneous, holistic appearance of emotion is a way for us to know something is important, without any need to know why. The emotion attracts and maintains us in the experience, and through this experience, it is possible to learn more about the idea within the emotion.

In the social or collective context, Durkheim (1912/1915) described emotional experience as collective effervescence giving rise to a sense of solidarity, comfort, and dependence within society. More recently the notion of Durkheim's collective effervescence was described as emotional energy (Collins 2004). Emotional energy is conceptualized as the glue binding social interaction. Like individual emotions it ebbs and flows as people seek out situations where emotional energy is present. Emotional energy is the collective energy unifying group members around shared objects. It becomes evident where embodied actions and conversation illustrate a heightened sense of arousal, attentiveness, and focus about an object as being shared by the group. These embodied actions tend to be synchronous such as eye contact, glances, head movement, body orientation, gestures, facial expressions, or verbalizations (Collins 2004).

In a sense emotional energy is observable through the social practices of the group, but it has a further dimension comprising the invisible or perceptual experience of emotional energy. This experience is internal to the members sharing the experience. It is the phenomenon of perception of emotion, the existence of which becomes visible in the group social practices (Durkheim 1912/1915). The reflexive property of social practices enables these internal emotions to be embodied, interpreted, and reflected to give rise to an outwardly visible, hearable, and shareable experience of emotional energy. This idea of emotional energy being internalized and acting on the social being from within the self is explained by Durkheim (1912/1915) in terms of moral force. For this reason the relationship between emotional energy and moral force is outlined below.

7.6 Emotional Energy and Moral Force

Durkheim (1912/1915) described the feeling of moral force as the emotion of *respect*. Moral force is the force driving peoples' actions as social beings. In a localized interaction, it is the force producing specific social practices between group members in response to each other. Like physical force people, including researchers, cannot see moral force. What researchers can see are the reflexive social interactions they may attribute to moral force. For example, consider a situation where students have a worksheet guiding them to observe an object. The situation below relates to the context of the earlier *beaker example*. As shown this example starts with a student with the pseudonym of Simon reading from the worksheet "what do you observe about movements in the contents of the beaker" at turn 1 below:

Transcript 7.2 Moral force and observation

1	Simon	what do you observe about movements in the contents of the beaker ((reading from the worksheet))
2	Shady	well we can't see much at the moment ((looking down into the beaker))
3	Max	wel:l: if we wait its gonna ((going to)) heat up: bubble
4	Shady	and their gunna ((going to)) go to the side eh ((places hand over top of beaker and raises hand))
5	Max	what do you reckons gonna ((going to)) happen to the balls when it starts boiling
6	Simon	awh no you have to observe what it is
7	Sam	its just sitting in the water

From a Durkheimian perspective, moral force is evident in the way Simon treats the worksheet instruction. In turn 1, Simon establishes *respect* for the worksheet by reading it and establishing the need for observation. The response to Simon's utterances from Shady at turn 2 is to look at the beaker and state that they "can't see much at the moment." This is followed by a series of speculations and predictions from Max (turns 3 and 5) and Shady (turn 4).

Moral force becomes evident at turn 6 when Simon utters "awh no you have to observe what it is." Simon's utterance at turn 6 follows his treatment of the worksheet instruction where he established the need for observation (turn 1). In turn 1, I determined in my study that Simon had treated the worksheet and its idea of observation with *respect*. In the situation where Shady and Max began to speculate instead of observing, Simon's sense of respect for the worksheet instruction may be said to have given rise to his response at turn 6. The Durkheimian (1912/1915) interpretation here is that turn 6 was made possible by a moral force energized by Simon's sense or feeling of respect. The emotion of respect was experienced through a moral force made visible by gestures and hearable by words. In my data, respect was also evident in the episode of interaction following turn 7 where Shady and Max resisted further speculation and practiced observation. This became a further example of moral force as they shared and enacted respect for the worksheet instructions.

Through this type of localized social practice, moral force has its origins in society. As a person accumulates experiences, moral force becomes internalized, and this internalized moral force shapes ideas about the self to define an individualized sense of self as a social being. Over time the moral forces acting on the sense of self become the accumulated experiences composing the mind. It is through this action of moral force and lived social experiences that Durkheim (1912/1915) explains the development of individual psychology. That is, moral force shapes the internal structure of the mind from its external experiences of society (Durkheim 1912/1915).

The internal-external dimensions of moral force arise by virtue of its social character. It originates through social practices, but it acts on and through individuals. As observed by Durkheim "society can exist only in individual minds and through them, it must penetrate and become organized inside us" (Durkheim 1912/1915 p. 157). In this way society becomes an essential part of peoples' sense of being. It is the basis of their sense of dualism as it raises people from being simple biological creatures and makes them part of something bigger than their biological being.

In my study, emotion was viewed as a collective, lived experience of emotional energy. The dimensions of emotional energy were taken as the holistic pulse of internalized feelings, coordinated embodied actions and sounds, and shared understandings or ideas. As a collective phenomenon, the emotional experience operates through the sense of social being generating feelings of respect. This emotional experience of respect is the source of energy for the moral forces guiding actions and becoming evident in social situations. These situations are visible and hearable through the embodied actions and ideas evident within social practices of the science classroom. In the following section, the notion of social practices is explained in terms of ethnomethods.

7.7 Social Practices as Ethnomethods

Social practices were defined earlier as the embodied motions and sounds of people when interacting with each other. In my study of the emotional dimension of analogical reasoning, I used a particular approach to social interaction by treating social practices as ethnomethods. Ethnomethods are observable as the taken-for-granted, everyday activities of social interaction, indigenous to particular localized social situations (Garfinkel 1967). By adopting these everyday activities as a resource for understanding the essence of social phenomena, ethnomethodology uses a bottom-up approach to accomplish an understanding of the lived order of everyday social practices (Pollner and Emerson 2002). The ethnomethodological approach to social interaction incorporates an epistemology that aims to know the essence of social phenomena (Garfinkel 1967). In this approach ethnomethods are treated as a resource for understanding how society's members accomplish their shared reality.

To illustrate ethnomethods and ethnomethodology, I refer back to Transcript 7.1 where Max, Simon, and I established a shared understanding about a concept as *circulation*. The initial ethnomethods in that situation were the sounds and gestures of Max, entraining Simon and I toward Max. In turn 1 of that episode of interaction, Max's ethnomethods accomplished the entrainment of our groups focused upon Max. His ethnomethods established or directed emotional energy, and in doing so, they situated Simon and I as observers of Max's performance. In turn 1, this reorientation of Simon and I upon Max was an accomplishment of Max's ethnomethods. Importantly Max would not have been aware of his ethnomethods or what they were accomplishing. If I had interviewed Max about these ethnomethods, he would not have understood my question. This is because ethnomethods are generally taken for granted.

To explain this further, if Simon and I had observed Max's sounds and motions from an outsiders' position, we may have seen an incoherent, linguistically challenged person, making ridiculous bodily actions. But instead, we accepted Max's ethnomethods as completely normal, and in doing so, we accounted for them by describing them as circulation. In essence we used Max's ethnomethods to make sense of the situation. We did this by putting his incoherent performance into the context of our previous interactions to create an account of Max that would make sense to all of us in that shared experience.

From a researcher's perspective, ethnomethodology uses Max's, taken-for-granted, ethnomethods to understand how the concept of circulation was co-constructed. By studying Max's ethnomethods, I arrived at a detailed study of the practices that produced the account of circulation as a form of knowledge about the beaker objects and Max. In this way, the ethnomethods are the taken-for-granted methods of the people, and they are also a resource for the researcher to understand how the topic of circulation was co-constructed by Max, Simon, and I.

7.8 Ethnomethodology and *Undramatic* Emotional Energy

The enactment of analogical reasoning with ethnomethods also gave rise to the coordinated, enacted movements of emotional energy. In the development of his concept of emotional energy, Randall Collins (2004) emphasized the empirical dimensions he considered to be observable and measureable. These included the coordinated actions of bodies evident through the alignment of posture, movement, eyes, voice, and facial expression. These methods have been utilized in previous studies of science classrooms (Milne and Otieno 2007). Collins (2004) also suggested self-report, and emotion diaries are an example of its application in emotion research (Bellocchi 2015).

In addition to these aspects identified as measurable, Collins (2004) discussed the nature of emotional energy from an ethnomethodological perspective. In contrast to the many examples Collins (2004) used to illustrate emotional energy, he emphasized how highly visible outbursts of collective emotion were not an essential feature of emotional energy. In fact, his conceptualization of emotional energy was an enduring emotion becoming visible in subtle or nuanced ways. Collins (2004) anticipated emotional energy being produced in covert, taken-for-granted ways such as the "undramatic... long-lasting... tones or moods that permeate social life" within our mundane reality (Collins 2004, p.106). For this reason, I have gone beyond the simple measures of overt instances of emotional energy and used the analytical techniques of ethnomethodology to identify and understand *undramatic* emotional energy. These are experiences of emotional energy giving us a sense that "nothing out of the ordinary is happening here" (Collins 2004, p.106).

One example of undramatic emotional energy drawn from my data was the practice of collective writing. The enactment of collective writing was a self-organizing practice, achieved through unspoken group consensus. In terms of emotional energy, the undramatic emotion of this mundane situation becomes visible through a sense of common purpose displayed through coordinated actions. These actions were the visible accomplishment of solidarity around a shared idea of collective writing about objects of inquiry. From a Durkheimian (1912/1915) perspective, I describe this solidarity as being sourced from a moral force driving coordinated actions toward a common purpose. While the sense or feeling of solidarity cannot be directly or empirically observed, it was visible through unspoken coordinated actions of the group. As these actions required no conversation to accomplish their coordination, the situation was interpreted in terms of a shared feeling of common purpose. This shared feeling was sufficient to apply moral force from within to produce practices of collective writing. Not only did this sense of solidarity make collective writing possible, but it established a social order in terms of roles such as who would scribe, who would speak, and who would observe in silence. In an ethnomethodological sense, this is the type of emotional energy Collins (2004) referred to as *undramatic emotional energy*.

To illustrate undramatic emotional energy, an example is used where students established embodied alignment and conceptual entrainment through the practice of collective writing. The group involved four students from the earlier beaker example, identified as Max, Shady, Simon, and Sam. Working together in a science inquiry lesson, they co-constructed a practice of collective writing. The vague idea of the group writing in a collective way was established early in the lesson when one of the students, Sam, stated "I'll get a pen." This was despite all group members having their own worksheets. He was then seen on video leaving the workbench and returning with a pen. Sam's adoption of this role was unchallenged and becomes observable through a single innocuous conversational turn. At the moment of Sam's statement, Simon had been reading the worksheet aloud to the group while Shady and Max were making tentative speculations about the inquiry. Sam's statement was quietly stated, and his action to get the pen was unchallenged. Some minutes after this action by Sam, Max began to dictate words and Sam began to write. Between a moment where Sam acquired a pen and the moment where Max began to dictate words, there had been no discussion about writing. The practice of collective writing was simply enacted by Sam and Max and tacitly accepted through the silence of Shady and Simon. The embodied actions of a speaker and a scribe become visible as and through the alignment of collective actions around the vague idea that the group would write something as a collective.

As the group stood around a workbench and observed their phenomenon of inquiry, Max began to speak in a style of speech, measured as being slower than his usual rate. This is shown in syllables per second (SPS) at the end of each turn, below. In addition the words he used were uttered in a selective, tentative tone. At the same time, Sam was making note of Max's utterances. This change in Max's style of speech was identified in my study as language practices contributing to the co-construction of objectivity, to be considered in more detail later. On this occasion, Max's language practices appeared as shown below:

Transcript 7.3 Language practices and self-prohibition

1	Max	so: uhm: we could note the ba::lls when their close together they stick together, when you get close to the edge its stuck to the edge(.) ((3.1SPS)) ((Max glances at Sam))
2	Max	say something like that, I don't know how you would word that(.) that's saying it like a ret*rd but ((4.1SPS)) ((Sam is writing as Max speaks))

These data show the speech rate in turn 1 at 3.1SPS and in turn 2 at 4.1SPS. In the first, slower utterances, Max is speaking on behalf of the group as part of their collective writing. These practices are objectifying the shared observations of the phenomenon. In the second, faster utterances, Max's speech rate increases as he speaks in his normal, *nonscientific* style. The glances to Sam direct Max's speech as something for inclusion in the collective notes. Sam responds by writing. In this situation, the vague idea of collective writing and the enactment of the idea are coordinated as an experience of undramatic emotional energy.

A further example of undramatic emotional energy was evident in an episode of interaction between me and a year 9 student with the pseudonym of Chris. Leading into this situation, Chris and I were sitting side by side. I was drawing a diagram of a carbon-14 nucleus on paper, and Chris sat with his right elbow on the desk, his hand supporting his chin, and his eyes entrained on the emergent diagram. I ask Chris "how many neutrons have we got" as shown at turn 1:

Transcript 7.4 How many neutrons?

1	Mr D	how many neutrons have we got
2	Chris	fourteen ((sitting with his right elbow on the desk, and his hand supporting his chin, eyes entrained on the diagram))
3	Mr D	well(.) we've got a MASS of 14(.) so if its fourteen minus six
4	Chris	eight ((Chris maintained his bodily alignment))
5	Mr D	eight(.) we've got one two three four five six seven EIGHT neutrons okay ((I draw 8 N's in circles near the protons))
6	Chris	yeah ((Chris maintained his bodily alignment))

Chris' response to my question was the utterance "fourteen" (turn 2). This answer was incorrect, but instead of telling Chris he was incorrect, I paused after uttering "well(.)" and then reframed the question, making it highly probable that Chris would utter the correct answer (see turn 3 above). Chris' response this time was the utterance "eight" (turn 4). I then mimicked Chris with "eight(.)" and then paused before counting and drawing each of the neutrons. Throughout these moments Chris maintained his gaze and his bodily alignment through to his final utterance of "yeah" as a form of agreement (Lambertz 2011).

The undramatic emotional energy in this situation is evident in the way that Chris and I maintained a shared gaze on the emerging diagram as evident by Chris' maintenance of his bodily alignment toward my gestures and the diagram. We were both entrained physically and conceptually on the carbon-14 nucleus being made visible on the paper. Most interesting about this situation was that Chris' did not lose emotional energy as he becomes aware of his incorrect answer (turn 2). The ethnomethod of my response may have contributed to the maintenance of emotional energy as I did not explicitly tell Chris his answer was wrong. By pausing and reframing the question, my ethnomethods situated Chris to remain focused on a new answer. My ethnomethods avoided social disjuncture and this retained the emotional energy as evident by Chris' continued focus. Chris remained almost motionless through the entire episode. He was certainly undramatic in these moments, yet he maintained a high level of shared focus with me, evident of undramatic emotional energy.

Undramatic emotional energy has not been reported in science classrooms. Catherine Milne and Tracey Otieno (2007) and Stacey Olitsky (2007) studied emotional energy in science classrooms but not in this undramatic form. Olitsky (2007) suggested researchers should anticipate undramatic emotional energy in science classrooms. This expectation is grounded on the common notion of science as an unemotional endeavor, despite its practice as a human endeavor. This anomaly has been noted throughout history (cf. Dewey 1895).

In my study I took this notion of undramatic emotional energy and explored it further in the context of Durkheim's (1912/1915) notion of respect. As noted in the earlier discussion, respect was observed by Durkheim as a fundamental emotional experience visible as enacted embodied practices attributed to moral force. In the following sections, I explain the relevance of respect in my study, but first, I illustrate how respect may be observed in different micro-social contexts.

7.9 Respect

Respect becomes visible across all areas of society in different ways. As an example of the importance of respect within contemporary Western culture, I cite two distinctively different examples. In the first example, respect is associated with the enactment of a sense of community within deprived inner-city communities in the United States (Patterson et al. 2011). In this sense respect gives rise to practices of acceptance of difference within groups and determines who is in the community and who is not.

In a totally different context, respect was applied to ideas as well as people. For example, in the context of medical practice in the United States, Larry Gruppen (2013) suggests knowing or not knowing concepts is not purely an intellectual state. Rather it is a powerfully emotional experience affecting judgment and professional practice. In this context respect is associated with an emotional experience around the inclusion or exclusion of concepts within medical practice. It is considered good medical practice to respect concepts from different disciplines. Yet this respect is not uncritical, as concepts must meet particular criteria to be included as relevant to particular situations of medical practice (Gruppen 2013). The inclusion or exclusion of concepts therefore becomes an accomplishment of the collective enactment of criteria as social practices, driven by a sense of respect.

These diverse examples of respect in different contexts illustrate the importance of respect across society. In the first example, respect was important for establishing community solidarity between people. In the second example, respect was important for establishing shared ideas across disciplinary boundaries. In my study of the science classroom, I illustrate the inseparability of particular enacted practices and emotional energy as respect. In school science I found respect as an important means for including and excluding ideas, forming a sense of kinship between group members, and in relation to shared ideas.

To illustrate this I outline a further interaction between Max, Simon, Shady, Sam, and I as the teacher. Prior to this occasion, the students had experienced a momentary loss of emotional energy about the objects of their inquiry. This was evident by their casual conversation with an adjacent group of students. As I arrived to join the group, I positioned myself at eye level to the objects on the workbench and asked "so what can you see here." The group responded by immediately realigning themselves with the physical objects on the workbench. Sam and Simon were opposite each other, and they both moved their heads lower, in unison, so that they could closely observe the objects. At the same time, Max turned toward the objects and started to speak on behalf of the group to explain their collective observations of the objects.

With this example I illustrate an accomplishment of both physical alignment and conceptual entrainment in the context of a shared experience of respect. As I arrived, these coordinated actions and the conceptual focus simply happened. Although I asked about the objects and looked at them, I did not request any particular actions from the group. The group could have responded to me by continuing their nonscientific conversation with the adjacent group. But they did not.

Based on the Durkheimian notion of respect discussed earlier, the actions of the group in my presence may be interpreted as a response to feelings of respect for me as the science teacher. This sense of respect maintained solidarity with me as the science teacher, and as such it reestablished solidarity within the group as they rallied around the objects. An interesting feature of this situation was the way respect for me as a science teacher was enacted in relation to the objects on the workbench. The emotional energy of respect for the teacher was transferred to the objects, unifying the group around the objects.

This simple example shows how the emotional energy of respect shapes the practices of a situation incorporating objects within the social order of a group. In this example the objects for which the group established respect were objects being used in the context of analogical reasoning. As such, this example illustrates the way in which respect operated upon enacted practices of analogical reasoning, to establish concepts as part of the social order of the group. But in this context, respect was observed as more than the shared inclusion of concepts into the group. As the following section shows, respect affected the status of people and concepts that contributed to how those objects became known.

7.10 Objectivity as Respect for Objects

The co-constructed status of objects and people is evident in my data through the enactment of objectivity as a way of establishing hierarchy within the group. This enactment of objectivity is evident through practices of self-prohibition involving language, physical distancing, and self-deprecation. Inseparable from these practices is evidence of emotional energy or respect toward the beaker objects. In this context where practices of objectivity and respect are inseparable, I have referred to objectivity as respect for objects. Evidence supporting this view of objectivity as an emotional sense of respect is visible through the practices of Max, Sam, Shady, and Simon. In this section I outline three of the more prominent practices.

The first of these were practices of language prohibition. Language prohibition established a sense of objectivity by treating the objects scientifically. Durkheim (1912/1915) considered language prohibition within religious practices as a way of making objects knowable as sacred objects. In my study language prohibition is evident in the way Max changed his language about the objects as the inquiry proceeded. In the initial episodes, Max referred to the objects using profane, everyday language. The objects consisted of a hot plate, water, some foam beads (balls), and some tea leaves. Early in the lesson, Max responded initially with "are we cooking tea" and by making comments about how he liked drinking his tea. In this way, Max was treating the objects in their familiar everyday sense. Later in the inquiry, as Max enacted the role of speaker for the group, Max's language about the objects was spoken in a more deliberate and tentative tone. As an example, in addition to the extract shown earlier, Max made the following statements about the objects as dictation to Sam (the scribe):

Transcript 7.5 Language practices and introduced concepts

1	Max	so by putting the tea bag into the beaker the water has been tinted ah: goldish tea-ish colour and you can see and it shows the convestion and the um the um movements of the water:: currents:currents as the currents of the water as it goes around

This statement by Max occurred quite late in the inquiry. Through this language Max now speaks of the tea in terms of an observation about concepts such as *convection*, *movements*, and *currents*. At this moment he is describing the tea as an object he has come to know through observation and social interaction. The observation was conducted via a further prohibition in the form of a physical spatial prohibition the group imposed upon themselves.

Like language prohibitions, Durkheim established that spatial prohibition was an important practice in religion that contributed to the sacred status of objects. In my study spatial prohibition was initially established through Simon's appeal for others to observe the objects from a distance. In support of Simon, Max also appealed for others to establish physical prohibition on two occasions. On one of these occasions, Max stated "just leave it(.) Shady leave (.) its what they want to do." In this statement, Max not only made an appeal to Shady for physical self-prohibition, he also established a sense of volition about the objects. Max's statement treated the objects with some form of status other than being simply water and tea leaves. This suggested they had some right or capacity to act on their own.

In this example, the group established practices prohibiting certain types of language in relation to the objects. This creates a sense of distance in an analytical sense as the language establishes conceptual knowledge of the objects beyond the previously existing everyday knowledge of those objects. This analytical space is concurrently reinforced by the physical space, established and maintained between the objects and the profane everyday bodies of group members. In Durkheim's social ontology, the human body is considered a profane everyday object. This view of the body is reinforced in my science classroom as group members attempt to observe objects from a distance, separate from their own bodies.

As this example illustrates, respect for the objects establishes practices of observation, physical, and analytical distancing. In addition this respect for the objects was more evident when Max suggested the objects be left alone as "it's what they want to do." Together these practices treated objects with a particular form of respect enabling them to become known scientifically as objects of inquiry. The experience of this particular form of respect for objects involves the enactment of a spatial arrangement, both physically and analytically. In this way respect for objects not only involves inclusion of objects into the group, but it also enables these objects to be transformed from everyday objects into sacred, scientific objects.

As an undramatic emotional energy, respect for the object is evident through these practices making visible objectivity in science. Physical space was created by physical prohibition and analytical space was created by language prohibition. In my study I establish a position suggesting emotional space was created by displacement of dramatic emotional energy, which was replaced with respect for objects in the form of objectivity. This was observable in my data as practices of prohibition were driven by a subdued, coordinated, focus on the scientific objects. My reference here to objectivity is therefore from two perspectives. One perspective views objectivity as a collection of practices evident through the ethnomethods and contributing to new ideas about the objects. The other perspective views objectivity as the experience of emotional energy of those practices I have called respect for the objects. In the context of science, by imposing practices of objectivity on the self, people experience a displacement of dramatic emotion with a replacement of undramatic emotional energy, in the form of respect for the object. It is this undramatic emotional energy I interpreted as the collective emotion of objectivity.

Finally, objectivity operates by establishing social status between people and objects, observable as a hierarchical spatialization of the social order. This is evident at a basic level when the group decides which objects will be the focus of the emotional experience and which objects will not. In the earlier example where I arrived in the group, the students immediately discarded profane concepts and immediately returned to concepts that were the focus of their emotion of objectivity. In this way objectivity establishes communities of in-groups and out-groups. Within the in-group, the emotion of objectivity drives objects to the highest level of the shared analytical space. This was evident in Max's utterances in Transcript 7.3, above. This example included a self-deprecating utterance of "say something like that, I don't know how you would word that(.) that's saying it like a ret*rd but." Through this utterance Max positions his social being as a lower status to the objects. This utterance suggests Max feels his words are too profane, or unworthy of describing the objects, thus situating himself within the hierarchy at a lower status in relation to the objects.

This issue of status was also evident in Max's second statement where he apportioned volition to the objects. In that statement self-prohibition of physical distance was justified by the rights of the object to move freely. Those rights were co-constructed by group members prohibiting their own spatial freedom. By prohibiting their own freedom so the objects could exercise freedom, group members were enacting a hierarchy, situating the objects as higher than themselves. Together these practices are indicative of a hierarchical distribution of objects and people within the group, observable as practices of objectivity and as respect for the objects.

7.11 Summary and Implications

In this chapter I illustrated the application of methodological tools to the analysis of data drawn from the lived experiences of science classrooms. I explained how *something more* may be known about an *established* topic such as analogical reasoning. To better understand this topic, I accessed the indigenous ethnomethods of student groups and used these as analytical resources.

This study has important implications for science education in terms of methodology, contributing to pedagogical theory and enacted pedagogical practices. Ethnomethodology has enabled me to capture living (in vivo) data about practices and emotional energy of the classroom, as experienced in everyday, taken-for-granted situations. This research approach is important as it adopts different ontological presumptions about being human. These different presumptions make possible different theoretical perspectives. The different perspectives I have described enable us as science education researchers to better explain student learning and pedagogical practices using ecologically valid empirical data. I suggest the social ontology and ethnomethodology described in this chapter may benefit other areas of science education research.

A significant benefit of ethnomethodology is the researcher's focus on how our shared sense of reality is co-constructed while leaving intact the sense of reality. For example, it is possible to think of analogical reasoning in terms of inter-domain structural relations while also appreciating the localized social practices and emotional experiences co-constructing the lived analogical situation. In this way ethnomethodology leaves the dominant theory intact while improving our understanding of how it is enacted in the localized situations where the society of the science classroom is continually reenacted and reproduced.

The power of ethnomethodological studies is the capacity to generate pedagogical findings (Garfinkel 1996). For science education researchers, these pedagogical findings may be translated to pedagogical theory. Out of this study, I have achieved improved understanding of student learning and teaching practices when using analogy in the science classroom. These improved understandings may impact pedagogical theory by connecting practical reasoning through embodied action, conversation, and emotional energy with the co-construction of shared conceptual understanding. This study has contributed to the definition of operational constructs such as conceptual entrainment, undramatic emotional energy, and respect. These constructs provide opportunities for future research and theoretical development.

Finally, the possibility for these operational constructs in my study to be observed in everyday science classrooms makes my pedagogical findings accessible to teachers. By understanding analogical reasoning from the perspective of observable social practices and shared emotional experiences, my pedagogical findings may be transferable to science teaching practices. The findings of this study should enable teachers to better understand how pedagogies involving social interaction and emotional experiences offer successful approaches to engaging students in learning science.

References

Bellocchi, A. (2015). Methods for sociological inquiry on emotion in educational settings. *Emotion Review, 7*, 151–156. doi:10.1177/1754073914554775.

Chi, M. T. H., Feltovich, P. J., & Glaser, R. (1981). Categorization and representation of physics problems by experts and novices. *Cognitive Science, 5*, 121–152.

Collins, R. (2004). *Interaction ritual chains*. Princeton: Princeton University Press.

Darwin, C. (1872/1965). *The expression of emotions in man and animals*. Chicago: University of Chicago Press.

Dewey, J. (1894). The theory of emotion: Emotional attitudes. *The Psychological Review, 1*, 553–569.

Dewey, J. (1895). The theory of emotion: The significance of emotions. *The Psychological Review, 2*, 13–32.

Durkheim, E. (1912/1915). *The elementary forms of the religious life* (J.W. Swain, Trans.). London: Allen & Unwin.

Durkheim, E. (2005). The dualism of human nature and its social conditions. *Durkheimian Studies, 11*, 35–45.

Garfinkel, H. (1967). *Studies in ethnomethodology*. Cambridge: Polity Press.

Garfinkel, H. (1996). Ethnomethodology's program. *Social Psychology Quarterly, 59*, 5–21.

Gentner, D. (1980). *Report no. 4451 for office of naval research and the defense advanced research projects agency: The structure of analogical models in science*. Cambridge, MA: Bolt, Beranek & Newman Inc.

Gentner, D. (1983). Structure-mapping: A theoretical framework for analogy. *Cognitive Science, 7*, 155–170. doi:10.1207/s15516709cog0702_3.

Gentner, D., & Markman, A. B. (1997). Structure mapping in analogy and similarity. *American Psychologist, 52*, 45–56.

Glynn, S. M., National Reading Research Center, A. G. A, & National Reading Research Center, C. P. M. D. (1994). *Teaching science with analogies: A strategy for teachers and textbook authors* (Reading Research Report No. 15). Athens: NRRC.

Gruppen, L. D. (2013). Humility and respect: Core values in medical education. *Medical Education, 48*, 53–58.

Harrison, A. G., & Coll, R. K. (2008). *Using analogies in middle and secondary science classrooms*. Thousand Oaks: Corwin Press.

Hume, D. (1777/2007). *An enquiry concerning human understanding*. (P. Millican, Ed.). Oxford: Oxford University Press.

Kant, I. (1787/1996). *Critique of pure reason*. (W. S. Pluhar, Ed.). USA: Hacket Publishing.

Lambertz, K. (2011). Back-channeling: The use of yeah and mm to portray engaged listenership. *Griffith Working Papers in Pragmatics and Intercultural Communication, 4*, 11–18.

Milne, C., & Otieno, T. (2007). Understanding engagement: Science demonstrations and emotional energy. *Science Education, 91*, 523–553.

Olitsky, S. (2007). Promoting student engagement in science: Interaction rituals and the pursuit of a community of practice. *Journal of Research in Science Teaching, 44*, 33–56. doi:10.1002/tea.20128.

Patterson, A., Cromby, J., Brown, S. D., Gross, H., & Locke, A. (2011). 'It all boils down to respect doesn't it?': Enacting a sense of community in a deprived inner city area. *Journal of Community and Applied Psychology, 21*, 342–357.

Pollner, M., & Emerson, R. M. (2002). Ethnomethodology and ethnography. In P. Atkinson, A. Coffey, S. Delamont, J. Lofland, & L. Lofland (Eds.), *Handbook of ethnography* (pp. 136–144). Thousand Oaks: Sage Publications.

Rawls, A. W. (2004). *Epistemology & practice: Durkheim's the elementary forms of religious life*. New York: Cambridge University Press.

Scherer, K. R. (2005). What are emotions? And how can they be measured? *Social Science Information, 44*, 695–729. doi:10.1177/0539018440582l6.

Silverman, D. (1998). *Harvey Sacks: Social science and conversation analysis*. Cambridge: Polity Press.

Treagust, D., Harrison, A., & Venville, G. (1998). Teaching science effectively with analogies: An approach for preservice and inservice teacher education. *Journal of Science Teacher Education, 9*(2), 85–101. doi:10.1023/a:1009423030880.

Turner, J. H. (2007). *Human emotions: A sociological theory*. New York: Routledge.

James P. Davis is a researcher and lecturer at the Queensland University of Technology, Brisbane, Australia. James has a background in qualitative and quantitative research in health and science education. His current research focuses on analogical reasoning in science teaching and learning, and the sociology of emotions in high school science, using interpretative research approaches such as ethnomethodology, phenomenology and micro-sociology.

Chapter 8
Our Neighborhood: A Place for Heightening Emotional Energy in Science Education

Kristin Cook and Gayle Buck

In this chapter, we present examples in which place-based pedagogy affected students' and their teachers' emotional energy in the science classroom. We incorporate findings from the teachers' efforts to implement place-based curricula. Findings from our work with these schools have indicated that allowing students to draw upon local referents to science adds an element to instruction on which teachers can build investigations alongside their students—offering participants opportunities to collectively and individually examine issues that affect their lives and giving them a holistic awareness and appreciation of how science connects to their community. Students' and teachers' emotional energy toward science learning also heightened feelings of connectedness with others and desire to do science. Moreover, we explore barriers to realizing the potential of place-based pedagogy in terms of fostering students' engagement with the community. With this chapter, we hope to provide science teacher educators who are interested in enhancing emotional engagement through place-based instruction with practical applications and theoretical underpinnings that could enhance their teaching and research.

8.1 Engagement in Terms of Emotional Energy

Research has linked student emotional engagement to the presence of affective indicators such as interest, identification with the classroom-based learning and community, attitudes, and motivation (Fredricks et al. 2004). Such markers can be more

K. Cook (✉)
Bellarmine University, Louisville, KY, USA
e-mail: kcook@bellarmine.edu

G. Buck
Indiana University, Bloomington, IN, USA

A. Bellocchi et al. (eds.), *Exploring Emotions, Aesthetics and Wellbeing in Science Education Research*, Cultural Studies of Science Education 13,
DOI 10.1007/978-3-319-43353-0_8

elusive to determine than student achievement scores (as in research on cognitive engagement) or time on task (as in research in behavioral engagement), so for this reason research on emotional engagement has suffered from a lack of methodological clarity. Additionally, it is problematic to attempt to isolate emotional engagement from the other domains of engagement; for example, Connell et al. (1994) found a combined scale of emotional and behavioral engagement predicted school performance in African-American youth, yet no determination was made as to the extent to what specific factors linked to engagement played a determinative role in learning or achievement. The associative aspects of emotional engagement with other constructs as well as the methodological imprecision make difficult an examination of it as a predictive or causal link to student learning and achievement.

However, a few studies of the connection between the affective domain and achievement have highlighted the significance of emotional engagement in enhancing learning for students. For example, Handelsman et al. (2005) found emotional engagement to be predictive of intrinsic outcomes associated with learning (e.g., valuing learning in its own right), while Ainley and Ainley (2011) found students' enjoyment of science positively predicted an interest in learning more about science topics. Although enjoyment does not equate to emotional engagement, there presumably is overlap between these constructs. Extending from a multitude of research that has pointed to positive emotional engagement associated specifically with strong teacher-student relationships, Sagayadevan and Jeyaraj (2012) found that undergraduate students who perceived a positive instructor-student relationship not only reported higher levels of emotional engagement but also showcased improved learning in the course.

The studies above have looked at emotional engagement mainly through the use of individuals' self-reported interview and survey data. Noting the limitations in these data, Olitsky and Milne (2012) presented empirical work that sought to explore the central aspects of the curriculum that led to students' emotional engagement as well as the collective engagement of the class—two dimensions of emotional engagement that had been otherwise hidden from survey data. Olitsky and Milne showcased work that argued individuals are a product of their social environment, and they assert the need to include the collective aspects of the classroom in their view of emotional engagement. In doing so, they recommended longer-term studies (rather than a pre- or post-snapshot of emotional engagement) that include classroom observations and interviews over time to look for patterns of collective engagement potentially influencing individual engagement.

Drawing from a sociological perspective on emotion, Bellocchi et al. (2014) described the emotional climate as determined by the collective conditions of emotional arousal. By combining interaction ritual theory (Collins 2004) and the sociology of emotions (Turner 2007), these researchers characterize the aspect of solidarity or group belonging as a condition for positive emotional climate in the classroom. By analyzing cogenerative dialogue, emotional climate ratings made by students, and video recordings of class interactions and accompanying facial expressions of secondary preservice teachers, researchers illuminated the complex relationship between positive emotional climate and high-quality learning experiences. More specifically, participants experienced emotional change and fluctuations throughout learning experiences considered high-quality learning experiences (i.e., science demonstrations, role-plays, group discussion). Bellocchi and Ritchie's (2015) work with eight-grade students also shows the changing emotions and emotional climate linked to classroom learning experiences. Learning experiences that were challenging led to negative and positive valences of emotions as precursors to feelings of pride and triumph emotions. In these and related works, researchers describe the emotional climate of the classroom as a collective emotional arousal, group identity formation, and a reduced sense of self.

Stemming from Collins' (2004) notion of *emotional energy* (EE), Olitsky and Milne also posit that EE is the "basis of why people engage in particular activities, join particular groups, or develop particular identities" (p. 14). The group solidarity that occurs as students engage emotionally in the classroom is characterized by:

> a build-up of mutual focus, the development of a common mood, an "entrainment," or coordination, of body movements and speech, shared experience between participants on both an emotional and cognitive level, and boundaries to outsiders. (p. 15)

We see that this focus on the collective, or group EE, as indicative of emotional engagement is resonant with Lave and Wenger's (1991) notion of communities of practice in that individuals express their movement into a community of practice (i.e., science learning environment) by showcasing enhanced EE (i.e., membership and shared experience). Beyond that, Olitsky and Milne point to the need for EE as a basis for cognitive and behavioral engagement—suggesting that other types of engagement build upon initial emotional energy. As Olitsky (2013) posits, "...people need to develop a sense of group membership and identification with the group in order to desire the acquisition of the relevant knowledge for participation" (p. 563).

Also emphasizing the importance of collective engagement, Seiler and Elmesky (2007) found that interest in science among urban African-American students increased due to emotions generated collectively from science activities that facilitated students enacting their cultural dispositions. Seiler and Elmesky show strong positive correlations between African-American students' science learning and classroom contexts that are communally oriented (as opposed to individualistic or competitive) to show that efforts made to engage the whole class in relevant science learning increased positive feelings and motivation to participate in science. Seiler and Elmesky argue that traditional science classrooms, with an emphasis on teacher-led questioning of individual students, inhibit students' engagement by devaluing communal dispositions in favor of individualistic success:

> Thus, in school science interactions most often truncate the expression of a core African American cultural dimension, namely communalism, and contribute to the perpetuation of the status quo in which most African American students remain unsuccessful in science classes and marginalized in society. (p. 397)

In their study, Seiler and Elmesky assert students' cultural capital is not always valued in traditional science classrooms; so students can perceive strong boundaries between their own knowledge, values, and dispositions and the cultural enactment of school science. Consistent with Olitsky and Milne's work with emotional energy, Seiler and Elmesky emphasize the importance of collective engagement by honoring students' social and cultural capital as primary in establishing the foundation for behavioral and cognitive engagement.

Our theoretical stance also aligns with Sewell's (1999) notion that one's cultural history as well as lived experiences leads to one's embodying of certain communal dispositions that affect learning and engagement. Although certainly not unique to one particular group, engagement in science is particularly enhanced for African-American students when it is communally oriented, that is, when the learning is centered on social relations and an awareness of shared responsibility. In our own work with urban African-American elementary students (Buck et al. 2014), we have found that the most influential aspects in our efforts to increase positive emotions toward science included (1) the establishment of collaborative activities, (2) a supportive lab teacher, and (3) inquiry-based experiences focused on local problems. All of these elements honored the communal and collective dispositions toward learning. In this way, we acknowledge that relational facets of learning experiences along with relevant curricula can foster opportunities for generating positive emotional energy in the science classroom.

8.2 Place-Based Pedagogy as a Context for Engagement

Research in science education has emphasized the importance of connecting students and teachers to place and has shown that cognitive and affective domains of science learning are supported through place-based pedagogy (Sobel 2005). In this context, place is more than an address or a plot of land; it involves the history, emotions, stories, culture, aesthetics, and social problems of the place (Sobel 2005). Important to our work, research has also indicated that students in high-poverty, urban contexts have been shown to benefit both academically and affectively in these domains from curriculum that engages place (Chinn 2007). Professional development efforts focused on place have also found that learning science in a local context provides teachers with both the rationale and agency to restructure their curricula toward field-based science learning, and such experiences enhance participation of underrepresented minorities by connecting the targeted science knowledge with issues relevant to their own lives (Nespor 2008).

There are many different approaches to place-based pedagogy; the purpose implicit in most of them is to engage students in meaningful science experiences that directly improve their quality of life (Gruenewald 2003). How a specific approach is structured differs greatly. Common principles include (a) taking advantage of local possibilities for curriculum development, (b) students as knowledge producers, (c) learning from community members as well as teachers, and (d) fostering a regard and appreciation for home community (Smith and Sobel 2010). Originally, the movement for place-based pedagogy emphasized a place-as-land approach with the focus on taking care of the local land (Karrow and Fazio 2010).

As time went on, however, place-based pedagogy evolved into a type of critical instruction with an emphasis on the underlying social, cultural, economic, and political issues that affect place. The inclusion of human culture brought with it discussions of conflicts inherent in dominant American culture (Gruenewald 2003). As a result, complications have surfaced as it quickly became more socially and politically charged (Karrow and Fazio 2010). In addition, the pedagogical approach necessary for this emerging understanding of place was different from anything teachers had experienced in their preparation programs. Place-based advocates seek an intra- and integrated curriculum, project-based learning, collaborative teaching and learning, and the extensive use of community resources (Sobel 2005). The new curriculum is necessarily interdisciplinary, even including many subjects not traditionally a part of a school's curriculum. All of these aspects of instruction differ from the common approach to science teaching and learning.

Science education research on place-based pedagogy has had, for the most part, conceptual change theoretical underpinnings. There is, however, a growing body of educational research that suggests science studied in one's neighborhood becomes a valuable venue for fostering not only students' conceptual understanding but also their connection to their place. An emerging research base is demonstrating that place-based pedagogy fosters stronger ties to the community, enhances students' appreciation for the natural world, and creates a heightened commitment to serve as contributing citizens (Gruenewald and Smith 2008). It is within this latter affective domain that we explore place in terms of *emotional energy* (Collins 2004). Drawing on this concept, which provides a theoretical foundation for why students choose to engage individually and collectively in learning activities and positions emotional engagement as a social as well as an individual construct, we explore the ways in which science via place-based pedagogy provides opportunities for emotional engagement with the local neighborhood for both students and their teachers. Throughout our narrative, we provide a review of our efforts in an urban school district designed to improve students' emotional engagement with science as well as to enhance teachers' capacity to deliver place-based instruction in an urban setting. We story the impacts on the teachers, students' connections to authentic science, and common themes that underlie the overall approach to teacher education.

Our work builds on research showing that curriculum which highlights personal relevance and honors students' sociocultural capital leads to increases in emotional connectivity to science (Seiler and Elmesky 2007). We concur with Seiler and Elmesky when they posit, "The theoretical lens of social capital is useful in understanding how communal practices become both resources for student participation in science activities and a means for accessing additional resources that afford science learning" (p. 404). Thus, it is the emotional experience that makes cognitive engagement possible. Here, our efforts aim to support teachers' development of curricula that emotionally engage students in science learning about local concerns. The research question guiding this exploratory study investigated the ways in which students and teachers emotionally engaged in the classroom during the implementation of place-based curricula.

8.3 Investigating Emotional Energy Through Place-Based Curricula

8.3.1 *Context*

Utilizing place-based pedagogy to situate our work with urban African-American youth provided an opportunity to investigate emotional energy, which we agree is foundational to cognitive and behavioral engagement in the science classroom. The case studies we share here took place in a large urban school district in the Midwest. The district was labeled as high needs by the state department of education, and many of the schools were not meeting annual yearly progress. According to the US Census Bureau School District Estimates for 2008, 7542 of the 20,290 youth in the school district are from families of poverty. In 2010, 67 % of the student population qualified for free/reduced priced meals. Students in this district also have an increased risk of academic underachievement, particularly in science (average pass rate for seventh-grade science was 23 %).

8.3.2 *Participants*

We explored emotional energy by focusing on a few classrooms with teachers committed to incorporating place-based pedagogy into their curricula. We asked for volunteers from among the teachers who had previously taken part in professional development workshops on designing place-based units to allow us to observe and film the implementation of their place-based unit, interview them as they planned and conducted the unit, collect students' work, and interview their students during and after instruction. Following IRB protocol of informed consent, our volunteers were two elementary teachers (from the same school within the district) and two middle school teachers (teaching science at two different schools within the district). The parents of student participants as well as the students also provided informed consent to be part of the research. Pseudonyms for all participants were used. Table 8.1 describes the place-based units the teachers had developed.

8.3.3 *Data Collection*

Data sources for this study included (a) pre-, mid-, and post-teacher interviews on the place-based unit implementation, (b) group interviews with student participants during and after the place-based units, (c) audiotapes and videotapes of the classroom place-based and professional development sessions, (d) written place-based unit plans including classroom artifacts from classroom implementation, (e) and field notes kept by the researchers during the study. The lesson plans, coupled with

Table 8.1 Classroom context and teacher units

Teacher *Pseudonyms used*	Students	Place as a focus of inquiry	Inquiry project
Mr. Hite, 10+ years	26 sixth-grade general science	*City*	Because of a dramatically increased cost of supplying power, our city has announced that it will be increasing the rates of energy. Your task is to generate a plan for the implementation of the most economical and easiest to put into service alternative energy plan that is available to the area
Male	All multigenerational African-Americans		
Caucasian			
Mrs. Morgan, 20+ years	24 eighth-grade students	*Large lake*	Your family is planning a family reunion at the lake and wants to know how the quality of the water will affect their swimming plans. Your task is to test the water of the lake to determine its quality and write a letter of recommendations to your family
Female African-American	All multigenerational African-Americans		
Mr. Franklin,10+ years	26 first-grade students	*Schoolyard*	Teachers at our school have received funding to "re-wild" (restore native plant and animal species) our schoolyard. The teachers need your help in developing a plan for the area
Male Caucasian	All multigenerational African-Americans		→First grade will focus on soil types for different conditions in the schoolyard (habitat for plants and animals, playground areas, etc.)
Mrs. Wills, 25+ years	24 third-grade students	*Schoolyard*	Teachers at our school have received funding to "re-wild" (restore native plant and animal species) our schoolyard. The teachers need your help in developing a plan for the area
Female African-American	All multigenerational African-Americans		→ Third grade will focus on the needs of plants and how to best grow different types of plants in the schoolyard.

observations of teachers implementing them and debriefing sessions during which teachers shared and reflected on what they had accomplished, allowed for identification of what teachers experienced as key events or opportunities for emotional engagement that were inspired by their teaching. The interviews of the students helped uncover students' emotional energy as they experienced place-based science in the classroom. The videos of the class sessions throughout the unit helped us gauge students' physical expression of their emotional investment in their learning (i.e., maintaining eye contact, leaning in, volunteering answers, smiling, etc.). Classroom documents served as secondary data sources. These included the place-based units, instructional plans, formative and summative assessments, and students' written work.

8.3.4 Data Analysis

This was a case study informed by phenomenological theory. The phenomenological orientation focused our efforts on revealing the participants' firsthand experience of the phenomenon (Merriam 1998). We used a multiple case study design based on criteria for case study research methods established by Yin (1994) to offer a rich understanding of the teachers' and students' experience. Data analysis was an ongoing activity (Merriam 1998) and consisted of a thorough review of all sources with notation of initial codes. A second reading of the data allowed for grouping and combining of specific codes into broader categories, enabling us to identify themes across cases. We used NVivo software for coding—reading and rereading the texts and highlighting in the body of the texts the sentences and passages that related to themes suggested by our research question—and made multiple analyses of the written and audio/video data to search systematically for emerging themes and patterns, as well as discrepant ideas. Any recurring themes that emerged were triangulated within and across data sources, with careful attention to maintaining an audit trail back to the original cases.

As a guide for coding to classify the ideas and events that the participants talked referenced, we first read the transcripts of interviews and group interactions, marked what seems of interest based on Wenger's (1998) concepts of "shared foci," "common mood," and "group membership," and developed low-inference codes directly from the data. Low-inference codes involve low amounts of abstraction and refer to information that is objective or accessible to multiple others (Carspecken 1996). Based on the low-inference codes, we identified the segments of the transcripts that seem most relevant to our research topic (emotional energy present in place-based learning) and formulated high-inference codes that represented the data more broadly by collapsing several low-inference codes into fewer high-inference codes. We then reviewed the data a second time, mining data for instances that countered or did not show evidence of emotional energy (positive or negative). High-inference codes thus represented the collective meaning of several low-inference codes while also including contextual factors such as social interaction patterns over time. This enabled us to glean the factors that students and teachers referenced as affordances or constraints to their emotional engagement with the place-based experience. When referenced, secondary data sources (classroom artifacts and researcher journals) were used to triangulate data interpretation. The data analysis thus consisted of an iterative and inductive process of analysis in order to formulate qualitative accounts. This resulted in the formulation of few exemplar accounts that underscored the analytic angles of the study, namely, the ways in which features of the place-based afforded or constrained the emotional energy in the classroom. Qualitative data gathered through this research and reported here were analyzed using the constant comparative methodology until researcher consensus was attained. During the coding process, we also looked for patterns of similarities and differences within and across the transcripts and noted which figures of speech the participants repeated and which ones they contrasted with others.

To understand the emotional energy of the students and the teachers, we used Collins (2004) ideas of interaction rituals (IRs) as a guide for coding ideas and events that may have enhanced the group's emotional energy. As a way to uncover evidence for group solidarity and mutually focused attention, we looked for moments within the curricula that seemed to engage the class as a whole. For example, instances in the classroom videos of head nodding, humor, eye contact, body orientation, overlapping speech, and the completion of each other's sentences pointed to, as Kenneth Tobin (2005) posited, synchrony that supported the emergence of emotions and student engagement. We also looked both in the videos and interviews for group attention linked to confidence, willingness to participate, membership in the group, and a shared experience. Collins' analysis of emotional energy, as referenced by Seiler and Elmesky (2007) below, helped guide us to look for moments in which:

> (1) two or more individuals are situated within the same physical space; (2) boundaries exist separating those involved in the interaction from others; (3) individuals have a shared focus upon an object or activity and become increasingly aware of each other's focus of attention; and (4) they share a common emotional experience or mood. (Seiler & Elmesky, p.401)

Through a careful analysis of the data, we noted instances that included and did not include evidence of positive or negative emotional energy. This resulted in the formulation of a few exemplar accounts from among our cases that underscored the analytic angles of the study, namely, the ways in which place-based inquiry afforded or constrained emotional energy in the classroom.

8.3.5 Validity

As a validity measure, agreement among data (convergence) as well as inconsistencies and contradictions (divergence) was found as a way to uncover new issues and interpretations. We observed each class and related activity over the course of the semester, which constitutes prolonged engagement. Both field notes and the thick record of each observation were transcribed using low-inference vocabulary to try to eliminate potential biases regarding what occurred in the setting. In addition, we also employed the use of negative case analysis, in which we noted instances that contrasted the reconstructed themes once they had emerged and explore explanations for the lack of fit.

A caveat to this study, as well as most qualitative studies, is that it focused on detailed interactions among a small number of participants. Thus, in qualitative inquiry, generalizability is not a function of sampling. Because individual experience is the unit of analysis, researchers are interested in selecting cases that provide a lot of detailed information about the topic of interest, rather than a representative group that will aim to provide insight to multiple populations. Judgments about transferability are based on information regarding the investigator, the setting, the methodology, the participants, and the nature of the relationships between the participant and the researchers.

8.4 Emotional Energy in Place-Based Curricula

The findings below draw from our multiple data sources from across our elementary and middle school classroom contexts. The emotional energy generated by the place-based curricula is presented first with an emphasis on the degree to which the units inspired a *shared foci* (i.e., noticing of place, extending learning, caring about details of classroom work), *common mood* (i.e., eagerly participating in activities, expressing ideas, asking questions, desiring to contribute), and *group membership* (i.e., sharing work, expressing communal interest/concern, sharing responsibility, insider knowledge), within the classroom community. In the second section, the potential impact of place-based curricula is explored in terms of inspiring and sustaining emotional energy.

8.5 Place-Based Curriculum Fostered Emotional Energy

In all of our cases, the place-based element of the curriculum enabled students to connect with their local surroundings (i.e., classroom, schoolyard, and/or nearby natural resource) to engage in a shared focus on science concepts and processes. Students in each classroom became interested in the science of their own backyard, showcasing not only their motivation to learn more but also their enthusiasm for engaging in scientific skills and practices. For example, in Mr. Franklin's first-grade classroom, students observed nature in their schoolyard to develop an appreciation of what existed there and how it interacted with the schoolyard environment as part of the ecosystem. Students recorded their observations in their science notebooks, detailing what they saw and describing what was unique about that place. For example, one student showcased her ownership for what she observed in the schoolyard but titling her reflection *My Special Tree*. In the drawing below, Angel describes the area under the oak tree as shady—an important observation on which she could later draw when asked to determine how to support best the plants and animals in that particular area of the schoolyard (Fig. 8.1).

My Special Tree

I like that people who cuts grass did not cut bushes off and I love the tall tree. There's not much of trash around the maple tree and I'd love to plant a tree I like how the tree's leaves look, smell, and fells the trash that was things like cans bottles and I saw bugs around the tree and a few days later we went to the since lab with my teacher we planted seeds my flower grew only a little bit it almost looked like a flower.

Fig. 8.1 Angel's drawing of an oak tree in the schoolyard and narrative describing its importance to her

In her narrative (orally provided and aided in transcription by Mr. Franklin), Angel notes the *bugs around the tree* and how the *people who cut grass did not cut bushes off* as she notices how humans and nonhuman animals interact with the tree. Her admiration for the tall tree (i.e., *love*) is further supported by her involvement in growing her own seeds in the science lab. This authentic connection to what was occurring in the science lab helped provide a context of care for her participation in science practices.

In Fig. 8.2, Angel's lab notes from her observations made in the schoolyard provided the foundation for her to develop a related experiment to test water flow through different types of soil in the science lab. As evidenced by her desire to conduct an experiment back in the lab (an idea she derived and connected to what she had observed in the schoolyard), her emotional connection to her schoolyard environment was connected directly to her development (and extension) of the inquiry and her participation in the science lab with her teacher.

A common mood was also evident in the classrooms. In Mrs. Morgan's eight-grade class, students investigated water quality at a nearby swimming hole in a large lake in their community. They excitedly discussed who might be potential polluters of the lake, often talking over one another to express their ideas. They enthusiastically researched water quality reports online, even asking for passes to go to the library to do more research on their projects during their study time; and they eagerly participated in data collection out at the lake despite the frigid winter temperatures (some students joyfully running into the cold lake water and getting their pants wet to collect water samples). As well, their engagement in the place-based units stimulated students to ask questions for future inquiries. For example, Breon said, "I learned that the water is not good. I mean 'cause if it's [pollutants] in the water that's in the beach that we go to, is it in our tap water at home?" Also, Devon asked if they could do follow-up tests on the air quality (in addition to their focus on water quality), wondering if there might be a connection between the water and the

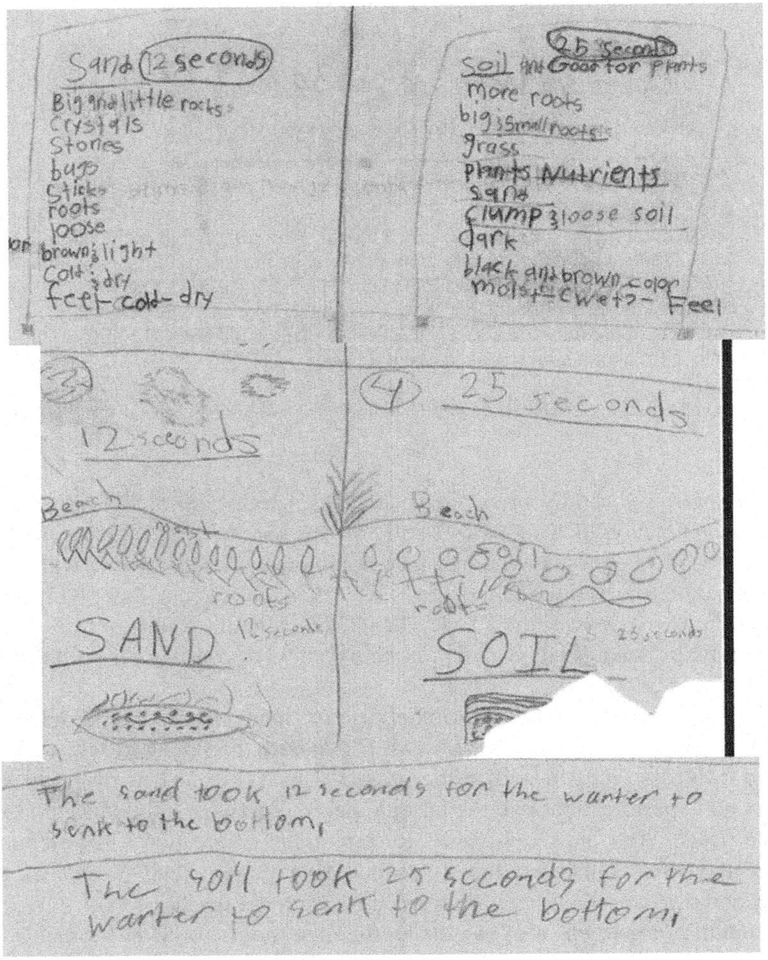

Fig. 8.2 Angel's observations of the schoolyard and lab notes

air pollution. This motivation to ask more questions and desire to engage in the processes of science (i.e., data collection and analysis) was evident in classroom.

The collaborative element of the units, centering on group data collection and analysis, encouraged students to share their findings with one another. In the picture below, Mr. Franklin's first graders excitedly show their drawings of the schoolyard to one another—eager to share what they found to be most interesting and unique about their area. As the teacher noted:

> The students have enjoyed the experience. They have been eager and curious when moving through the lessons. All students were actively engaged and asking question that pertained to the topic. There was very little behavior problems to contend with and everyone stayed on task. The different levels of questions made it interesting.

This positive emotion regarding the place-based unit also extended to the teachers, who were inspired by the enthusiasm of the students to explore the schoolyard. For example, Mrs. Wills noted, "Each class was exploring the same area [of the schoolyard], but did [the exploration of the schoolyard] in a different way. I like the fact that the pictures showed things that each [student] found important…makes me want to go back and explore more with my class." The communal exploration and resulting whole class discussions invoked a mood of wanting to be involved in the science investigation as evidenced by teacher reflections during and after the learning experience.

Students also showcased feelings of inclusion in a group through their exploration of place. In all of our classrooms, students quickly became entrained with joining others to solve a problem in their environment (i.e., asking others to join them in explorations, showing their work to each other, discussing their ideas about the issues in the schoolyard). As a third grader in Mrs. Wills's class noted, litter in the schoolyard was an area of concern:

> My class explored the playground area at Academy. I don't like how the tree branches are sticking out. I don't like how candy wrapers are around the tree. I don't like how the weeds are around the tree. What can we do to make this area look better.

Rather than pointing to others to deal with this issue of litter, Lovice asked, "What we can do to make this area look better?" With an emphasis on "we," he alluded to the shared responsibility that the class and the teacher had now that they've noticed and begun to explore this problem. His statement could be taken to mean that he saw the class as having a membership to a special group that shared allegiance to the schoolyard and on which the onus was placed to improve local conditions.

In Mr. Hite's place-based unit on alternative energy forms, a sixth-grade student remarked, "I learned that usually people always want to talk about wind and solar but like tides and biomass and garbage and human driven… I have never heard about that stuff since we started this." Students acknowledged they were learning aspects of science that changed their knowledge on important topics—a knowledge that not many others had and they now shared with their classmates and a teacher.

The degree to which students wished to connect their classroom inquiries to what was important to them (as described below, their school and their city) was also indicative of their emotional connection to their science learning experience. In Mr. Hite's unit on alternative energy forms, students were asked to modify their own energy-consumption behaviors to notice how that might have an effect on their energy usage readings at home. In the transcript below of our interviews with the students, a group of them share their experience with this aspect of the unit. The extract from the interviews shows not only how the project impacted them but also how they wished to extend their understanding through their care for the school (Extract 8.1).

Extract 8.1 Interview with Mr. Hite's students about collecting data on their energy usage in their homes

Silas	We had to read our electric meters for what about a week? I believe- or two. He had us read it for like a week and then we had to try and then we tried to turn off the lights and stuff and then we went back and read it again. To see if we saved energy or not.
DJ	Or like that- the people that me and my mom live with- they already chastise us about leaving the TV on, leaving the light on, going to sleep with the TV on, so uh, like I was doing my meter readings they were kinda low anyway cause we were…
Cy	Mine was 7
Researcher	Do you know why yours was so high?
Cy	Uh, yeah cause we do leave the lights on a lot and we turn on the air conditioning and the fan and like that. But now we just use the fan.
Researcher	So, this [the project] actually made a difference in your household?
Cy	*[Grunts]*
Anton	I mean like when he [Mr. Hite] have us do like our houses and whatever, it would be interesting to do the school- you know? Even though it is fittin' to close, I still like to know.

Here, Anton's interest in collecting data on his own school connected to the value he placed on the school, a building that had been condemned by the state and was set for closure in the coming year. Students in Anton's school talked to us about their sadness over the school shutting down, as they had been attending it for many years and claimed they had positive emotions with regard to their time there. We thought it was interesting that Anton showed his care for the school by wanting to include it in their study—a study he thought was important.

Similarly, Mr. Hite's students began thinking about how their place-based unit could be important for considering improvements in their city. As the students researched the advantages and disadvantages of various types of energy and the feasibility of alternative forms of energy for their city, they considered not just what energy might be plausible but also how that change in infrastructure might improve others' perceptions of their city. In the exemplar below, students discuss how using biomass as an energy source (as opposed to tidal energy as one student suggests) might help with enhancing the aesthetics of their city (Extract 8.2):

Above, students debate the best approach for their city to take to reduce their energy usage—making specific considerations about their place and what would work best. In their discussion that centered on a shared focus topic, they employ the use of humor as well as their desire to improve their city (both showing the common mood among their group).

In both of the above examples, students illustrated how the place-based element to their science learning stimulated a shared focus in their desire to participate in science process skills as well as common mood in extending their learning to what they valued as important. Their membership in a classroom community among their peers and teacher was evident as they referenced insider knowledge and ownership in doing something about the issue they were exploring. In the following section, we

Extract 8.2 Interview with Mr. Hite's students about improving environmental conditions in their city

DJ	Actually it is low and high tides, but we couldn't use the Lake because the difference between the Lake's tides is like 2 or 3 inches. That is not going to produce enough energy to do nothing.
Cy	I think biomass [is better] cause we have a lot of garbage and we could go ahead and clean up the city and...
Anton	*[Interrupting]* It is the reason why our city stinks.
Cy	That and I- but I mean with tides it has been raining a lot last year and all that rain and all the flooding and mosquitoes coming in that could have been change the water.
Anton	I told my dad that I was going to look into different ways to conserve energy in the house. He decided one way to conserve energy was to kick my brother out *[All laugh]*

explore how the latter seemed to be a critical aspect of the curriculum that fostered emotional energy that emerged from our classroom data.

8.6 Potential Impact as a Critical Element to Emotional Energy

Students wanted their learning to make an impact outside of the classroom, to realize change, through their learning experiences in science (i.e., Cy's desire to clean the city by using biomass for power, Lovice's desire to make the schoolyard look better). On several occasions, students noted their excitement to do "real" science and participate in "actual" scientific exploration that may lead to authentic change. Locating the inquiry project in students' place afforded the opportunity to have a voice in potential change, so students quickly realized their learning in science class might have the potential to impact something positively in their area. Rather than creating a project around a contrived or made-up issue, teachers wanted students to participate in and ultimately have the potential to influence a real problem in their area.

In the unit centered on re-wilding the schoolyard, the first and third graders took photographs of what they perceived to be important areas of the schoolyard that needed improvement or consideration. They also wrote narratives (some first graders orally stated their narrative and it was transcribed by Mr. Franklin) describing what was present in the photo and why they thought it was important to the ecology of the schoolyard. The photos and narratives were prominently displayed in the school for all other students and teachers to see, and the principal was personally invited to see the students' work. The students were thrilled to see their work on the wall in front of the library, stopping to look at it and point it out to their parents and friends as they walked into school. The principal also visited their classrooms to thank them for their work and applaud them for thinking so deeply about the welfare of the schoolyard. In this vein, locating the study of science in local matters of

Fig. 8.3 Students' photos and narratives about re-wilding the schoolyard displayed on library wall

interest to students enabled them to bring a voice to others about what they had learned (Fig. 8.3).

In contrast to the aforementioned successful units in achieving a connection place, not all of the teachers' units emphasized this culminating communicative aspect of bringing the classroom learning to a broader audience despite a sincere interest in doing so on the part of the students and the teacher. When asked about his experience with Mrs. Morgan's unit on water quality, Davon claimed:

> I think it's important because a lot of people don't get interested in school like this. So like if you make it fun, you make something that's like really an issue, they'll want to do it more. They'll like pay more attention because it is really something they can affect them.

Even Mrs. Morgan herself was interested in doing something to effect positive change in the water quality in the area. She herself had worked in the nearby steel mill plants before her career as a teacher and believed that if they could connect to the steel mill industry representatives, their unit might make an impact or start a dialogue. Although she attempted on several occasions to contact someone at the mill, she was unsuccessful in establishing a dialogue with anyone who could speak on their behalf. As she reported, "The guy told us from the steel mill—'No problems here,' but we don't believe him." Here, she illustrates her frustration and desire to do something more to help improve the water quality in the community. Because there was nothing structured into the unit that would help students connect their learning and understanding beyond that of the classroom experience (i.e., no dialogue with other interested parties), students were left only to guess as to what was causing the pollution in the lake.

Similarly, Mr. Hite's students were highly motivated to make positive change in their community after learning about energy overconsumption and the potential of

Extract 8.3 Interview with Mr. Hite's students about extending their project to community action

Tyreece	And uh most people think that our city is like a trash or ghetto but it can have. Like we have people in our community that actually is good citizens and willing to help our city out. We could be as good as we want to be. Like our city- people thinking it is bad. We just gotta come together.
Deja	Then you see that like the air pollution thing- that is messing up our atmosphere so like doing other energy sources and not polluting. And tell them like for real, you can see these streets and all these other kinds of stuff and it is just not right. Maybe if we write a letter to the mayor- I tell them this should be this and maybe we can get other people to help out in our community. To get our city back the way it was, whenever that was...
Researcher	So you'd even like to take this project a step further and actually connect it to other community members?
Deja	Yeah, I think we should ask everybody to do this- join on. Maybe we should do like a petition and instead of just writing a letter and saying we want this done or that done maybe we should just all come together and we all just do it. Instead of just write a letter cause most people don't even listen to that. Go strikes. Build projects- do stuff like that.

different energy alternatives. Their enthusiasm in the following plea to get others involved underscores how much students came to care about their environment (Extract 8.3):

When Mr. Hite's unit culminated with the students' building a solar cooker—an undoubtedly fun event with food and merriment—the students invited other grade levels at the school to come enjoy hotdogs cooked with solar radiation. However, when discussing the unit, which ended with no connection to positive change in the community (other than the potential impact students might have had on their family members' energy usage), students expressed their desire to do more with the information they had learned. Deja's idea to write a letter or start a petition indicated her feeling of shared responsibility (i.e., what *we* should do); the group had to improve conditions in the community.

8.7 Conclusion

Overall, the experiences reported in this study of students and teachers engaged in place-based pedagogy enhanced our understanding of the potential of this instructional approach for heightening emotional energy in science education. More specifically, we (1) witnessed the emotional connections that resulted from the place-based experiences, (2) realized how much the students valued solidarity-building interactions, and (3) observed the unrealized potential of place-based pedagogy in terms of fostering students' engagement with the community. These understandings are discussed below.

First, we were able to witness the emotional connections that resulted from the place-based experiences. We witnessed Angel lovingly describing her tree and how she can take care of it, Breon and Devon expressing concern about the water in their community and wanting to know more, and Anton and his classmates wanting to learn more about their condemned school. Additionally, we witnessed this increased emotional engagement as the students quickly became involved in asking questions to further their classroom inquiry. We were pleasantly surprised at the enthusiasm with which the students probed the teacher and each other about the topics, as well as what they could do to protect their place. These were science lessons, but the emotions that they stirred did so much more than address the content standards. Our findings support the work of Seiler and Elmesky (2007) by further demonstrating how young urban African-American students show increased emotional connectivity to science as a result of highlighting personal relevance during learning experiences. We observed such emotional connectivity when the discussions focused on their schoolyard, school, lake, and neighborhood. As a result, we agree with Tan (2009) who argued that science, technology, and environmental education have been taught as neutral and technical subjects in abstract, decontextualized ways. He suggested place-based education as a means through which some of these challenges can be faced. Place-based education seeks to help communities through employing students in solving community problems. Place-based education differs from conventional text and classroom-based education in that it frames students' local community as one of the primary resources for learning. Thus, place-based education promotes learning that is rooted in what is local—the unique history, environment, culture, economy, literature, and art of a particular place, that is, in students' own "place" or immediate schoolyard, neighborhood, town, or community. We certainly found this to be true in these classrooms.

Second, we came to realize how much the students valued solidarity-building interactions. The elementary students were very excited to share their work and pictures with parents, friends, and school personnel. Eighth-grade students, such as Deja, expressed a desire to talk to their friends and community members in an effort to get them involved in the water quality of the local lake, as well as getting others involved through the use petitions, projects, and letters. The sixth graders shared how they talked to family and friends about conserving energy. Drawing on Collins' (2004) work, Olitsky and Milne (2012) noted, "People are EE seekers, choosing courses of action based on their anticipation of the emotional payoff from participation in solidarity-building interaction rituals" (p. 14). The students we followed quickly and naturally sought to engage with others about science-related issues. Unfortunately, they were not able to take part in solidarity-building exercises very often or for very long—which leads to our third understanding.

Third, our findings revealed unrealized potential of place-based pedagogy in terms of fostering students' engagement with the community. Place-based education focuses on the concept of place in relationship to education, which leads to avenues that are exceptionally rich and creative of exploration especially when coupled with a social justice perspective (Aikenhead 2006). Schroder (2006) outlined how schools can focus on both inquiry and action in local, regional, and global space, studying the culture of nearby places with the study of culture in distant places and shifting the discourse of accountability to that of place-based accountability. In our study, we witnessed that the students did experience a shared foci, common mood, and group membership but attempts to be heard or initiate involvement within the community fell short due to lack of response from the community (i.e., steel mill, local environmental agencies) or failure to build this into the curriculum. In these cases, the students were left to try to make sense of their data, but did not get the opportunity to dialogue with interested parties and generate discussion about potential solutions to the pollution problems. The inherent value of the dialogue was stifled when the teacher only got a closed-ended response of "No problems here…" or did not get a response at all when she attempted to foster a discussion between her students and their community. This lack of response greatly inhibited the students' knowledge, skills, and experiences. This lack of response was a barrier to realizing the full potential of place-based pedagogy. It is a barrier we, as science teacher educators, believe we can help teachers and future teachers address.

8.7.1 Implications

We argue here that the local neighborhood is a valuable venue for fostering an understanding and appreciation of science concepts. Taken together, the cases reported here provided us with understandings of affordances and obstacles to place-based attempts in fostering diverse students' emotional energy toward science and their local environment. We now turn the attention back to our own work as teacher educators to describe our role in furthering these efforts.

As noted earlier, we are science teacher educators who have worked with the teachers in these schools to design place-based curricula. Bianchini et al. (2003) suggested that teacher education programs need to focus more attention on highlighting the connection between science and local interests and issues, enhancing engagement of students in using science toward social and political ends, and empowering teachers to be reflective of how good practice can exist within the culture of high stakes testing. Teachers should be supported in their endeavor to consolidate standards within place-based teaching so as to maximize the content embedded in these long-term units. Given our findings, we now conclude that our work with current and future urban teachers needs to focus more on supporting them in connecting their classrooms to their community while still supporting the content and integration of standards. While our case studies revealed great promise in that the place-based teaching with relevant inquiry about their place fostered the emotional energy of the students, if we wish to meet the goals of solidarity building, it is essential that these place-based units support student voice in the community. Students should dialogue with their community members about their findings and discuss possibilities for enriching the community through their understanding of science. We further assert that this type of engagement would necessitate a need for students to understand deeply the meaning of their data so as to enable fruitful dialogue with community members, enable students to pursue their own inquiries to further their understanding of the issue, and allow for a deeper understanding of the complexity of local environmental issues as multiple voices would be heard.

To accommodate our new understanding, our teacher education experiences will now include a focus on fostering connections with local stakeholders who may be interested in the students' place-based units. As former teachers and current science teacher educators, we see the potential for the research practice of photovoice (Wang and Burris 1994) to be a useful pedagogical tool with which teachers can actively engage their students in community-based socioscientific inquiry (Cook and Quigley 2013). Participants collaborate on the reasons for and use of their pictures and reflections to showcase relevant issues and generate dialogue with community members and policy makers who may be in a position to mobilize change. Students' lives in the classroom and outside of it are often directed and structured by decisions made by others. Those in positions of power to change the community may not reflect what change is most desired by students. Ultimately, students are able to participate in a process that may lead to unforeseen social change. A student-guided focus of scientific issues could enable teachers further to gain knowledge about the larger social context of the lives of their students (Cook and Buck 2010).

With regard to the strengths of place-based curricula in terms of fostering emotional engagement to the neighborhood, our work supports our current commitments to preparing teachers to address cultural congruence in their lessons. As illustrated in these case studies, by allowing the students to explore place-based questions through opportunities embedded within learning units, the environment became more congruent between home and school. This congruence is a crucial part of bridging any disconnect between home and school science that often silences marginalized groups (Calabrese Barton et al. 2008). Cultural identity also has a significant impact on students' achievement and attitude toward school (Helms and Cook 1999). Catsambis (1995) showed that African-American students' attitudes are often positive despite low achievement. It is suggested that this attitude-achievement paradox may be related to factors external to the classroom (e.g., family, community). Furthering our understanding of such related factors is an increasing amount of writing attending to African-American student identities in relation to their attitudes toward science and science education. Kane (2011) showed how three dimensions of science identity (competence, performance, and recognition) shaped how third-grade students constructed their notion of "self." From work such as this, we have come to understand that in order to improve students' attitudes toward science, we need to be concerned about the development of strong science identities. Efforts to characterize the pedagogical improvements that would better support the formation of positive science identities are increasing (Simpson and Parons 2008). Although not as vast as the research on gender, research on cultural identity has clearly demonstrated that this identity does make a difference in science education, and a substantial number of studies are also available to guide reform efforts in regard to improving young African students' attitudes toward science. Preparing our teachers to put the students' neighborhood in the center of their instruction and implementing a science education that values all students' experiences prepares them to develop a more inclusive practice.

Through these case studies, we came to better understand the potential of this instructional approach for heightening emotional energy in science education. By turning a critical eye back on our own practice, we are now better positioned to prepare teachers to make the neighborhood a valuable venue for fostering not only students' conceptual understanding but also a connection to their place.

References

Aikenhead, G. S. (2006). Border crossing into the subculture of science. *Studies in Science Education, 27*, 1–52.

Ainley, M., & Ainley, J. (2011). Student engagement with science in early adolescence: The contribution of enjoyment to students' continuing interest in learning about science. *Contemporary Educational Psychology, 36*, 4–12. doi:10.1016/j.cedpsych.2010.08.001.

Bellocchi, A., & Ritchie, S. (2015). "I was proud of myself that I didn't give up and I did it": Experiences of pride and triumph in learning science. *Science Education, 99*, 638–668. doi:10.1002/sce.21159.

Bellocchi, A., Ritchie, S. M., Tobin, K., King, D., Sandhu, M., & Henderson, S. (2014). Emotional climate and high quality learning experiences in science teacher education. *Journal of Research in Science Teaching, 51*, 1301–1325. doi:10.1002/tea.21170.

Bianchini, J. A., Johnston, C. C., Oram, S. Y., & Cavazos, L. M. (2003). Learning to teach science in contemporary and equitable ways: The successes and struggles of first-year science teachers. *Science Education, 87*, 419–443.

Buck, G., Cook, K., Quigley, C., Prince, P., & Lucas, Y. (2014). Seeking to improve young African American girls' attitudes toward science: A participatory action research study. *The Elementary School Journal, 114*, 431–453. doi:10.1086/674419.

Calabrese Barton, A., Tan, E., & Rivet, A. (2008). Creating hybrid spaces for engaging school science among urban middle school girls. *American Educational Research Journal, 45*, 68–103. doi:10.3102/0002831207308641.

Carspecken, P. (1996). *Critical ethnography in educational research: A theoretical and practical guide*. New York/London: Routledge.

Catsambis, S. (1995). Gender, race, ethnicity, and science education in the middle grades. *Journal of Research in Science Teaching, 32*, 243–257.

Chinn, P. (2007). Decolonizing methodologies and indigenous knowledge: The role of culture, place and personal experience in professional development. *Journal of Research in Science Teaching, 44*, 1247–1268. doi:10.1002/tea.20192.

Collins, R. (2004). *Interaction ritual chains*. Oxford: Princeton University Press. doi:http://dx.doi.org/10.5944/empiria.10.2005.1054; http://dx.doi.org/10.5944/empiria.10.2005.1054

Connell, J. P., Spencer, M. B., & Aber, J. L. (1994). Educational risk and resilience in African-American youth: Context, self, action, and outcomes in School. *Child Development, 65*(2), 493–506. doi:10.1111/j.1467-8624.1994.tb00765.x.

Cook, K., & Buck, G. (2010). Listening to the learners: Proposing the tool of photovoice for engaging students in community-based socioscientific inquiry. *Science Scope, 33*, 35–39.

Cook, K., & Quigley, C. (2013). Connecting to our community: Utilizing photovoice as a pedagogical tool to connect college students to science. *International Journal of Environmental and Science Education, 8*((2), 339–357. doi:10.12973/ijese.2013.205a

Fredricks, J. A., Blumenfeld, P. C., & Paris, A. H. (2004). School engagement: Potential of the concept, state of the evidence. *Review of Educational Research, 74*(1), 59–109. doi:10.3102/00346543074001059.

Gruenewald, D. A. (2003). Foundations of place: A multidisciplinary framework for place-conscious education. *American Educational Research Journal, 40*, 619–654.

Gruenewald, D. A., & Smith, G. A. (Eds.). (2008). *Place-based education in the global age*. New York: Taylor & Francis.

Handelsman, M. M., Briggs, W. L., Sullivan, N., & Towler, A. (2005). A measure of college student course engagement. *Journal of Educational Research, 98*(3), 184–191. doi:10.3200/JOER.98.3.184-192.

Helms, J., & Cook, D. (1999). *Using race and culture in counseling and psychotherapy*. Needham Heights: Allyn & Bacon.

Kane, J. (2011). Young African American children constructing academic identities in an urban science classroom. *Science Education, 96*, 457–487.

Karrow, D., & Fazio, X. (2010). Educating-within-place: Citizen science, and ecojustice. *Cultural Studies and Environmentalism, 3*(2), 193–214.

Lave, J., & Wenger, E. (1991). *Situated learning. Legitimate peripheral participation*. Cambridge: University of Cambridge Press.

Merriam, S. (1998). *Qualitative research and case study applications in education*. San Francisco: Jossey-Bass Publishers.

Nespor, J. (2008). Education and place: A review essay. *Educational Theory, 58*(4), 475–489. doi:10.1111/j.1741-5446.2008.00301.x.

Olitsky, S. (2013). We teach as we are taught: Exploring the potential for emotional climate to enhance elementary science preservice teacher education. *Cultural Studies of Science Education, 8*(3), 561–570.

Olitsky, S., & Milne, C. (2012). Understanding engagement in science education: The psychological and the social. In B. J. Fraser, K. Tobin, & C. McRobbie (Eds.), *Second international handbook of science education*. Dordrecht: Springer.

Sagayadevan, V., & Jeyaraj, S. (2012). The role of emotional engagement in lecturer-student interaction and the impact on academic outcomes of student achievement and learning. *Journal of the Scholarship of Teaching and Learning, 12*(3), 1–30.

Schroder, B. (2006). Native science, intercultural education and place-conscious education: An Ecuadorian example. *Educational Studies, 32*, 307–317.

Seiler, G., & Elmesky, R. (2007). The role of communal practices in the generation of capital and emotional energy among urban African American students in science classrooms. *Teachers College Record, 109*(2), 391–419.

Sewell, W. H. (1999). The concept(s) of culture. In V. E. Bonnell & L. Hunt (Eds.), *Beyond the cultural turn: New directions in the study of society and culture* (pp. 35–61). Berkeley: University of California Press.

Simpson, J., & Parons, E. (2008). African American perspective and informal science educational experiences. *Science Education, 93*, 293–321.

Smith, G. A., & Sobel, D. (2010). *Place- and community-based education in schools*. New York: Routledge.

Sobel, D. (2005). *Place based education: connecting classrooms and communities* (2nd ed.). Great Barrington: The Orion Society.

Tan, M. (2009). Science teacher activism: The case of environmental education. *Journal for Activist Science & Technology Education, 1*, 32–43.

Tobin, K. (2005). Urban science as culturally and socially adaptive practice. In K. Tobin, R. Elmesky, & G. Seiler (Eds.), *Improving urban science education: New roles for teachers, students and researchers* (pp. 21–42). Boulder: Rowman and Littlefield.

Turner, J. H. (2007). *Human emotions: A sociological theory*. London: Routledge.

Wang, C., & Burris, M. (1994). Empowerment through photo novella: Portraits of participation. *Health Education Quarterly, 21*, 171–186.

Wenger, E. (1998). *Communities of practice: Learning, meaning, and identity*. New York: Cambridge University Press.

Yin, R. (1994). *Case study research: Design and methods* (2nd ed.). Thousand Oaks: Sage Publishing.

Kristin Cook is an Assistant Professor of Science Education at Bellarmine University. She received her Ph.D. at Indiana University in Curriculum and Instruction, specializing in Science Education and Environmental Science. Kristin's research focuses on engaging students and preservice teachers with the community of science through exploration of socioscientific issues.

Gayle Buck is a Professor of Science Education at Indiana University Bloomington. Her scholarship and teaching foregrounds relationships across learner diversity and science. Her research focuses on student populations traditionally underserved by science education and neglected epistemological assumptions in science teaching and learning.

Chapter 9
Emotions: Connecting with the Missing Body

Liv Kondrup Kristensen and Kathrin Otrel-Cass

9.1 Introduction

This chapter presents a framework for investigating emotions in science classrooms based on ideas described by the French philosopher Maurice Merleau-Ponty (1908–1961) who wrote about the intertwined relationship between the body and the mind. This framework will further be exemplified by presenting the analysis of a group of year 8 Danish primary school students engaging in a physics task. We are interested in Merleau-Ponty's ideas because he opposed the classical Descartian view that the body's function is purely biological. Merleau-Ponty proposed the concept of "the lived body" (*le corps propre*) arguing that human perception is not solely located in people's minds, but stems from people's experiences with and of things, such as vision or the experience of touch and being touched. These experiences *of* the body open up the world *for* the body to make sense of. Merleau-Ponty explained that it is only through embodiment that it is possible to sense, perceive, and experience this world.

This chapter presents how these ontological assumptions can be used to analyze emotions in classroom interaction. We facilitate this intention by building on the works proposed by Marjorie Goodwin, Asta Cekaite, and Charles Goodwin (2012) who consider emotions as situated practice to examine bodily stance and talk. We begin by presenting background to the ideas of Merleau-Ponty, starting with his theorizing of consciousness and perception. We then outline how we apply Merleau-Ponty's thinking about the body and mind to examine our own case of a science lesson in a Danish primary school.

L.K. Kristensen (✉)
Centre for Social Educational Studies, University College Zealand, Roskilde, Denmark
e-mail: likr@ucsj.dk

K. Otrel-Cass
Department of Learning and Philosophy, Aalborg University, Aalborg, Denmark

A. Bellocchi et al. (eds.), *Exploring Emotions, Aesthetics and Wellbeing in Science Education Research*, Cultural Studies of Science Education 13,
DOI 10.1007/978-3-319-43353-0_9

9.2 Introducing Maurice Merleau-Ponty: Consciousness and Perception

The notion that body and mind are split was made famous by French philosopher and mathematician Rene Descartes (1596–1650). The Cartesian split idea supported the tenets of rationalism and empiricism and regarded the body as an object that does not affect thinking about or rationalizing the world. To Merleau-Ponty, such a position failed to address what it means to be in the world and that the world is phenomenally experienced. He regarded the body as the anchorage in the world that our existence is bound to and which consequently shapes how we experience the world. His first publication was *The Structure of Behavior* (1942), followed by *Phenomenology of Perception* (1945), *Sense and Non-Sense* (1964a), and *The Visible and the Invisible* (1964b) that was posthumously published. This interest in the body made him also known as the philosopher of the body.

For Merleau-Ponty the body represents an equivocal existence between object and subject, for example, when we are experiencing the sensations of touch: when we touch one body part (the right hand) with the other (the left hand). When the right hand grasps the left hand, it is the subject of action, while the left hand is the object. But as soon as the left hand feels the right, it becomes the subject, which senses the right as an object (Merleau-Ponty 2012). With this example, Merleau-Ponty draws attention to how consciousness and perception are intertwined. The hand is part of the sensing body that is continuously making experiences. Merleau-Ponty explains that we *are* our bodies and that it is through the body that we experience the world. The physical dimension of the body defines the texture and the place in the world.

Just as the lived body makes it possible to experience the world, the lived body is always expressing something, even when the body tries to remain quiet or refrain from visible interaction. Then, quietness and inaction are expressions in themselves. This means that expressions are inseparable from experiences. We make sense of others' expressions and our own by use of our experiences. As Thøgersen (2014) writes: "We are entangled or intertwined in a flow of living relations in which expression come to happen because we respond to the experience of the immediate surroundings" (p. 25). Building on the understanding that expressions *are* experiences that are made visible to others, then, what we are experiencing are our emotions made visible.

In the following section, we discuss Merleau-Ponty's understanding of emotions further.

9.2.1 Merleau-Ponty on Emotions

To Merleau-Ponty emotions are part of our existence. They shape how the world comes to matter to us (Thøgersen 2014, p. 26), and emotional meaning becomes visible as a way of taking up a position in the lived world. Emotions are a way of relating to the world, because when an event is experienced, it triggers certain emotions. In other words, the event has an emotional atmosphere, which links to the world and shapes how the world exists for the person who experienced the event. Emotional atmosphere is not an inaccessible private space, but signifies the interactions between the public surface of the body situated in a given environment and the inner parts of the body. This means that deep feelings are connected with the inner body and the outer life and affect our perceptions and physical responses with others. Merleau-Ponty recognized that these responses vary for people. He wrote in his book *Phenomenology of Perception*: "But in fact, the gesticulations of anger or love are not the same for a Japanese person and a Western person… the difference between gesticulations covers over a difference between the emotions themselves" (2012, p. 195). In other words, behavior and emotions are connected and shape the manner in which we meet a situation and live it. Feelings and emotions can also be of different quality, and true feelings like love may come with more intensity from within as well as from the outside. Merleau-Ponty regarded emotional experiences as a specific kind of consciousness that we carry with us and cannot shake off. He described that we do not possess emotions but that emotions possess us. Understanding emotions as a kind of practical consciousness (Merleau-Ponty 1998) that enables people to act and react in social interactions has implications for how we view learning in schools. Since we adopt the position that learning is socially situated in human interactions, the very (emotional) character of the interaction affects how people make sense of a given situation and thus also what learning may be afforded from that situation.

Maurice Merleau-Ponty became known for his philosophical arguments on the role that perception plays in making sense of the world. With our particular interest in emotions, we selected three concepts from Merleau-Ponty's phenomenological perspectives, which we found particularly useful for the analysis of emotions. First, we will present the notion of the habitual body and continue with the idea that the body is defined through its permanency, and finally we will introduce the argument that emotions constitute styles of conduct.

9.2.2 *The Habitual Body: The Body's Relationship with the Environment*

Defined habitats such as classrooms or schools are environments that produce and reproduce particular ways to engage and behave. For example, coming to school, entering the building, walking to the classroom, and sitting down at the desk are a rhythm of actions. Upon engaging in these repetitive events, a body memorizes a set of ways of acting, and this is what Merleau-Ponty refers to as the habitual body (2012). These habits are embedded in the body and exist when we engage in familiar situations. Richard Shusterman (2012, p. 39) explains that "…in most of our activities, attention is and needs to be primarily directed …to the objects of our environment." The present body is not to be understood to be different to the habitual; it is simply our body in this very moment that is engaged in activities in which it draws on embodied knowledge that is already embedded in the body – the habitual body. We draw on our habitual body for the many things we do without thinking, such as walking. It is a skill that is available to the present body but does not intrude on the present body. Merleau-Ponty argues that skills that are learnt will in time become part of the habitual body and may be used by the present body.

As noted above the habitual body is shaped by its relationship to a particular milieu or environment. Merleau-Ponty stresses the importance of the milieu by stating:

> Insofar as I inhabit a "physical world," where consistent "stimuli" and typical situations are discovered – and not merely the historical world in which situations are never comparable – my life is made up of rhythms that do not have their reason in what I have chosen to be, but rather have their condition in the banal milieu that surrounds me. A margin of almost impersonal existence thus appears around our personal existence, which, so to speak, is taken for granted, and to which I entrust the care of keeping me alive. Around the human world that each of us has fashioned, there appears a general world to which we must first belong in order to be able to enclose ourselves within a particular milieu of a love or an ambition. (Merleau-Ponty 2012, p. 86)

The "margin of almost impersonal existence" (ibid) Merleau-Ponty refers to above shapes people's engagement in the world, imposes certain ways of being, and privileges particular actions over others. This means that understanding how people experience and respond to the environment requires connecting the habitual body with the physical world in which it carries out its operations.

9.2.3 The Body Defined Through Its Permanency

The second important aspect we would like to explore for our analysis is the idea that bodies pertain a particular permanence that form the basis for our perception of the world (Merleau-Ponty 2012, p. 93). The permanence of the body refers to the idea that we cannot remove ourselves from our bodies. It means that the body is a constant condition of our existence, unlike other physical objects, such as a chair, which we may leave at any given time. Merleau-Ponty (2012) writes "To say that my body is always there for me is to say that it is never truly in front of me, that I cannot spread it out under my gaze, that it remains on the margins of all my perceptions, and that it is with me" (p. 93). For Merleau-Ponty there is distinctiveness to the embodied life of a person because a person has history, language, and culture. This constitutes uniqueness and permanency that go beyond a short glimpse of embodied performance. Merleau-Ponty (1964b, p. 245) regards the body also as one whole organ that is to be seen by others and experienced by oneself. It is only through this visibility that reflexivity or reciprocity is possible. However, Merleau-Ponty also explains that the experience of one's own body is personal and not like the experience we have with other objects that can be moved away and cannot be experienced when they have gone. In the school setting or any lived environment, this concept of permanency applies to any bodily experiences one might have sitting at a desk, putting up your hand, talking to your friends, and looking to the teacher or at the black/whiteboard. Those embodied experiences make the experience "classroom" a personal and embodied one.

9.2.4 Emotions as Styles of Conduct

The third aspect we would like to explore for our analysis is the role of emotions on our behavior or styles of conduct as Merleau-Ponty (1964a) refers to it. Merleau-Ponty's thinking presupposes that the perception of the world (or how we experience and respond to the world) is constituted by the lived body and is situated in the world. This means that for Merleau-Ponty perception is not an inner (cognitive) event but a sophisticated mode of engaging with the world.

> If I try to study love or hate purely from inner observation, I will find very little to describe: a few pangs, a few heart throbs – in short, trite agitations which do not reveal the essence of love or hate. [...] We must reject the prejudice which makes 'inner realities' out of love, hate or anger, leaving them accessible to one single witness: the person who feels them. Anger, shame, hate and love are not psychic facts hidden at the bottom of another's consciousness: they are types of behaviour or forms of conduct which are visible from the outside. They exist on this face or in those gestures, not hidden behind them. (Merleau-Ponty 1964b, p. 52)

Emotions impinge on our being at all times and contribute to the kinds of actions that the perceiver sees. The way in which we conduct ourselves when engaging in the world are expressions of emotion, and this implies that all behavior is emotional. Merleau-Ponty (1964b) explains that emotions are styles of conduct that are visible from the outside. He writes that through the visibility of the body, we anticipate and respond to each other in such ways that we do not even have to look at ourselves but receive information by sensing how others perceive our body. This in turn leads to adjustments and responses produced through and with our body. Merleau-Ponty describes this also as "reflective redoubling" (Merleau-Ponty 1964b, p. 249), the way how the body receives and responds to its own perception.

When we explore the interactions between the students in our example, we are exploring how their emotions can be detected when they are expressed through the body. This view of emotions distances itself from the idea that emotions are a purely mental state, which only the person who displays emotions has access to. On the contrary, Merleau-Ponty indicates that a person is often unaware of their emotions and that these can only be discerned through reflections following an (emotional) event. Even then, embodied emotional expressions such as an increased pitch in the tone of voice may mean several things – love, hate, and fear – the details of the emotional spectrum can only be defined by examining the context (Crossley 1995). To understand embodied practice, it is important to unpack the body's "relational and interactive dimensions" (O'Loughlin 1998, p. 281).

To articulate this tacit dimension of human interaction may be achieved by locating emotions neither exclusively in the mind nor in the body but in the entanglement of engagement and interactions with a world that is complex (O'Loughlin 1998). Merleau-Ponty conceptualizes emotions as an atmosphere, which is intangible and obscure because it surrounds, encapsulates, and guides actions and behavior. When exploring the role of emotions in classroom interactions, we are interested not in categorizing disjointed emotional displays but rather to go into a dialogue with our empirical data to explore how the visible interactions privilege certain meanings. This means that we attempt to explore and explain embodied displays in school settings from multiple angles, giving our data multiple voices in an attempt to come closer to explaining the complexity of people's emotions.

To move now from the theoretical to the applied approach of Merleau-Ponty's ideas, in the next section we explain interaction analysis and how to approach the detailed examinations of bodily stance.

9.3 Examining Emotions Through Bodily Stance

Human emotions are often thought of as responses to changes in the surroundings in which people operate. Charles and Marjorie Goodwin (2000), linguistic anthropologists, regard the practices of individuals as socially situated and examine emotions as constructed social phenomena. In an article on emotions and bodily stance, they wrote together with Asta Cekaite that emotions are "mediated by culturally variable display rules and made visible on the body of the actor expressing the emotion" (2012, p. 12). The interesting angle in their mode of investigating emotions is that they take note of the bodily expressions of social players, as well as traceable speech features such as pitch and volume (prosody), and "the unfolding flow of action in interaction" (p. 17). To highlight these aspects in transcripts, they use conversation analysis (CA) and line drawings. CA is typically used to notate the social interaction between different actors. The term conversation analysis is somewhat misleading because CA takes note of more than talk. Based on Garfinkel (1967) and other's work, this method is used to transcribe social interaction, outlining both verbal expressions and nonverbal behavior, indicating pauses, emphasis, or nonverbal sounds. Here we have used Gail Jefferson's (2004) transcript symbols to prepare the selected transcripts in our examples for analysis (for overview of notations used in this article, see Table 1 in the Appendix). We also used line drawings of selected situations and combined them with transcripts to emphasize bodily stance, which Goodwin and colleagues described using arrows to show either congruent (double headed arrow) or oppositional alignments (a vertical line at the end) (p. 22). Through this, it is possible to show bodily (mis)alignments between actors (see example in Fig. 9.1).

Goodwin and colleagues noticed that the traditional and important contributions of investigating emotions, notably by Charles Darwin (1872/1998) but also in more recent and contemporary works by Paul Ekman (e.g., 1982; Ekman and Friesen

Fig. 9.1 Emotional stance taking – an example of conversation analysis and line drawing, including alignment arrow

1.	Mira:	Who should r:un?
2.		Oh no you (.) run
3.		↑Adi runs
4.		I don't want ↓ to run
		(Well it's really fast)

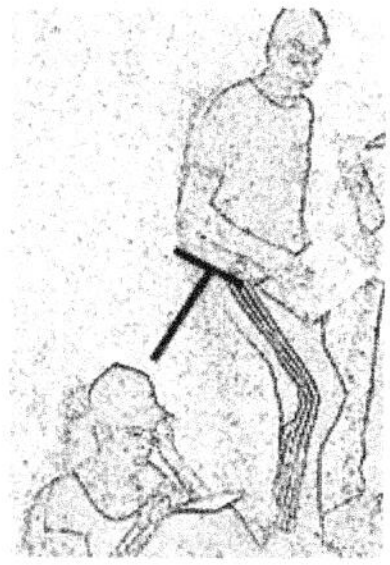

2003), focused on the close examinations of facial expressions only. They found that this kind of analysis ignored to examine the environment around those (faces) because "one cannot detect the environment when studying expressions" (2012, p. 16–17). They wrote that "emotional expressions be investigated within an environment of unfolding action being constituted in part through orientation to the bodies and actions of others" (p. 21). Goodwin et al. expanded the systematic investigation of emotions from facial expressions to taking careful note of interacting bodies, including alignment, stances, and orientations. We found this approach useful because this allowed us to take note of the entire body and enactment of emotions. However, we wanted to expand this approach further by taking note of the material world/environment/milieu to explore how "meanings and materials are coming together in everyday practices" (Orlikowski 2007, p. 13). In our analysis, we connect the analysis of bodily orientation or stance with the different ways of movement of different actors (students) in specific spaces, to show how these are linked to emotions during science activities at school. Our premise is that an environment's spatial qualities shape how it is being experienced (Boone 2013) and that the conditions of the physical environment are in direct connection with how emotions play out or are moderated. We take note in our analysis of the physical environment to explore also where and how it shapes social interactions and emotional responses.

9.4 The Investigation

Next we will describe the context in which we collected our observations and how the data sources were collected. The activity involved a year 8 physics class exploring the concept "light and sound". The teacher organized group activities and the students had to conduct several experiments. We focus on a group of students who were working on an experiment about the Doppler effect. The Doppler effect is the change in frequency of a wave for an observer who is moving relative to its source, such as may be experienced when a siren approaches, passes, and recedes from an observer. What the observer should experience is a higher frequency during the approach, identical at the moment that it passes by, and lower frequency when it recedes.

The episodes analyzed and presented here involve a group of six students. The students received instructions from their teacher how to carry out the experiment. The students were asked to perform the experiment in the hallway just outside the physics lab. The hallway is about 50 m long and marked with lines every 5 m, resembling a running track in a track and field stadium. There were several classrooms along the hallway and a large section of windows facing the schoolyard. Prior to using the hallway, the teacher had made sure that there were no exams in the adjacent classrooms not to disturb anyone (pre-interview with the teacher).

In the Doppler effect experiment, the teacher asked the students to use a sound source that made a constant high pitch, such as from an app on a smartphone. Then, they positioned a student at the end of the hallway with the sound source, and the rest of the group had to stand midway and evaluate how the sound changed as the student with the sound source ran from one to the other end of the hallway. The students had to do five trial runs at full speed, each student running at least once. They had to record the change in sound from when they could hear it from the end of the hallway, then again, when a runner was moving the sound source toward them, and again when the runner was passing by, and finally when the runner was moving away. The students had to read background information in their textbooks. Their task was to produce a video that showed and explained the experiment. Their results had to be posted on the class website, on a page featuring the sound and light unit. The learning objectives in this activity were (1) to let the students experience the Doppler effect and (2) to give them an idea of what the Doppler effect "was all about" (interview with teacher).

The distance markers on the hallway floor were added recently when the school adopted a sports profile, which involved a renovation of the school facilities to promote more physical activity. The adoption of the sports profile included that school policy mandated the deliberate infusion of bodily activity in the learning activities across all subjects. As a response to this recent initiative, by Danish policymakers, part of the teacher's plan was therefore to include physical activity into the experiment by having the students run.

The data collection included approximately 8 h of video-recorded observations, field notes, pre- and post-interviews with the teacher, group interviews with the students, and video stimulated recall interviews with selected students.

Goodwin et al. (2012) wrote that "the display of emotions is a situated practice entailed in a speaker's performance of affective stance through intonation, gesture, and body posture" (p. 16). These qualities provide the entry point into the analysis of our first vignette, where we will illustrate how emotions played out in the observed episodes. We then apply Merleau-Ponty's ontological assumptions (the habitual body, permanency, and styles of conduct) to examine when the students engage with each other in the setting of the hallway over the task of conducting a Doppler effect experiment.

9.5 Vignette 1: Risky Running

In this first vignette, we illustrate the interaction of Mira, a 15-year-old girl, with her group members when they talked about who should be running in their experiment. She made her intentions clear through her embodied stance and by that we mean the way in which she used gestures, talk, and intonation.

Figure 9.2 shows how emotions played out in this vignette. The group, three boys and two girls, had just left the classroom and entered the hallway just outside the classroom. Mira sat down immediately and pressed her back against the wall of the hallway. Then she looked up and explained the others that she did not want to run, and then she looked down again (see lines 1–2 in Fig. 9.2). Next, Mira looked back up at Alfons, who was standing next to her with his back leaning against the wall. Alfons discussed with Mira who should run and suggested that Hai (who was standing next to Alfons) should not run (see lines 3–4). Mira agreed and suggested that Adi, who was walking up to them from behind the camera, should be the runner (see lines 5–11).

In lines 1 and 2, Mira expressed her refusal to run both verbally and nonverbally. She explained that she did not want to run (line 1), through her nonverbal actions (turning her face away and covering her eyes with her hand so that her face is not directly visible to the others). Mira spoke in a low pitch (200Hz) when stating that she did not want to run. Prosodic features such as pitch, that is, a lowering or raising of the voice, communicate emotions (Frick 1985). Pitch reinforces verbal content, and although there are cultural variations, basic emotions are described to show universal patterns. Low pitch with soft loudness and slow tempo can be interpreted as boredom or grief/sadness (Scherer 1986; Ververidis and Kotropoulos 2006). Mira's low pitch can be interpreted together with her gesturing and talk as being unenthusiastic about the idea to run. When Alfons suggested that Hai should run (line 3), Mira quickly responded by stating that Adi should run (line 10). She emphasized the name "Adi" to signify this suggestion as a better option. Mira's interacting body showed congruent alignment marking and oppositional positioning (Goodwin et al. 2012). At this point in the unfolding situation, the intonation of her voice changed from a low to a high pitch (400Hz). High pitch, fast tempo, and increased loudness can indicate happiness/joy, confidence, activation, or anger (Scherer 1986, 1994) and can be interpreted as increased excitement and in Mira's case a change in her emotional state.

9.5.1 Applying Merleau-Ponty to "Risky Running"

When the students came out of their classroom, Mira sat down immediately, and Alfons aligned with her by sitting down in the same position. Hai was standing next to Mira and Alfons, also leaning against the wall. Adi was walking back and forth in the hallway, interacting with those passing by, such as the girl he was talking to

01 Mira **Jeg gider ikke ↓ løbe** *((looks down and touches her forehead with her hand, addressing her talk to no one in particular))*
I don't want to run

02 (Altså det vildt hurtigt)
Well it's really fast

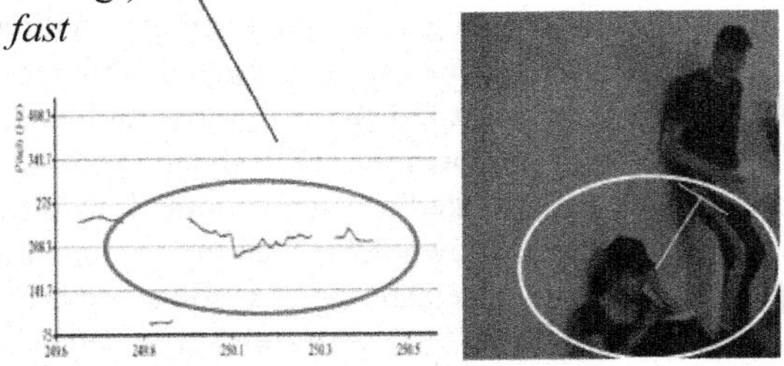

03 Alfons **Hai** løber
Hai should runs

04 Du løb heller ikke i går= *((addressed to Hai))*
You didn't run yesterday

05 Mira =Adi du skal løbe *((Alfons looks up at Adi and smiles))*
Adi you must run

06 Adi **E::::j**
No

07 Adi Hai så **er det dig** der løber ((adressed to Hai))
Hai then it's you who'll run

08 Alfons **Du løb ikke engang derude mand** *((shakes his head))*
You didn't even run out there man

09 Adi Jeg ved det Alfons
I know it Alfons

10 Mira **Ne:j ↑ Hai er langsom** *((Alfons sits down next to Mira))*
No. Hai is slow

11 **Hai** er langsom
Hai is slow

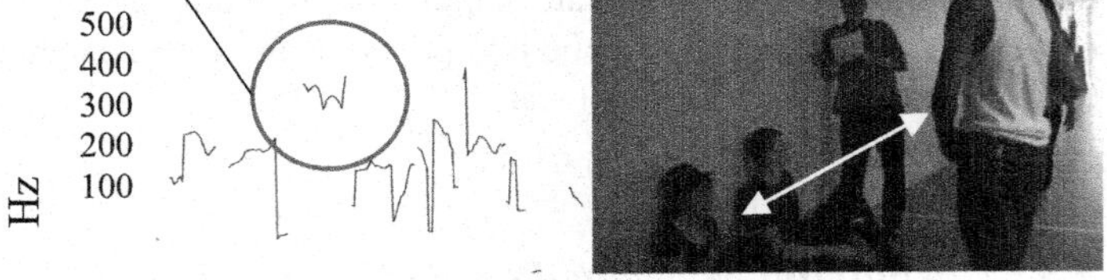

Fig. 9.2 Deciding who should run

Fig. 9.3 The group inhabiting the hallway

in Fig. 9.3. Adi inhabits the hallway as an environment he possesses which is indicated by his confidence moving around and stopping those who pass through. While the other three keep to the margins, the walls.

Taking note of Mira's enactment of seemingly making herself "smaller," we asked her during an interview why she did not want to run.

Interviewer: Hvad siger du Mira? .
What do you say Mira?
Mira: Det var fordi vi skulle optages, og det er sådan ikke ((makes a face))… Jeg gider ikke løbe og blive optaget og sådan og så bliver det vist…
It was because we had to record ourselves, and it's like… I don't want to run and be recorded and for it to be shown…
Interviewer: Også når det kun er mig der ser det?
Also when it's just me watching it?
Mira: Nej men ***vi*** skulle jo optage det med video
No but **we** *had to record it on video*
Interviewer: Det er rigtig, I skulle jo optage hinanden. Det gider du ikke at andre ser?
That's right, you had to record each other. You don't want others to watch this?
Mira: Nej.
No
Interviewer: Hvorfor ikke?
Why not?
Mira: Jeg kan ikke lide det.
I don't like it.
Interviewer: Jamen er der forskel på den måde du løber og den måde de andre løber?
But is there a difference in the way you run and the way the others run?
Mira: Jeg ved det ikke. Jeg tror bare ikke rigtig… Det er sådan lidt ubehageligt
I don't know. I just don't think… It is kind of uncomfortable.

In the interview, Mira confirmed our analysis that she did not want to run, both by telling us in words but also by frowning her face, which is a sign of disgust (Ekman in Darwin, 1872/1998). The hallway with its confines and closeness and also the video recording meant for Mira that her embodied display would be not only visible to everyone being there but also permanent if she was recorded running. She referred to being uncomfortable and not liking it. From her responses both during the observation and when she was prompted again later by viewing the scene on video we take that sitting down (Mira never moved away from this position during most of this activity) provided safety and comfort to her.

Merleau-Ponty explains that the feeling of familiarity stems from the habitual body that "knows" from experiencing things over and over. Sitting down, feeling the back of the wall, is a feeling Mira is used to and feels at the time most comfortable with. Mira's habitual body knows also how to run, yet immersed in the environment of the hallway, the task of the activity, and her reflection on being video recorded, she decided as she also told us in the interview that this is not an attractive option. This is what Merleau-Ponty describes as reflective redoubling. Mira's present body, faced by the task at hand, had to make a decision to break from the "norm," the habit (how to "be" in the hallway, running in front of others, being filmed). Merleau-Ponty explains that when we inhabit an environment, it shapes how we perceive our body to be perceived by others. When Mira referred in the interview to her feeling uncomfortable in relation to running and being filmed, she demonstrated her consciousness of how others might perceive her embodied performance.

9.6 Vignette 2: The Confident Runner

This vignette presents Adi, a boy of 16 years old, before, during, and after he sprints down the hallway. He is confident and likes to run, which he embodies in several different manners supported by gestures, talk, and intonation.

The group made the decision that Adi should run and not Hai. Adi picked up Hai's phone, which they were using to make a sound as part of the Doppler experiment, and he walked down to the end of the hallway (see lines 1–2 in Fig. 9.4). On the way, he stopped briefly, made some wriggling movements, adjusted his pants, and looked back to the others, smiling (see lines 3–4). When a girl passed, he stopped for a short embrace before jogging to the end of the hallway. In the meantime, the group was joined by another student, Lucy, and discussed how to record the volume changes when a sound source passes. Hai suggested that they should use a minus to symbolize a "dark" sound and a plus to symbolize a "light" sound. Adi called out to them and asked if he should time the run, and Hai and Alfons replied that it was not the time that they needed to record but that he should play the sound on the phone while running (see lines 5–11). Adi sprinted down the hallway while the group watched. Lucy, Mira, and Alfons smiled as Adi ran. As soon as Adi

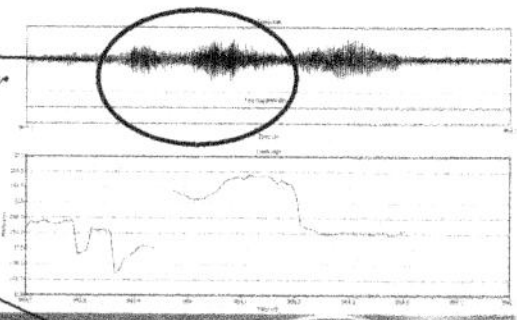

01 Adi	>Lad **mig gøre det** så< *((walks over to Hai and takes his phone))*
	Let me do it then
02 Alfons	**Ja: I** skal skiftes
	Yes you have to switch
03 Adi	Yes
	Yes
04 Hai	**Behold den op ved den linje Adi**
	((Adi adjusts his pants he looks back and smiles))
	Keep it at that line Adi
04 Adi	Ja Hai
	Yes Hai
05 Hai	Ja **bare** begynd
	Yes just start
06 Adi	(hvor mål)
	Where goal
07 Mira	Hai du skal løbe
	Hai you have to run
08 Alfons	**Ja-**
	Yes
09 Hai	**Den skal ikke måle ti:den**
	We are not supposed to measure time
10 Alfons	Vi skal ↑bruge **lyden**
	We need the sound
11 Hai	Ja kom nu bare *((Adi starts to sprint down the hallway with the phone in his hand))*
	Yes come on
12 Adi	>Hallo<
	Hallo
13 Adi	>Kameraet kunne sikkert ikke engang fange mig<
	The camera probably coudn't even catch me
14 Adi	>er vi enige om det?<
	do we agree on that?
15 Adi	Jeg **fløj** siger jeg til dig ↑ *((walks back to the group with his arms out, while he makes a take-off sound))*
	I tell you I flew.
16 Adi	A::::h. Jeg **ku'** mærke vinden i mit hår
	Aaaah. I could feel the wind in my hair

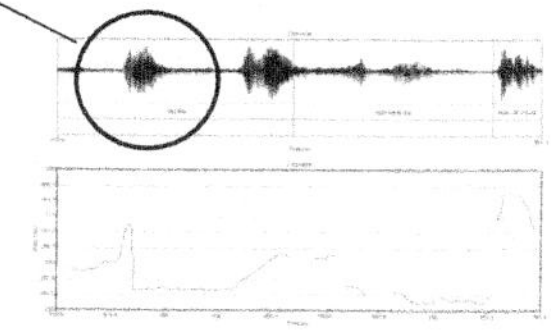

Fig. 9.4 Running

passed by, Hai started to evaluate the sound recording. Adi walked back to the group and told the researcher how fast he was (see lines 12–16).

In line 1, Adi stated he should run and supported this intention by an increase in pitch (nearly 260 Hz) on the phrase "Lad **mig gøre det** så / *Let me do it then*." High pitch, increased volume, and fast tempo are indicative of confidence (Scherer 1986; Ververidis and Kotropoulos 2006). Adi's high pitch combined with his dialogue with Hai was further supported by his nonverbal actions (taking Hai's phone without asking). This behavior indicated his confidence in performing. When Adi decided to run, there was to begin with some confusion about what he had to do (see lines 4–11). Adi focused on how far he had to run and who would time his run. The group corrected him that time did not matter, only the recording of the changing sound patterns. With their backs against the wall, the group watched Adi run. In turning their heads to follow his run they aligned their bodies congruently, which Goodwin et al. (2012, p. 21–22) write, indicates a mutual stance toward Adi's running. After the sprint Adi approached the researcher (see lines 12–15) to discuss how fast he was. Adi spoke with a higher pitch in line 15 and used the word "flew" to describe his running. He further emphasized his sensation by putting his arms out to the sides, gesturing that he was flying. Adi described this in line 16 when he exclaimed "A::::h," which Goodwin et al. (2012, citing Scherer 1994, p. 170) describe as an emotive interjection, the "very brief, discrete, nonverbal expressions of affect in both face and voice."

In a later interview, Adi elaborates on why he wanted to be the runner rather than Hai. He explained that it was important that someone who is fast should run. Adi said that he sees himself as someone who enjoys running and who is fit for running. He was asked what made him try and give all he could when he was running and he replied:

01	Adi:	Jah. Du ved, jeg skal jo lige vide hvor hurtig jeg er. Forstår du.
02		Det er ligesom en 60 m hele vejen ned *Yeah. You know, I have to know how fast I am. You see. It's like a 60 m all the way down*
03	Interviewer:	Okay *Okay*
04	Adi:	Så jeg vil jo gerne lige måle mig selv *So I really want to measure myself*
05	Interviewer:	Hvorfor vil du gerne måle dig selv? *Why do you want to measure yourself?*
06	Adi:	For at gøre det bedre næste gang *To do it better next time*

Adi explained that he takes pride in his body. He used to train for soccer four times a week before he suffered a knee injury. At the time of the recording, he had not started rehabilitation yet because he was still in pain, but this assignment provided an opportunity to "test" his knee. The lines in the hallway also acted as an incitement for testing his knee (see lines 1–2). Adi was further motivated to sprint because he was being video recorded. He stated on several occasions in the interview that he liked being recorded and that he would not have been so keen on sprinting if the camera had not been there. Adi also explained that he felt happy because it was summer time and he was able to wear a singlet and hereby show off his biceps. He said that he did not like the cold because it affects his mood. The summertime, combined with the ability to show off his arms and interact with the passing students and the group, seemed to have a positive effect on how he felt about being in the hallway, which according to Adi can be dull at times. He explained that he was able to enjoy himself in a setting that was different from the "normal" classroom environment:

01	Interviewer:	Er det sjovere at være ude i gangen end i klasselokalet? *Is it more fun to be out in the hallway as opposed to the classroom?*
02	Adi:	Jah, meget. Nej ikke kun, ikke selve gangen. Der kan også være
03		kedeligt på gangen. Det er bare hvem du er sammen med og
04		hvad du laver. Sådan noget der (referer til aktiviteten), som ikke
05		har så meget med timen at gøre, hvor jeg bare skal sidde og kig-
06		ge, kan jeg godt lide. Og når jeg er sammen med de der (hans
07		gruppe), så er det også godt nok. *Yeah a lot. No not just, not the hallway in itself. It can also be boring in the hallway. It is all about who you are with and what you are doing. This thing (refers to the activity) that doesn't have much to do with class, where I just have to sit and watch, I like it. And when I am with those guys (the group), then that's okay.*

9.6.1 Applying Merleau-Ponty to "The Confident Runner"

Adi's habitual body presented itself in a different manner compared to Mira. Adi was comfortable with running in front of others (and on video) as he was used to perform competitively. He was jovial about being recorded and explained in the interview that if you make a mistake you can just redo the video recording. His way of inhabiting the hallway space, for example, by spreading his arms after the run and stating that he was flying showed a familiarity with situations that entail physical performance in public. Different to Mira Adi was positive about his body's performance. Adi sensed how his body and performance may be received, and in anticipation of a positive response, he was comfortable and looking forward to running, which is what Merleau-Ponty describes as "reflective redoubling" (1964b, p. 249).

Being watched and also recorded was for him a positive reinforcement and made him feel confident about putting his present body into this situation.

When Merleau-Ponty talks about the idea of permanency, he explains that the body is at the margin of all our perceptions and that we cannot remove ourselves from our bodies. In Adi's case the sports culture and history of a knee injury that reside in his body provided him with a particular perspective about the task of running. From Adi's perspective, being able to run fast and the right way was of importance. An excerpt from a later interview with Adi illustrates this, where, after watching selected sequences of the video observation, he laughed when Hai ran. When asked why he laughed, Adi responded:

01	Interviewer:	Hvorfor griner du? *Why are you laughing?*
02	Adi:	Fordi han (Hai) er så doven mand *Because he (Hai) is so lazy man*
03	Interviewer:	Er han doven? *Is he lazy?*
04	Adi:	Ja, Hai. Han bliver ved med at snakke, lad os bare tage den her (run from halfway) *Yes, Hai. He keeps on talking, lets just take it here*
05	Interviewer:	Okay. Men er han doven hvis han mener at I ikke behøves at lø
06		be længere? Hvorfor skal han løbe hele længden? *Okay. But is he lazy if he means that you do not have to run further? Why does he have to run the entire distance?*
07	Adi:	Nej, men det er bare han orker ikke det der. Han orker ikke at rø
08		re sig. *No, but it is just that he cannot be bothered to do it.* *He doesn't have the strength to move about.*
09	Interviewer:	Okay. *Okay.*
10	Adi:	Så siger jeg bare at han er doven. Kom i gang! *So I just say that he is lazy. Get going!*

Adi refers back to his embodied culture and history, for example, when he talks about his involvement in competitive sports. The hallway environment, in particular the lines on the floor, the assignment, and the video recordings, privileged sprinting for him as a sensible action he felt comfortable with and was good at. This position is so strong within Adi that he was observed sprinting whenever there was a possibility but also when he urged Hai to sprint and bantered him when he did not perform as well as Adi.

9.7 Emotional Atmosphere

The examples above show two very different reactions to the task of running in this particular setting. While Mira refused to run, Adi took pride in running. In other words, within the same group of students, there were different modes of engaging with the task presented.

According to Merleau-Ponty, emotions are visible in our behavior, as they form an atmosphere, which privileges some actions over others. However, Merleau-Ponty distinguishes between flat affective responses and deep emotional expressions (Cataldi 1993) because emotions, he argues, are only experienced at a deep level. In the case of Mira, sitting with her back against the wall was her preferred response over running. Although Mira was capable of running, being situated in a hallway with many students passing, cameras recording their actions, and asked to run fast, this was a situation that created feelings of discomfort. In the interview, she stated that she felt *uncomfortable* about being recorded while running. She did not comment on what it felt like sitting on the ground, because this was a superficial experience that did not require any deeper reflection. However, the necessity to run while being recorded created negative inner feelings. Mira decided to "walk away" from this task by sitting down. Her vocabulary of distancing herself includes phrases like "I don't want to," "I don't like it," and "it makes me feel uncomfortable." Merleau-Ponty explains in *Phenomenology of Perception* that words are not used without meaning. He classifies this further by stating that:

> Our view of man will remain superficial so long as we fail to go back to that origin, so long as we fail to find, beneath the chatter of words, the primordial silence, and so long as we do not describe the action which breaks the silence. The spoken word is a gesture, and its meaning the world. (2012, p. 184)

Merleau-Ponty reminds how important it is to examine the details of what has been observed words, gestures, and embodiment.

Returning to Mira's example, the emotions she reflected on were observable and shaped the mode of engagement between her and the other students. Her emotional response was prompted by the nature of the task, the environment, and artifacts, including the video cameras, the other students, the public hallway, the meter marks, and the assignment.

In the case of Adi, the artifacts and the surrounding had the opposite effect: He willingly sprinted and displayed enjoyment and commitment when performing in public. His body posture and embodiment while being in the hallway (walking in the middle of the hallway, talking and engaging with all who walked by) was also a display of his positive stance toward the activity and the others. While Mira's body posture combined with the prosodic indicators in her voice for negative emotions when asked to run indicated her embodied affective refusal to the task at hand.

9.8 Discussion

The aim of this chapter was to present an argument on how to take note and analyze emotions as they played out in our example of a science activity at school. We were using theoretical ideas proposed by the philosopher Maurice Merleau-Ponty. Leaning on methodological approaches that analyzed talk and embodiment and considering the physical environment, two worked vignettes were presented to show how to identify and analyze feelings and emotions. The examples were from a group of Danish primary school students who were tasked to conduct physics experiments about the Doppler effect that involved a student running while the others were to record and observe the runner passing by with a sound source.

We found through observations, interviews, and analysis of embodiment in given situations, as well as taking note of pitch, volume, or intonation when talking (prosody) that we were able to identify how different students responded to a task they were given by their teacher.

All students were capable "runners" in the experiment, but one objected to doing this task because it made her feel uncomfortable (Mira). Another one was rejected by his peers because they did not perceive him to be a good runner (Hai), while another one had the support of his peers and the task itself aligned with him seeing himself being a physical performer (Adi). The activity was not about being able to run fast or well, but the deliberate public display paired with recording the activity itself on video foregrounded physical performance.

The school hallway was a particular environment that framed this group activity. The inspiration to consider the environment to a greater extent was adopted from Merleau-Ponty. In *Phenomenology of Perception*, he identifies how the body enables the perception of the physical environment to make it possible to experience what is "there." The body allows grasping an object and holding it but also to move around that object and gaze upon it. Through our bodily immersion, the environment is being perceived. This embodied perception means that objects and the environment exist as *lived* in that moment but also have a past memory that is embedded in the body, which also shapes the anticipation of perception. In our examples the students lived in the environment "hallway" that they perceived in specific ways with the constituents' video cameras, other group members and students, or the marking on the hallway floor shaping the perception of what is or might be perceived.

Taking into account gestures, facial expressions, prosody, and embodiment means that we came closer to an interpretation of how the young people in our examples felt. The theoretical space that Merleau-Ponty's opens up is that there are dimensions beyond the content of talk that show how people respond and feel in a given situation. Merleau-Ponty's idea of the lived body highlights that the body makes it possible to experience reality. Following this line of thinking implies that examining the students' experiences requires the consideration and analysis of the embodied experiences.

The affective dimension of the body presents what Merleau-Ponty describes as the emotional atmosphere that shapes our perceptions with the world. Perception he explains is a constituent of the lived body that is situated in the world. According to Merleau-Ponty, perception is not an "inner" show but forms an opening onto the world. The ways in which people conduct themselves in the world are expressions of emotions or styles of conduct that are visible from the outside (Merleau-Ponty 1964a).

The vignettes we presented unpacked different behaviors that were displayed by the students as they made them visible through their styles of conduct. Drawing on analytical tools that take note of embodied conduct such as conversation analysis, prosodic analysis, and descriptions of bodily stance helped us to identify individual responses or emotional atmospheres.

9.9 Conclusion

In summing up, we find that making sense of the body and how oneself and others perceive its abilities is a useful way of understanding the connection between how one behaves and responds emotionally to a given situation. Merleau-Ponty's attention to the bodily sensations, and how they shape the decision-making process of actions and interactions, implies that emotions are grounded in the body that carries memory, attitudes, and feelings.

While it may be argued that Merleau-Ponty described emotions in his works as an explicit concept only vaguely, we see this as strength that forces us to go beyond categorization and into a dialogue with the empirical data to explore how visible (emotional) actions privilege certain meanings.

Taking this theoretical point of departure has analytical implications because it requires that we identify how people respond and interact through their whole bodies. This means also that, when trying to understand interactions based on the notion of the lived body, it is important to take note of interactions in context, the milieu or environment, because this is where responses are formed.

Analyzing emotions this way means also that certain actions/relations are privileged over others. The cultural processes and settings they are embedded in shape this process. In the case of our examples, the Danish school, the science education context, the task, the hallway, and the students formed all part of this cultural setting.

Reflecting on this analysis, we find that exploring the role of emotions in classroom interactions is insightful, not necessarily to categorize emotional displays but rather to go into a dialogue with the empirical data to explore how the visible interactions privilege certain meanings.

9.10 Appendix: Transcript Conventions

The conversations captured in the presented vignettes have been transcribed using Gail Jefferson's transcript symbols (2004). Translations are provided for each line. Bodily stance has been identified using Goodwin et al.'s description of interacting bodies (2012).

::	Lengthening of the preceding sound
( )	Item in doubt
(0.5)	Silence in seconds
> <	Enclosed text is said more quickly
< >	Enclosed text is said more slowly
h or hhh	An audible aspiration
.h or .hhh	an audible inhalation
↑	Onset of noticeable pitch rise
↓	Onset of noticeable pitch fall
=	The equal sign shows that there is no discernible pause between two speakers' turns or, if put between two sounds within a single speakers' turn, shows that they run together
Bold	Bold sounds are louder
-	A dash shows a sharp cut-off

References

Boone, G. E. (2013). *Emotion, community development, and the physical environment: An experimental investigation of measurements*. Theses and dissertations–Community and Leadership Development. Paper 10.

Cataldi, S. L. (1993). *Emotion, depth, and flesh: A study of sensitive space: Reflections on Merleau-Ponty's philosophy of embodiment*. Albany: SUNY Press.

Crossley, N. (1995). Merleau-Ponty, the elusive body and carnal sociology. *Body and Society, 1*, 43.

Darwin, C. (1998). *The expression of the emotions in man and animals* (Introduction, afterword and commentaries by Paul Ekman). Oxford: Oxford University Press. (Original work published in 1872).

Ekman, P. (1982). An argument for basic emotions. *Cognition & Emotions, 6*((2–4), 169–200. doi:10.1080/02699939208411068.

Ekman, P., & Friesen, W. V. (2003). *Unmasking the face: A guide to recognizing emotions from facial clues*. Cambridge, MA: Malor Books.

Frick, R. W. (1985). Communicating emotion: The role of prosodic features. *Psychological Bulletin, 97*(3), 412–429. http://dx.doi.org/10.1037/0033-2909.97.3.412

Garfinkel, A. (1967). *Studies in ethnomethodology*. Englewood Cliffs: Prentice-Hall.

Goodwin, M. H., & Goodwin, C. (2000). Emotion within situated activity. In *Communication: An arena of development* (pp. 33–53). Stamford: Ablex.

Goodwin, M. H., Cekaite, A., & Goodwin, C. (2012). Emotion as stance. In A. Peräkylä & M. L. Sorjonen (Eds.), *Emotion in interaction*. Oxford: Oxford University Press.

Jefferson, G. (2004). Glossary of transcript symbols with an Introduction. In G. H. Lerner (Ed.), *Conversation analysis: Studies from the first generation* (pp. 13–23). Philadelphia: John Benjamins.
Merleau-Ponty, M. (1964a). *Sense and non-sense*. Evanston: Northwestern University Press.
Merleau-Ponty, M. (1964b). *The visible and the invisible*. Evanston: Northwestern University Press.
Merleau-Ponty, M. (1998). *Structure of behaviour*. Pittsburg: Duquesne University Press.
Merleau-Ponty, M. (2012). *Phenomenology of perception*. Oxon: Routledge.
O'Loughlin, M. (1998). Paying attention to bodies in education: Theoretical resources and practical suggestions. *Educational Philosophy and Theory, 30*(3), 275–297. doi:10.1111/j.1469-5812.1998.tb00328.x
Orlikowski, W. J. (2007). Sociomaterial practices: Exploring technology at work. *Organization Studies, 28*(9), 1435–1448. http://doi.org/10.1177/0170840607081138
Scherer, K. R. (1986). Vocal affect expression: A review and a model for future research. *Psychological bulletin, 99*(2), 143. http://dx.doi.org/10.1037/0033-2909.99.2.143
Scherer, K. R. (1994). Affect bursts. In S. H. M. van Goozen, N. E. van de Poll, & J. A. Sergeant (Eds.), *Emotions: Essays on emotion theory* (pp. 161–193). Hillsdale: Lawrence Erlbaum.
Scherer, K. R. (1995). Expression of emotion in voice and music. *Journal of voice, 9*(3), 235–248. doi:http://dx.doi.org/10.1016/S0892-1997(05)80231-0
Shusterman, R. (2012). *Thinking through the body: Essays in somaesthetics*. Cambridge: Cambridge University Press.
Thøgersen, U. (2014). The embodied emotionality of everyday work life: Merleau-Ponty and the emotional atmosphere of our existence. *Philosophy of Management, 13*(2), 19–31. doi:10.5840/pom20141329.
Ververidis, D., & Kotropoulos, C. (2006). Emotional speech recognition: Resources, features, and methods. *Speech Communication, 48*(9), 1162–1181. doi:10.1016/j.specom.2006.04.003.

Liv Kondrup Kristensen is a PhD student at the Department of Health Science and Technology at Aalborg University. Her research is classroom-based, in which she explore different ways of capturing the role of the body in learning process. At the same time she is an associate professor at the University College of Zealand, where she teaches health related courses in the Social Education program.

Kathrin Otrel-Cass is professor MSO (with special responsibilities) in visual ethnography at the Department of Learning and Philosophy at Aalborg University. She is an experienced science and technology education researcher primarily in classroom-based settings with a particular interest in the study of interactions and what this means for teaching practices and critical education research.

Chapter 10
Interest and Emotions in Science Education

Morten Rask Petersen and Niels Bonderup Dohn

10.1 Introduction

Over the last four decades, a great deal of educational and psychological research was devoted to the clarification of cognitive processes involved in students' learning in science. Although these efforts have come a long way to explain the unique nature of student learning, they also demonstrated how little is known about the influence of affective variables (Pekrun 2006). Affective variables, such as enjoyment, are central for learning, because they guide cognition, assist people to select among beliefs, arrange priorities among goals, determine access to memories, and provide heuristics that influence reasoning, judgment, and planning (Oatley 2001).

In this chapter, we take a psychological perspective on emotions and interest in science education. We begin by exploring topic and epistemic emotions, that is, emotions triggered by the characteristics of the learning content as well as triggered by the engagement with that content. Then we unfold the concept of interest and link it to emotions and learning.

Affective variables are central to emotions and are physiologically bound to subsystems of the limbic system (or paleomammalian brain) (Pekrun 2006). Emotions are defined as multicomponent, coordinated processes of psychological subsystems including affective, cognitive, motivational, expressive, and peripheral physiological processes. The term emotion refers to a collection of responses triggered from parts of the brain to the body and from parts of the brain to other parts of the brain,

M.R. Petersen (✉)
University of Southern Denmark, Odense, Denmark
e-mail: mrask@sdu.dk

N.B. Dohn
Aarhus University, Aarhus, Denmark
e-mail: dohn@edu.au.dk

A. Bellocchi et al. (eds.), *Exploring Emotions, Aesthetics and Wellbeing in Science Education Research*, Cultural Studies of Science Education 13,
DOI 10.1007/978-3-319-43353-0_10

using both neural and humoral routes. The end result of the collection of such responses is an *emotional state* (Damasio 1998).

Such emotional states are individual and can be caused by prior experiences. For instance, Öhman et al. (2001) found that people who fear spiders and snakes had higher attention toward pictures where some contained their specific object of fear. So just seeing a picture of a spider or a snake could cause an emotional state of increased attention.

Students experience different emotions during science lessons in school. These emotions can be positive (e.g., enjoyment, pride, and hope) or negative (e.g., anxiety, anger, and boredom), and they can be frequent and intense. Some of these emotions are brought into the classroom from life outside school, but most originate within classroom settings (Pekrun 2006).

Emotions are of educational importance for two reasons. First, emotions can affect students' interest, engagement, and achievement, as well as the social climate in classrooms (Pekrun et al. 2002). Second, emotions are central to psychological health and well-being, implying that they should be regarded as important educational outcomes in themselves, independent of their functional relevance (Fredrickson 1998). Both reasons can impact students' learning. In this chapter, we focus on the former, that is, on how emotions can affect students' interest in science. Interest refers to a psychological state which has both affective and cognitive components: it includes emotions and valuing of disciplinary content (e.g., biology), as well as the perception of having and being able to develop knowledge about that content (Renninger and Hidi 2011).

Emotions that relate specifically to academic learning and classroom instruction are defined as *academic emotions*. Students' academic emotions are related to studying, test taking, and attending class. According to Reinhard Pekrun (2006), academic emotions can be categorized into four groups:

1. *Achievement emotions* relate to achievement activities and to success and failure resulting from these activities. Examples are enjoyment of learning, hope and pride related to success, and anxiety and shame related to failure. Achievement emotions are pervasive in academic settings, especially when the importance of success and failure is made clear to students.
2. *Social emotions* relate to teachers and peers in the classroom, such as sympathy, compassion, admiration, contempt, envy, anger, or social anxiety. These emotions are especially important in teacher/student interaction and in group learning.
3. *Epistemic emotions* are emotions triggered by cognitive problems, such as surprise about a new task; curiosity, confusion, and frustration about obstacles; and delight when the problem is solved. Epistemic emotions are especially important in learning with new, non-routine tasks.
4. *Topic emotions* pertain to the topics presented in lessons. Examples are enjoyment of performing inquiry-based or hands-on activities (Palmer 2009) and disgust when dealing with heart dissections (Holstermann et al. 2012).

Activating positive topic and epistemic emotions, such as enjoyment of learning the material, is assumed to trigger interest, strengthen intrinsic motivation, and deactivate negative emotions, such as boredom (Pekrun 2006). In the following, we discuss the conception of interest and its relation to emotions.

10.2 Interest and Emotions

Interest is defined as a positive cognitive and affective experience that directs attention to, and focuses it on, the activity or task at hand (Rheinberg 2008). Interest is content specific, such that it is always directed toward an object, activity, field of knowledge, or goal. When persons experience interest, their actions acquire an intrinsic quality; they are driven by enjoyment rather than external reasons (Krapp 2002). However, it is important to note that interest cannot be equated with "enjoyment while learning." Enjoyment can occur for many reasons, and interest is only one of these as we will outline in the section of different domains of interest that is presented later. Interest creates the urge to explore, take in new information and experiences, and expand the self in the process. It is commonly assumed that promoting interest increases students' intrinsic motivation to learn and the number of learning strategies they use to do so (Hidi and Renninger 2006).

Interest is an important affective variable that is associated with focused attention, higher cognitive functioning, and learning (Ainley et al. 2002). The experience of being interested is characterized as an optimal state that combines positive emotions, e.g., feelings of immediate enjoyment, and positive cognitive qualities, e.g., striving for meaningful goals (Rathunde and Csikszentmihalyi 1993). In recognizing the strong affective component of interest, many researchers went so far as to argue that interest is a basic emotion (e.g., Izard 1977). Suzanne Hidi (2006) suggests that if we only consider the moment in which the psychological state of interest is triggered, interest may be appropriately considered as an emotion. However, as interest develops and is maintained, both affect and cognition contribute to the experience. Furthermore, the relative strength of the two components of affect and cognition change over time, with cognition having an increasing presence as interest develops (Hidi and Renninger 2006).

Topics of interest are generally characterized as those that students perceive to be personally meaningful. In science classrooms, interest is often associated with a particular science topic (e.g., Hoffmann 2002). It may be that a student enjoys learning about science and attending science class; however, the student may experience boredom if the content is not perceived to be meaningful.

In the following section we first give a brief introduction to the traditional view of the interplay between learning as the cognitive variable and interest representing the affective variable. Then we explore the concepts of affective variables, emotions, and interest development where we gather and marshal arguments for another perspective on the interplay between learning and interest. This theoretical outline will then be exemplified through our empirical studies on interest development in upper secondary biology classes. Finally we will discuss the implications of our theoretical and empirical approach as well as its applicability beyond science education research.

10.2.1 Interest as a Facilitator for Learning

Interest has a long history in educational research. For example, early educators such as Johann Friedrich Herbart (1965) and John Dewey (1913) pointed to the importance of interest for supporting learning. They noted the importance of interest in encouraging effort, focused attention, and persistence to understand content and that the design of tasks was likely to promote learners' interest in content to be learned.

Interest is conceptualized by most researchers as a phenomenon, which emerges from an individual's interaction with his or her environment (e.g., Krapp et al. 1992). The etymology of the term *inter-esse*, "between being," points in the same direction (Dewey 1913). Interest is defined as a positively cognitive (involving knowledge and experience) and affective experience (involving positive feelings and appreciation) that directs attention to the activity or object at hand (Rheinberg 2008). Interest is characterized by focused attention and engagement, and feelings of pleasure, happiness, and well-being are typical emotional aspects of interest-based activities (Schiefele 1991). Suzanne Hidi and Judy Harackiewicz (2000) argue that students, who are interested in a particular subject, exhibit greater attention, are more persistent, feel greater joy, and learn more than students who do not have this interest.

The close connection between interest and learning is seen by many as self-evident; the more interest a student has in a particular topic, the more willing he or she is to learn about that topic. Alternatively, those who have little interest in a discipline tend to learn less. More specifically, interest affects goal setting and learning strategies in ways that make it a particularly relevant variable for improving educational practice (Hidi and Renninger 2006). It is generally assumed that interest is a motivational variable for learning: it induces learners to persist with a task, even if it is a difficult one; it focuses their attention on the task, and it produces positive affect regarding the task, and the result of this is learning.

10.3 Interest Development in Theoretical Perspectives

When looking at interest as a theoretical topic, it is common to see interest appearing in three distinct domains involving a subject, an object, and an action. These domains are (i) dispositions, (ii) interestingness, and (iii) the psychological state of interest (Krapp et al. 1992). In the following section we will outline these three domains. In doing this we are also referring to their relations to their affective variables. After this short outline of each of the domains, we will combine the three domains with *learning* in order to describe some of the interplay between cognition and affection in learning and interest development.

10.3.1 Interest as Disposition

We all have dispositions for getting involved in and with different kinds of objects and activities. When interest is understood to be a disposition, it is therefore a domain involving a subjective perception, recognition, and interpretation of an object combined with the will to act upon it. Whereas the first part involves cognition, the latter involves affection and thereby goes beyond cognition. The will to act is in other words a matter of affectional involvement with the perceived object. The affectional involvement is therefore motivated but not necessarily as an interest in the object. Involvement with the object could be either a unified or a divided activity (Dewey 1913). When the activity is divided, it is not the object in itself that is of interest but instead the fulfillment of others' goals, e.g., the teachers while a unified activity is an activity for the sake of the object. So even though both kinds of activities lead to the fulfillment of a goal, there is a great difference in the affective state involved in those different forms of activity. Although divided activity calls for willpower and maybe fear of the consequences for not fulfilling the goal, a unified activity involves joy and positive expectations towards goal fulfillment. It is in the unified activity that we find an interest in things, and it is therefore in dispositions that lead to recognition of possible unified activities that we find interest as a disposition. Such positive expectation of goal fulfillment and positive emotions of understanding can turn into pleasure and joy in working with the specific topic. Klaus Fiedler (2000) provides a model – the dual-force model – for combining learning and affective variables during learning. One of the key components in this model is that it builds on our internal long-term interpretations of emotional stages. The model refers to the constructivist thoughts of Jean Piaget in that it sees the involved affectional variables as negative when doing accommodation, while the affectional variables are positive while doing assimilation. Positive emotions and feelings entail that there is a general drive where the individual investigates and becomes more creative with the task while there is an aversion or frustration in the negative affective variables that forces the individual to focus more on getting the task right and finding structure. This model is useful in seeing the understanding of a topic as a trigger for positive emotions.

In the dispositions we again find a division into two subcategories, namely, general biological dispositions and individually experienced dispositions. It is outside the scope of this chapter to outline fully the differences, but in short the general biological are characteristics of an object that draw the attention of the individual toward the object. Such characteristics are referred to as interestingness (see next section), but they are characteristics that in general draw the attention of people, and therefore they are not tied up with the experiences and values of the individual (Wolf et al. 2014). In contrast, there are the individually experienced dispositions. Such dispositions are based on prior experience with an object similar to or related to the current object of interest. As previously stated, the difference between these dispositions lies in the personal value of engaging with the object. Dewey (1913) stated that the individual seeks growth, and if prior experience involves interacting with an object leading to growth, then there is a disposition for engaging with such an object when faced with it again. Such a growth of mind stimulates enjoyment, while experiences without growth on the other hand stimulate boredom and are not likely to create dispositions for further engagement. Such an understanding of individual dispositions towards interest are founded on individually valuable experiences involving both cognitive growth *and* positive affective experiences of value. So one claim made here is that it is the personal value of learning that constitutes the emotional input for the individual to experience positive feelings towards learning in science. This implies that that object of interest contains some kind of substance that is recognizable for cognitive growth in the individual. This is the basis of the interestingness of the object.

10.3.2 Interest as Interestingness

The specific characteristic of an object, that makes it interesting, is referred to as *interestingness*. When students are handling live earthworms in biology lessons, for example, some may find the activity interesting. Interestingness may here refer to the hands-on aspect of the activity or the "sliminess" of the worms. Students have dispositions for different perceptions and interpretations of some objects, while others are of a more general nature. The concept of interestingness was first examined by Roger Schank (1979) who found that an object can contain three different factors to draw the attention of a subject: firstly if one's expectations were not fulfilled, secondly if some information was missing, and thirdly if the object included distinct themes like death, danger, power, or sex. Suzanne Hidi and William Baird (1986) describe the first two factors as context dependent, while the last is described as cross contextual content.

Because the object is given in a context – and in school within the context of teaching – it is also given that the interestingness of the object is the only thing that an outsider can form and manipulate to have the subject to draw attention toward it. But because these categories are of general interestingness, they also draw upon the general biological disposition from above. This means that there are factors that can draw a person's attention toward an object, but this is not necessarily the object that is more interesting from a personal view. There is only limited personal growth involved in interacting with the object, and drawing on such general interestingness, in the terminology of Dewey (1913), is called *sugarcoating* or, in contemporary usage from game-based learning, chocolate-covered broccoli (Habgood and Ainsworth 2011). It is not the object in itself that has interest but instead a general interestingness outside the actual object, and by covering the object up in either sugar or chocolate, we distract students from seeing and valuing the object and instead digesting the coating, and by doing so we actually introduce the students to a divided activity. As Dewey (1913) states: "When things have to be *made* interesting, it is because interest itself is wanting" (p. 11). And interest in itself is not of value. The value connected to the individually experienced dispositions lies within the specific object and its possibilities to stimulate personal growth. It is therefore not enough to focus on the interestingness of an object to create fruitful affective stages in learning. In order to create a context containing interestingness, there has to be congruence between the object of intended interest and the object of attraction. If not, then the attraction is drawn in other directions, and we will be back at the divided activity described above.

In order to maintain interest, there has to be to growth in perception and valuing of the object. This means that not only does interest work as a facilitator for learning, it also operates in the other direction where learning is working as a facilitator of interest. Here, it is not enough to discuss if the object has some kind of interestingness. It is also necessary to look at the quality of such interestingness and if this interestingness is embedded in the object or introduced as a covering layer designed to coax students into swallowing it.

10.3.3 Interest as a Psychological State

Interest can also be a psychological state. Traditionally this state is divided in two parts, namely, (a) situational interest and (b) actualized individual interest (Krapp et al. 1992). Originally situational interest represented interest initiated from the interestingness of the object – including some of the general and cross contextual factors from above – while actualized individual interest was initiated from individual dispositions to recognize content from prior experiences in the current object. Even though there has been substantial research focus on situational interest, the concept of actualized, individual interest is almost neglected. This could be due to the fact that the differences between the two, when observed in research, are not clear. In fact they may be so similar that Paul Silvia (2006) proposed that the concept should be removed from research on interest because it was a pure theoretical construct that was not measurable in practice. Although research has been focusing on situational interest, it seems hard to find a coherent approach from where to understand and investigate interest as a psychological state (Renninger and Hidi 2011).

The distinction between situational interest and actualized individual interest is, in our context, seen as fruitful in that it reveals a distinction between the different qualities of interest and interestingness.

Interest as a psychological state implies that there has been a connection between the subject and the object. Within some research literature, this connection is often referred to as a *catch* (Mitchell 1993) or a *triggering* (Hidi and Baird 1986). The distinction between these two approaches lies in the direction of the connection where *catch* implies that the object has sufficient interestingness to catch the attention of the subject, whereas *triggering* implies that the subject through dispositions encounter the object due to recall of prior experiences. Within research in science education, there has during the last decades been a focus on triggering as the connection between the object and the subject (Renninger and Bachrach 2015). Again referring to the quality of the interest, a catch may be seen mostly referring to the general biological dispositions, while triggering is mostly an outcome of individual experienced dispositions. In order to constitute further the psychological state of interest, there must be a hold of the attention and possible engagement from the triggering. Here, Suzanne Hidi and Ann Renninger (2006) provide a model dividing this holding in two main categories, namely, (i) situational interest and (ii) individual interest. Situational interest is further divided into two subcategories of triggered situational interest and maintained situational interest. Both phases have in common that there is a need for outside support from, e.g., a teacher to keep focus on the object and finding characteristics worth working with. Individual interest is also divided into two subcategories which are emerging individual interest and well-developed individual interest. In these phases the drive for engaging with the object primarily comes from the individual and thereby from the individual experienced dispositions.

Even though Silvia (2006) makes a valid point about the practical investigations of interest, it seems that there is a need for further distinction between concepts within the domains of interest as a psychological state. In the following section, we will look at the interconnections between these three domains and link these connections toward emotions.

10.4 Connecting the Domains of Interest

There is a clear connection between going from the domain of dispositions and the domain of interestingness towards the domain of interest as a psychological state, which is also described in research literature (Krapp 2007). What is not that clear is the reverse connection from interest as a psychological state toward interest as dispositions or as interestingness. The triggering/catch and the hold of an actualized interest give a direction of a somewhat momentary state of interesting interaction between the subject and the object, but it does not give an explanation for the development of interest and the role of an actualized interest in this development. If the psychological state of interest is actualized and held, we claim there can be a combined effect of the four-phase model of interest development (Hidi and Renninger 2006) and the personal growth of mind (Dewey 1913) involving both cognitive development and affective well-being.

The affective dimension may be negative to begin with, such as frustration or avoidance of a difficult task. In order to contribute to the development of interest, this negative affect must be turned into a positive which is what happens when difficulties are overcome (Fiedler 2000). This can happen either as an "aha experience" with a sudden positive feeling (Bechara and Damasio 2005) or as a more slow understanding with a change in long-term emotions (Linnenbrink and Pintrich 2004). In fact it is in the overcoming of difficulties that the positive emotions can help to establish and develop an interest. As Dewey (1913) states:

> It is not too much to say that a normal person *demands* a certain amount of difficulty to surmount in order that he may have a full and vivid sense of what he is about, and hence have a lively interest in what he is doing. (Dewey 1913, p. 52)

This is yet another argument for reversing the connection between interest and affection. It is through the learning and the cognitive gain that the individual creates positive feelings toward an object. So our claim here is not only that the triggering/catch and the hold parts direct interest from the subject and the object toward a common domain of an active psychological state of interest. When this state is actualized, it also feeds back to the other two domains in that the object is opened up and deeper layers of understanding become accessible while at the same time the active interaction with the object and the new dimension opening up is valued by the subject thereby strengthening the individual's experienced dispositions for further engagement.

To sum up the theoretical approach, we do not accept the common approach that interest is a facilitator for learning (Krapp 2002). Learning is also a facilitator for interest development.

Patricia Alexander (2004) proposed the Model of Domain Learning (MDL), which describes interest development in parallel to a person's increasing academic competence. Alexander et al. (1994a) have shown that levels of individual interest and domain knowledge are highly correlated. Both situational and individual interests are included in discussions of the role of interest in the MDL. The MDL specifies stages of individual expertise development (and of concurrent interest development). Alexander suggested that the final stage of expertise is only reached after high school and that the stages of the MDL are sequential and irreversible. As a consequence, if a person is an expert, then he or she has an individual interest for the subject matter domain. This also means that individual interest is not present before reaching expertise. The model has met critique in that Hidi and Renninger (2006) claim that individual interest can be present in much earlier stages and in younger students than postgraduate high school students. We here agree with the critique but also keep in mind that the MDL does not have the intention to describe interest development but rather to present a clearer view of learning thereby having a focus on cognition and not on affective variables.

We found underpinning arguments for the view that learning may facilitate interest development through research on interest development in science education. In the following section, we will show how students experience the affectional side of leaning. This is done through two cases where the empirical work thereby will underpin our theoretical statements on the directions and interplay between cognition and affection.

10.5 Cases

To illustrate our theoretical statements, we here present two cases from upper secondary biology education. The combination of the two cases addresses the point taken in the earlier theoretical discussion that learning can be a facilitator for interest development.

10.5.1 *Learning to be Interested: The Case of Animal Physiology*

The first case concerns an undergraduate biology course in animal physiology (Dohn et al. 2009). Students' situational interest was investigated by observation, informal conversational interviews, and a questionnaire. Students described what had caught their interest in previous lessons (lectures, theoretical exercises, and laboratory exercises) and described why they found the situations interesting. The aim was to explore students' perceptions about sources of interest.

Knowledge was a very important source of interest in this study. Knowledge-based interest was identified at two levels: (1) *aha experiences* and (2) *background knowledge*.

When a student is "stuck" on a problem, he or she sometimes achieves a clear and sudden solution through insight – the so-called *aha experience*. Aha experience refers here to a knowledge-based interest that is triggered by a sudden and unexpected flash of insight. For example, a student stated that "it was really fascinating when I suddenly realized how muscles function."

The experience of being interested seems to be the consequence, rather than the cause, of the intellectual activity involved in resolving some issue. An explanation of why aha experiences can trigger interest must be sought in closely related variables like optimal challenge, novelty, and optimal discrepancy between input and cognitive structure (Deci 1992). The aha experience is situated in the context of problem-solving. As such, the first step toward promoting the aha phenomenon is to present students with interesting and challenging problems.

One of the most common interview responses was that interest emerged by acquiring knowledge in physiological processes and how these processes are expressed in different living animals in comparison with human beings. The responses refer to a knowledge-based interest which is generated due to relevant background knowledge. This category has much in common with aha experience but is much more persistent and of more individual character.

In a study by Patricia Alexander and Karen Murphy (1998), the undergraduates demonstrated significant growth in domain knowledge and were more personally interested in that domain, which suggest that knowledge and interest should be significantly and positively correlated.

According to Alexander et al. (1994b), there appears to be a reciprocal relationship between knowledge of a domain and interest in the domain. That is, we pursue learning about things we are interested in, and the more we know about something, the more we become interested in it. Previous research suggests that background knowledge is related to both individual and situational interest, even though knowledge appears to be related more strongly to individual interest (Bergin 1999). In a study by David Palmer (2009), learning (i.e., the acquisition of domain knowledge) was found to be the most important source of situational interest among K-9 science students.

From an educational point of view, the major challenge is how educators can help students in the acquisition of domain knowledge and thus interest. Unfortunately, background knowledge is a factor which is difficult to change, because it is predominantly an individual variable. As Dewey cautioned decades ago (1913), transient (i.e., situational) interest alone will not sustain learning, and such sustained learning is requisite for proficiency in any complex domain. Thus, abstract, demanding exposition will need to be carefully anchored to the goals and long-term interest of students (Alexander et al. 1994a). Findings from studies of interest suggest that educators can help students sustain their attention toward tasks even when these tasks are challenging. This could either mean providing support so that students may experience a triggering of situational interest or providing feedback which allows them to sustain their attention, generate their own questions, and select resources which promote problem-solving and strategy generation (Hidi and Harackiewicz 2000; Schraw and Lehman 2001).

10.5.2 Learning to be Interested: The Case of Simulating Natural Selection

As shown in the above case on animal physiology, it was found that knowledge and new learning could be what the students found interesting. Another study on this topic was approached more directly (Petersen 2012). In Morten Petersen's (2012) study, students at upper secondary level were doing a simulation of natural selection. All students had read and had lessons within the area of natural selection and evolution beforehand, and the simulation was therefore not introducing new knowledge but instead consolidating prior knowledge.

The approach in the study was to test the development of conceptual understanding of core concepts of natural selection and evolution addressed through the simulation. Students who showed development in the direction of a more scientific understanding of either one of these concepts were then interviewed afterward specifically addressing their experiences of learning and understanding these concepts and their emotions in doing so. The target group of students informing the project was therefore limited to students who showed signs of having cognitive expansion of the prior understanding.

In this exploration of interest development through learning, it became clear that every student interviewed in the beginning found it hard to cope with the simulation of natural selection with Lego® bricks which was the material for modeling (Christensen-Dalsgaard and Kanneworff 2009). When the students started working with the simulation, they saw the bricks for what they were – bricks. But after working with the simulation, almost every student changed their perception from seeing a pile of bricks to seeing small animals. This change of perception is what we argue for as an essential part of interest development in the theoretical approach to interestingness and the interplay between interest as a psychological state and interest as interestingness.

In the follow-up interviews, students connected this change of perception to the understanding of not only the exercise but also the natural selection in general. Even though the students had had lessons in natural selection before forehand, almost all of the students experienced a deeper understanding of the concepts involved in natural selection and especially the interplay between these concepts. It is through this change of perception that we find not only a shift in emotional state from slight frustration and lack of understanding toward understanding and enjoyment but also an unfolding of the object – in this case the bricks – helping the students to become aware of their personal growth through this deeper understanding.

When discussing the research setup with colleagues, a common critique was that because almost every child has had good experiences with Lego® bricks, it would be the bricks that caught the students' interest and not the biological simulation of natural selection. In other words the bricks would have a functioned as sugarcoating or chocolate covering and thereby lowered the quality of the perceived interest for students. When confronting students with this critique in the interviews, it turned out that many actually disassociated from the bricks as they felt that they were too old to use toys in upper secondary school. This disassociation thereby confirms to us that when interest is present, it is also present in a quality that could be seen as an actualized individual interest. This means that when we find interest, it is due to that biological domain knowledge in the simulation and not due to the recall of childhood play with bricks.

As seen in the former case, students experience new knowledge in different ways as either aha experiences or as background knowledge. In the case of natural selection, this is also the case, but some of the students report that they actually had an aha experience but that this experience did not happen during the exercise but when they were working with their written protocol of the exercise that they had to hand in. This indicates that even though the students were capable of showing progression in their conceptual understanding from the pretest to the posttest, this new knowledge did not reveal itself to the students as explicit knowledge until they again had to work with the concepts and knowledge of the simulation.

Such a delay of understanding and revealing of knowledge is also found by Kevin Pugh (2011) in the investigation of transformative experiences, and actually the simulation of natural selection led many students toward such transformative experiences. The understanding of the concepts involved in natural selection and their interplay did not only open up as deeper knowledge. It was also valued knowledge for the students making some of them see their everyday life from a new perspective. So the findings of the study indicate that from the actualized interest there is a feedback to both the object as deeper understanding and to the subject and thereby the dispositions in valuing this new deeper understanding.

We thereby find empirical evidence that the process of learning and interest development in science education is not a unidirectional process but can be viewed upon as an interdependent process similar to Alexander's (2004) MDL, but also taking the critique from Hidi and Renninger (2006) into account in that we find an interest well founded in individual dispositions even though students may not be experts in the domain of biology and evolution.

10.6 Perspectives of Focusing on Interest Development

In this chapter we argued that the connection between affective variables in interest development and cognitive processes in learning is not a unidirectional connection but rather a connection that can be both *interest as a facilitator for learning* and *learning as a facilitator for interest* development. Through our theoretical outline we have shown that even though the common approach to the connection between interest and learning is that interest is a facilitator for learning, we also find arguments in the research literature for the opposite approach seeing learning as a facilitator for interest development.

Throughout our argument we have also discussed the quality of the interest in the sense that the object of interest has a need for an inbound possibility of unfolding itself to the individual through learning and understanding instead of the object being covered in stimuli for drawing attention toward the object. In order for an object to entail true interestingness, the object has to offer possibilities for personal growth and valuing of this object. If there is no learning and personal growth during an activity, the individual may hold the activity for a while, but the activity will not contribute to strengthening a future engagement with similar content.

Through our empirical work we found that not only do university students in biology find it interesting to learn through either aha experiences or through background knowledge in the case of animal physiology. Students in the case of natural selection actually find that the deeper understanding of the interrelation between previously learned concepts is the core in awakening their interest. The empirical work thereby goes hand in hand with our theoretical outline in questioning the unidirectional connection between interest and learning.

Through our theoretical and empirical argumentation, we highlighted learning as having the qualities of understanding, cognitive growth, value, and transformative experiences. So in order to go beyond cognition in learning, this learning has to be a deep learning going from learning *about* biology toward a learning *of* biology (Scardamalia and Bereiter 2006). Such deeper learning is more time demanding than an introduction to topics without this deeper understanding and thereby leaving teachers with a challenge toward fulfilling a standard curriculum. From our previous arguments with reference to the dual-force model (Fiedler 2000), we can state that such a lack of time for getting to understand the topic may in the worst case introduce students to a long-term emotional state of negative emotions without having the time at hand for students to follow the emotional progression in understanding toward a more positive emotional state. Such an experience is not likely to enhance the development of more enduring interest. In fact an experience of that kind would be capable of drawing students' attention in other directions other than science in that they experience the lack of personal growth.

A critique of such arguments can be that we thereby are postulating that a person cannot be interested in something unless this involves deep learning and understanding. We do not mean to exclude short-term positive engagement with objects as not being interesting to the person but in order to contribute to the personal growth and the valuing of the content the interaction needs to be over a period long enough for the object to open up toward seeing new dimensions of it. Here, the short-term engagement can be placed as situational interest that can be held through general biological dispositions, and covering the objects in such interestingness is a distraction from an object not unfolding itself.

In this chapter, we have concerned ourselves with biology students in upper secondary school and at university. It should be kept in mind that, to the extent that the results are statistically generalizable, they might well hold only for similar populations. It is likely that different emotional and motivational dimensions would emerge with different populations and in different subjects. Replicating this study with different populations, not least ones with younger students, would be an important next step. Despite these limitations we would like to pinpoint our main claim in this paper once again. We have argued that not only can interest be a facilitator for learning. It also operates in the opposite direction in that learning can be a facilitator for interest development. For a practical implementation of our claim, a teacher is therefore not in need of focusing on interest in itself for getting students motivated and interested. A focus on learning and understanding of content knowledge can also be a way for students to get a positive affective experience with learning and thereby go beyond cognition.

References

Ainley, M., Hidi, S., & Berndorff, D. (2002). Interest, learning, and the psychological processes that mediate their relationship. *Journal of Educational Psychology, 94*(3), 545.

Alexander, P. A. (2004). A model of domain learning: Reinterpreting expertise as a multidimensional, multistage process. In D. Y. Dai & R. J. Sternberg (Eds.), *Motivation, emotion, and cognition: Integrative perspectives on intellectual functioning and development* (pp. 273–298). Mahwah: Lawrence Erlbaum Associates, Inc.

Alexander, P. A., & Murphy, P. K. (1998). Profiling the differences in students' knowledge, interest, and strategic processing. *Journal of Educational Psychology, 90*(3), 435–447.

Alexander, P. A., Kulikowich, J. M., & Jetton, T. L. (1994a). The role of subject-matter knowledge and interest in the processing of linear and nonlinear texts. *Review of Educational Research, 64*(2), 201–252.

Alexander, P. A., Kulikowich, J. M., & Schulze, S. K. (1994b). How subject-matter knowledge affects recall and interest. *American Educational Research Journal, 31*(2), 313–337.

Bechara, A., & Damasio, A. R. (2005). The somatic marker hypothesis: A neural theory of economic decision. *Games and Economic Behavior, 52*(2), 336–372.
Bergin, D. A. (1999). Influences on classroom interest. *Educational Psychologist, 34*(2), 87–98.
Christensen-Dalsgaard, J., & Kanneworff, M. (2009). Evolution in Lego®: A physical simulation of adaptation by natural selection. *Evolution: Education and Outreach, 2*(3), 518–526.
Damasio, A. R. (1998). Emotion in the perspective of an integrated nervous system. *Brain Research Reviews, 26*(2), 83–86.
Deci, E. E. (1992). The relation of interest to the motivation of behavior: A self-determination theory perspective. In K. A. Renninger, S. Hidi, & A. Krapp (Eds.), *The role of interest in learning and development* (pp. 43–70). Hillsdale: Lawrence Erlbaum Associates, Inc.
Dewey, J. (1913). *Interest and effort in education*. Boston: Houghton, Mifflin and Company.
Dohn, N. B., Madsen, P., & Malte, H. (2009). The situational interest of undergraduate students in zoophysiology. *Advances in Physiology Education, 33*(3), 196–201.
Fiedler, K. (2000). Toward an integrative account of affect and cognition phenomena using the BIAS computer algorithm. In *Feeling and thinking: The role of affect in social cognition* (p. 223). New York: Cambridge University Press.
Fredrickson, B. L. (1998). What good are positive emotions? *Review of General Psychology, 2*(3), 300.
Habgood, M. J., & Ainsworth, S. E. (2011). Motivating children to learn effectively: Exploring the value of intrinsic integration in educational games. *The Journal of the Learning Sciences, 20*(2), 169–206.
Herbart, J. F. (1965). Pädagogisches Gutachten über Schulklassen und deren Umwandlung (1818). In W. Asmus (Ed.), *Johann Friedrich Herbart – Pädagogische Schriften* (Vol. 3, pp. 89–128). Düsseldorf: Küpper.
Hidi, S. (2006). Interest: A unique motivational variable. *Educational Research Review, 1*(2), 69–82.
Hidi, S., & Baird, W. (1986). Interestingness—A neglected variable in discourse processing. *Cognitive Science, 10*(2), 179–194.
Hidi, S., & Harackiewicz, J. M. (2000). Motivating the academically unmotivated: A critical issue for the 21st century. *Review of Educational Research, 70*(2), 151–179.
Hidi, S., & Renninger, K. A. (2006). The four-phase model of interest development. *Educational Psychologist, 41*(2), 111–127.
Hoffmann, L. (2002). Promoting girls' interest and achievement in physics classes for beginners. *Learning and Instruction, 12*(4), 447–465.
Holstermann, N., Ainley, M., Grube, D., Roick, T., & Bögeholz, S. (2012). The specific relationship between disgust and interest: Relevance during biology class dissections and gender differences. *Learning and Instruction, 22*(3), 185–192. doi:10.1016/j.learninstruc.2011.10.005.
Izard, C. E. (1977). *Human emotions*. New York: Plenum Press.
Krapp, A. (2002). Structural and dynamic aspects of interest development: Theoretical considerations from an ontogenetic perspective. *Learning and Instruction, 12*(4), 383–409.
Krapp, A. (2007). An educational–Psychological conceptualisation of interest. *International Journal for Educational and Vocational Guidance, 7*(1), 5–21.
Krapp, A., Hidi, S., & Renninger, K. A. (1992). *Interest, learning, and development*. Hilldale: Lawrece Erlbaum.
Linnenbrink, E. A., & Pintrich, P. R. (2004). Role of affect in cognitive processing in academic contexts. In *Motivation, emotion, and cognition: Integrative perspectives on intellectual functioning and development* (pp. 57–87). Mahwah: Lawrence Erlbaum Associates.
Mitchell, M. (1993). Situational interest: Its multifaceted structure in the secondary school mathematics classroom. *Journal of Educational Psychology, 85*(3), 424.
Oatley, K. (2001). Emotion in cognition. In N. J. Smelser & P. Baltes (Eds.), *International encyclopaedia of the social and behavioral sciences* (pp. 4440–4444). Oxford: Pergamon.
Öhman, A., Flykt, A., & Esteves, F. (2001). Emotion drives attention: Detecting the snake in the grass. *Journal of Experimental Psychology: General, 130*(3), 466.

Palmer, D. H. (2009). Student interest generated during an inquiry skills lesson. *Journal of Research in Science Teaching, 46*(2), 147–165.
Pekrun, R. (2006). The control-value theory of achievement emotions: Assumptions, corollaries, and implications for educational research and practice. *Educational Psychology Review, 18*(4), 315–341.
Pekrun, R., Goetz, T., Titz, W., & Perry, R. P. (2002). Academic emotions in students' self-regulated learning and achievement: A program of qualitative and quantitative research. *Educational Psychologist, 37*(2), 91–105.
Petersen, M. R. (2012). *Interest development in science through learning progression – An investigation of the interplay between change of concepts and interest development in high school biology classes (in Danish).* Ph.D. thesis, University of Southern Denmark, Centre for Science and Mathematics Education.
Pugh, K. J. (2011). Transformative experience: An integrative construct in the spirit of Deweyan pragmatism. *Educational Psychologist, 46*(2), 107–121.
Rathunde, K., & Csikszentmihalyi, M. (1993). Undivided interest and the growth of talent: A longitudinal study of adolescents. *Journal of Youth and Adolescence, 22*(4), 385–405.
Renninger, K. A., & Bachrach, J. E. (2015). Studying triggers for interest and engagement using observational methods. *Educational Psychologist, 50*(1), 58–69.
Renninger, K. A., & Hidi, S. (2011). Revisiting the conceptualization, measurement, and generation of interest. *Educational Psychologist, 46*(3), 168–184.
Rheinberg, F. (2008). Intrinsic motivation and flow. In J. Heckhausen & H. Heckhausen (Eds.), *Motivation as action* (pp. 323–348). New York: Cambridge University Press.
Scardamalia, M., & Bereiter, C. (2006). Knowledge building: Theory, pedagogy, and technology. In K. Sawyer (Ed.), *Cambridge handbook of the learning sciences* (pp. 97–118). New York: Cambridge University Press.
Schank, R. C. (1979). Interestingness: Controlling inferences. *Artificial Intelligence, 12*(3), 273–297.
Schiefele, U. (1991). Interest, learning, and motivation. *Educational Psychologist, 26*(3), 299–323.
Schraw, G., & Lehman, S. (2001). Situational interest: A review of the literature and directions for future research. *Educational Psychology Review, 13*, 23–52.
Silvia, P. J. (2006). *Exploring the psychology of interest.* Cary: Oxford University Press.
Wolf, K., Bach, A.-M., & Waitz, T. (2014). Media–chemistry–interest? Identifying the types of students chemistry-related media eeception. *Eurasian Journal of Physics and Chemistry Education, 6*(1), 76–87.

Morten Rask Petersen is an Assistant Professor in Science Education. His special field of interest is the theoretical foundation of interest development in science education.

Niels Bonderup Dohn is an Associate Professor in Science Education. His special field of interest is methodological approaches towards measuring interest development in science education.

Chapter 11
Science and the Arts: Curriculum Integration, Learning, and Emotions in Schools

Tatiana Chemi

11.1 Introduction

This chapter presents several perspectives on the meeting of the arts with science. Partly referencing my previous work on arts integration (Chemi 2014) and artistic creativity (Chemi et al. 2015) and partly using an original empirical case of "tinkering pedagogy," I intend to unfold the relevance of and complexity behind the meeting between art and science. The purpose is to uncover the complexity of arts integration practices and to offer a nuanced view of the relevance of the arts for learning and teaching science, considering affects as the foundation of learning processes. I will discuss conceptual differences and similarities in artistic and scientific approaches, using the concrete example of Danish schools that are developing a cross-disciplinary approach to the arts and sciences. The case reported here is described against the background of qualitative research completed in 2014. Data sources included semi-structured interviews with students and teachers; informal conversations with teachers, artists, and school leaders; observations of learning expeditions; participation in facilitated school meetings (made visible by teachers' qualitative assessment); and document analysis (teaching plans, educational material, students' self-reports). As a supplement to the qualitative data, I also used a short survey, which included quantitative items and was developed as an internal evaluation tool (Fredens and Fredens 2014).

T. Chemi (✉)
Department of Learning and Philosophy, Aalborg University,
Kroghstræde 3, Aalborg, Denmark
e-mail: tc@learning.aau.dk

A. Bellocchi et al. (eds.), *Exploring Emotions, Aesthetics and Wellbeing in Science Education Research*, Cultural Studies of Science Education 13,
DOI 10.1007/978-3-319-43353-0_11

11.2 Opposites?

Art and science have often been opposed to each other throughout the history of Western cultures (Wilson 2010). During the early European Enlightenment, the artist's creativity was supposed to adjust to domain-agreed rules in order to be accepted. On one hand, the academic discourse was able to systematize creative works of science and arts, seen as "twin aspects of the same 'scientific' interest" (Goldwater and Treves 1976, p. 13) and give them a "professional" or specialized audience. On the other hand, artists and creators in general were presented with new beliefs that dispensed with classical and medieval philosophies and valued instead scientific truth and reasoning.

Periods of holistic views of human knowledge and expression, like the Renaissance, which cultivated the balanced unity of science and arts, were overruled during the Romantic era. In the Romantic schism between science and art, scientists were seen to be wise representatives of rationality, logic, and practical scientific thought (Runco and Albert 2010). Artists were spontaneous, mad people, prey to their feelings, and characterized by genius and extraordinary talent. The Romantic stereotype prevented the following centuries from looking at artists as navigating through *both* feelings *and* rational thinking, both openness and rules, and both individual and dialogical processes, as studies focused on artistic creativity have shown (Chemi et al. 2015). Systemic (Csikszentmihalyi 1999) and sociocultural perspectives (Connery et al. 2010) on artistic creativity offer an alternative to the dualistic separation of emotions and cognition proposing instead a holistic view that combines emotions and cognition and individual and society. The heavy heritage of Romanticism has influenced to the same extent both scientific and artistic practices. According to Damasio (1999), this has, for instance, seriously restrained the sciences from approaching areas such as emotions and feelings, thought to be irrational, and exclusive domains of the arts. It has often kept the arts, on their part, from taking an interest in scientific inquiry or applications.

This discourse and attitude radically change with the cultural avant-gardes throughout the whole nineteenth century, when artists played with theoretical concepts, to the point of founding the movement of conceptual art and stimulating a new reunion of the arts with science (Wilson 2010). Over the last 20 years, increasing collaborations between scientists and artists have changed the dualistic paradigms, offering a different ontology for human expression and creativity. After all, both scientists and artists share a commitment to creative processes.

Dichotomist conceptualizations miss the fact that both science and art focus on phenomena and make use of experimental approaches. Yet they differ from each other, because science seeks to test hypotheses and scientific analysis is targeted at disseminating conclusions, whereas art prefers to create and shape, without necessarily achieving and showing conclusions. Artworks might be the product that conclude a creative process, but they tend to be open ended and leading to novel questions.

However similar and jointly contributing to a human gestalt, the arts and sciences can still unfold their episteme differently. What is specific to artistic practices is already implicit in the very word "art." It is therefore essential to address the implications behind this word. The modern use of the word "art" is located in the Latin *ars*, meaning art, skill, or craft (Morwood 2005). Most European languages borrow the Latin lexeme: English and French *art* and Italian, Spanish, and Portuguese *arte*. Others adopt a semantically similar root, originally coming from the Proto-Germanic *kunnana*, "to know how to," and German, Danish, Dutch, and Norwegian translate it as *kunst* and Swedish as *konst*. A peculiar difference in origin but not in meaning is represented by the Greek *τεχνη* ("tekne"), meaning, once again, skill or craft. The latter is most significantly the origin of the modern word *technique* and its derivatives, *technical* and *technology*. Both in Germanic and Romance languages, art is related to the semantic field of practical skills.

In Western cultures, art refers to what is made by human beings by applying skills, crafts, and capabilities, and which is at the same time meaningful, meaning generating, and to be shared with others. Still today, in spite of the many challenges of avant-garde experiments of cross contamination, such as the readymade (the use of objects and artifacts) or happenings, art is in our linguistic perception something linked to the tradition of "handicraft," something made with the hands or whole body, which at the same time is meaningful and is to be shared in and with a given community. The artistic process is accomplished when a maker produces "something" (a work of art, a performance, an event, a display), which is purposely made for a receiver, to be perceived, understood, and decoded. The natural sciences, on the other hand, tend to be defined as the study and observation of natural phenomena, as opposed to culture (i.e., made by humans).

Eisner (1981) describes the differences between art and science in terms of fundamental characteristics and not as oppositions. He makes the distinction between old views of scientific and artistic manifestations into a systematic taxonomy with the following elements:

1. Forms of representation: science is based on codification and literal meanings; the arts are based on evocation and metaphor.
2. Criteria for appraisal: science's main method of appraisal is validity and lack of bias; in the arts, credibility is extended also to the verisimilar, and bias can be positive.
3. Points of focus: in science, what is manifest and the observable are objects of analysis; in the arts, the acts of empathy and imagination are the focus.
4. The nature of generalization: statistical or probabilistic generalization in science emerges out of randomness; in the arts the belief is that a case can apply to other cases.
5. The role of form: in the arts form is content.
6. Degree of license allowed: in science no license is allowed; the arts expect artistic license.
7. Interest in prediction and control: classical and positivist science is based on naturalistic-like generalization; the arts tend to qualitative explication.

8. The source of data: standards are the source of scientific data, with the risk that the hard to see is overlooked; in the arts, the main source of data is the investigator.
9. The basis of knowing: classical science is basically positivistic; the arts allow and expect methodological pluralism and the inclusion of emotions in the epistemological process.
10. Ultimate aims: in science, it is the discovery of truth; in the arts, it is the creation of meaning.

Eisner's lifelong work focused on recognizing the differences between the arts and science and on building paradigms of integration that could bring the two together. To be fair to different tendencies in the philosophy of sciences (Kukla 2000), I must mention that constructivist perspectives are being considered as well in natural sciences as in humanities. One of the consequences of the constructivist turn is the possibility of moving beyond positivism in all kind of sciences. But what kind of paradigm should be used when the arts and science actively collaborate with each other? What happens in paradigms that unite the arts and sciences? And more importantly in the present chapter, what are the implications of Eisner's ideas in pedagogical experiments that attempt to integrate the arts and sciences?

11.3 The Arts and Sciences in Learning

Many questions arise when arts and science meet and when art meets pedagogy. How can art contribute to the individual's optimal learning and self-development? What is the role of artists when they collaborate with schools? To what extent can the meeting with schools enrich artistic exploration? Further questions emerge when the arts meet school subjects such as the sciences that are, only apparently, very distant from aesthetic approaches. This meeting requires closer consideration, together with focused attention on its emotional and motivational components.

A number of research findings have shown that aesthetic, cultural, and artistic experiences can help in creating and maintaining meaningful engagement with learning (Chemi 2014), whether this takes place within formal or informal learning environments (Chemi and Kastberg 2015). These studies show that artistic experiences can greatly facilitate dedication, self-development, and learning and that what has a major effect on learning and creative expression is when children and young people experience great commitment and passion for what they do (McLellan et al. 2012). Through the arts, individuals can express emotional experiences that are strongly linked to learning and development. According to neuroscience (Immordino-Yang and Damasio 2007) and embodied cognition (Johnson 2007), individuals learn against a background of affective processes. In this chapter, emotions are seen, as in Damasio (1999), in an embodied perspective, and aesthetics is here approached as the engagement of the senses and appreciation of beauty. In the last 20 years in the pedagogical field, there has been increasing interest in the learning potential implicit in bridging the arts and sciences and toward the role of emotions in learning processes (Lund and Chemi 2015).

In order to operationalize the unity of cognition, body, and emotions in learning processes, several innovative educational designs are based on the integration of the arts and sciences (Marshall and Donahue 2014). These educational designs, which aim at and often succeed in achieving better learning, social skills, and well-being, are victims of several dilemmas. On the one hand, the arts are viewed with instrumental interest, with the focused purpose of exacting as much numeracy and literacy out of individuals and groups as possible. On the other hand, in spite of the general notion that the arts foster unique learning potential, funding for arts projects and teaching hours in the arts are reduced at all levels of education.

However, new trends in education are emerging, such as "the integration of arts education in the promotion of science, technology, engineering and mathematics (STEM) education" (Winner et al. 2013, p. 27). Korea and the USA, for example, are adding the arts (or arts and design) to the STEM group of disciplines, creating the new acronym "STEAM." Arguments for this integration are the expectations of specific outputs, such as students' self-confidence, creativity, and innovation. At national level, these results are directly linked to a political discourse that encourages all activities that might bring economic growth to the nation. The truth of these claims, already challenged in the past (Winner and Cooper 2000), is problematic and poses the question of whether learning in the arts transfers to non-arts domains. There is no consensus among scholars on the answer to this question. Evidence has not yet demonstrated causality but only assessed correlation. In other words, participation in the arts correlate with high or improved academic performance, but there is no clear evidence that improvement in school is due to art experiences. Learning in the arts seems to be correlated to learning in non-arts subjects, but the reason and the cause of this are still uncertain. Moreover, it might be unproductive for the arts to relegate them to an ancillary position in the educational hierarchy. Are students expected to engage in the arts only because they will learn STEM subjects? Or rather because the learning implied in artistic experiences is unique and offers distinctive developmental opportunities?

Arts-integrated projects are based on the arts but not arts enriched. In other words, art is both cultivated as a subject in itself, and it is integrated with other school subjects with the purpose of facilitating learning across other disciplines (Koppman 2010). Authentically integrated school projects or learning environments are built upon an optimal balance among subjects, insofar that all the subjects are present and effective to the learning process. Integration in the arts needs to be meaningful and not just being entertainment or serving an ancillary role to other academic works (e.g., singing the alphabet song). Students in arts integrated programs are stimulated in their thinking and questioning explicitly and purposely (Willner 2010). As Willner (2010) highlights, it might not be enough to establish cross-disciplinary projects, but arts-integrated activities raise the need for thinking about education differently, in terms of critical thinking, and scientific questioning. As other studies also emphasize (Deasy 2002), the matter of learning outputs in arts-integrated projects or environments is closely related to the educational approach to these activities. In other words, better learning or learning in itself is not an immediate consequence of integrating the arts with non-arts subjects (Fiske 1999). Rather arts-integrated environments can offer the preconditions for deep learning to occur. This must be planned, nurtured, and developed in educational cultures that value the holistic development of individuals and make it possible.

The arts/science partnership is, in the context of arts integration, a case of exceptional interest. Given the domains of art and science, projects that integrate the two subjects hold learning potential at several levels: learning outputs might gain from engagement and pleasure, schools might become the change they envision by practicing a holistic paradigm of education, and societies might gain not only economically but also socially and culturally by engaging with well-balanced and developed citizens. These ideal outcomes are based on the belief that art and science share common values and practices. Horvath and Marshall (2014) define the learning achieved in the natural sciences as similar to the sense of wonder to be found in the arts. They maintain that the curiosity and wonder that initiate participation in both art and science begin with experiences. Similarly to the arts, science in school meets learners with questioning, interrogating, need for understanding, and with an open attitude toward healthy skepticism, scrutiny, and critical thinking (Horvath and Marshall 2014). When integrating art and science, educators should keep in mind similarities and divergences, but mostly they should look at art in a novel way, as discovery and research.

11.4 Art Making Is Discovery and Research

One of the main findings in Chemi et al. (2015) is that the interviewed professional artists seemed to perceive artistic creation *as research* and not just as a formal exercise. Discovery as the process, exploration, research, discovery as pleasure and artistic research were recurrently mentioned in the interviews. The artist's compositional method is based on the conceiving of art making as a form of research, a learning project that starts by asking: "what do I know?" When artists contribute with their way of thinking and behaving to school projects, they also bring about a methodology that is defined by curiosity, by experimentation and inquiry, by problem solving, by pragmatism, and by sensitivity to changes.

Asking and formulating questions are fundamental steps of any research process (Denzin and Lincoln 2005). The intersection between science and art is characterized by the willingness of engaging in a learning journey and the commitment to experiences that can change the way we think by means of deep observations in the world, systematic connections, interpretations, and thinking (Horvath and Marshall 2014). Artistic research is characterized by the central role of heuristics and by a multitude of investigative directions (Vickery and King 2012), which can be also one of the methods used in the sciences for solving problems. Artists, like scientists, approach the task at hand by considering its context and responding to it, but they also address more philosophical questions about their own need for the task and methodological questions. However, artistic questioning is often about finding new questions to ask. Artistic research adjusts working methods to the given task as it goes. It depends on what one is doing or is expected to do, and it varies depending on the project at hand. Scientific and artistic methodologies have been successfully compared before (Weisberg 1993), showing structural resemblances between their procedures. Both domains are characterized by creative drives and a variety of searching directions or trajectories (Sternberg 2003, p. 100).

Artistic cognitive processes are described in a variety of ways (Chemi et al. 2015) and differently compared to scientific creativity. Creative processes tend to be relational and collective in both domains. Creative people in both fields seem to share processes that unfold not as successive stages, clearly and orderly occurring through time, but as overlapping and interconnected waves flowing in complex, serendipitous patterns (Sawyer et al. 2003). Like artistic methods scientific methods can be of very diverse nature: experimentation, observation, hypothesis testing, trial and error, or intuition (Horvath and Marshall 2014). Both creative practices consist in systematic experimentation and in finding new problems and/or solutions that are meaningful to others. Any artistic process or product or performance is an original interpretation of some sort, which is meaningful to someone or appropriate to a given situation. Modern science tends to adopt similar strategies, but with some fundamental differences that characterize and distinguish the scientific domain from the artistic one. The functions of the arts are more complex and diverse than the mere creation of something new with value – as creativity is consensually defined (Feist 2010). In other words, the arts often serve simultaneously different functions or combinations of functions even beyond creativity: creation of beauty, feeling of sociality, well-being, cognitive challenges, problem solving, provocation and rebellion, and aesthetic pleasure. Systematic, professional artistic practice not only includes creative processes but also accepts and expects acts of creativity, acts of novelty and appropriateness, acts of recreation and renewal (Dewey 2005, p. 113), or acts of redefinition of creativity. In the present chapter, I look at the arts as an epistemological or methodological approach: a way of inquiring, questioning, and making sense of the world. In this sense, the arts can generate knowledge and encourage learning with its tools, approaches, and mind-set. In order to look at empirical evidence of this, in the following section, I will address a Danish case that exemplifies the educational considerations and learning outputs of art/science integration.

11.5 Tinkering Pedagogy

Since 2011 in Jutland, the northern region of Denmark, two state schools have engaged in a pedagogical approach that integrates science and art in a common, experiential, real-life, cultural journey. The educational approach is defined by the schools as "Vandringspædagogik" (VP), which could be translated as wandering pedagogy, or better, tinkering pedagogy, because it is based on school trips to cultural and natural sites. VP is an approach that looks at learning as a dynamic enterprise that can be achieved between formal (i.e., school-based) and nonformal learning environments (i.e., the learning that occurs during leisure or spare time experiences). Learning happens and flourishes in environments that are complex, engaging, diverse, and experiential. It means that students are involved in expeditions that allow them to move their own learning outside of school. In these out-of-school settings, they meet experts in cultural topics or nature in arts or natural science museums, visit natural sites of special interest, and talk with artists. Learning sessions are partially designed by the team of teachers beforehand, but not entirely. This allows, according to the teachers interviewed, for experimentation, improvisation, and flexibility. Students pack and carry in their rucksack their own supplies for the journey but also, metaphorically, they carry a rucksack of cognitive and emotional resources. Teachers lead them and are together with them on their journey. The teachers' team carefully plans the trip beforehand, taking decisions on which cultural activities to experience, in terms of accommodation (often publicly funded shelters), length of activities, and transportation (often by bike). In the cultural meetings, the journey often takes place from the school to the nonformal learning environment (nature, museum, town center), but other journeys are of the opposite character, the nonformal learning environment comes to the students in school, for instance, in the form of an expert, cultural entrepreneur, or artist who visits the school and sets up a project.

The VP approach was previously evaluated in 2012 (BMMK 2013), but my analysis, presented in this chapter, supplements this earlier evaluation with new empirical data and the search for commonalities in the different forms of learning expeditions, as well as unusual characteristics of or responses to the learning experiences. Although I looked at a variety of learning expeditions, the examples I will present are mostly taken from the expeditions I personally observed. I focused on the elements of science/art integration and on the students'/teachers' emotional responses during the learning experiences. I hypothesized that learning in integrated learning environments stimulates emotions as part of learning processes. My hermeneutic perspective on emotions and well-being in educational contexts is social, psychological, and pragmatic. My analysis led to three central concepts of VP: (1) tinkering, (2) movement, and (3) learning.

11.5.1 Tinkering

During the learning expeditions, the students seem to be involved in serious and systematic tinkering. One of the projects observed aimed to engage a fifth grade class in learning about Moclay (a special kind of clay found in the area of Fur, Denmark) from scientific perspectives (guided visit to a natural science museum and Moclay digging site) and artistic perspectives (meeting with storyteller and visual artist who paints using Moclay). The expedition was prepared with thorough introductory work at school and was targeted at the production of a child-friendly tourist film about the area visited. Even though the trip was planned (by teachers), it left time and space for experimentation and tinkering, to the extent that learning could happen serendipitously. An example of this might be the students' amazement regarding the silent and black-and-white movies they were watching in preparation for the trip. The teachers had collected old documentaries on the Moclay excavations and trade in the area and, as a preparation for the outdoor activities, they showed to the students chosen clips of these old films. Attention to film media and its history was not in the program, but the students' amazement about the lack of sound or color directed the class toward this theme. Students were able to question, wonder about, and exchange their previous knowledge of these earlier forms of cinematic production.

Time and opportunity for tinkering are systematically included in the learning expeditions. This allows for surprises and focusing on emerging interests and is strictly connected to movement as pedagogical tool and perspective.

Where the arts contribute the most in this process is in their strategies of expression, which can extend to ontological and epistemological differences. The phenomenon of the arts has a tangible quality (it can be seen, heard, touched, smelled, and so on). But it also contains an intangible element (meanings, emotions, understanding, and the like) and an ineffable one, the latter being what is perceived and understood but struggles for expression through language. Symbolic and intended elements are fundamental in the arts, but not necessarily in science. The symbolic dimension adds a structured narrative to perception, allowing the extraordinary to emerge from the ordinary and the ordinary from the extraordinary. The intentionality of the symbolic or metaphoric expression is implicit in the arts, but not necessarily in the sciences, which tend to be direct, transparent, and "propositional." Propositional language, according to Susanne Langer (1953), is the verbal means of expression that makes use of everyday language, as distinguished from presentational language, the symbolic means of expression of the arts. The former might be insufficient in order to convey artistic experiences; for the language of art, presentational language stretches its epistemological reach by means of metaphor.

Art is meaningful in ways that written and spoken languages cannot be, because it adds to everyday communication a wider, more systematic, and purposeful practice of embodied expression of meanings and emotions, after all "if all meanings could be adequately expressed by words, the arts of painting and music would not exist" (Dewey 2005, p. 77).

Serendipitous experiences during the tinkering journeys are encompassed in both science and art. What the arts add on to the scientific approach is the possibility of expressing the experience by symbolic and narrative means.

11.5.2 Movement

Movement in VP unfolds physically in space and with the body, but also psycho-emotionally (the cognitive and emotional backpack), for example, students bike or walk through the chosen area and challenge their previous knowledge on specific topics. They are physically moving from school to the museums, natural, or cultural sites, but also they are moving toward (and are being moved by) the meeting with collaborating artists. These educational approaches differ from the school routine, to the extent that they enable students to "see the world as it is" (girl, fifth grade). Cognitively the students "move" into unexplored environments by means of questioning, exploration, wonder, and humor. The latter is able to establish relational and cognitive dynamics that are novel and surprising. Teachers, artists, or museum guides often introduce humor to the dissemination of knowledge, in order to make the classic lecture more engaging. Students seem to understand this element of VP as fundamental. They describe VP as something active, something one does in nature, that involves movement. Learning activities are something that happens in "fresh air" and involves actively making, seeing, and testing. They believe that this contributes to a better learning: "you understand it way better" (girl, fifth grade). They understand the expeditions as something other than school. They describe school as sitting inside and being told what to do, as less effective than expeditions that happen outside, with rather more active tasks, and happier teachers. Students report that the teachers' more relaxed and joyful approach is contagious to them. A fifth grader outlines the difference between school and expeditions in terms of learning quality: "it is as if one thinks more, you learn better when you come outside; it is not the same as just hearing about it." Learning seems to be more effective and creative ("there are more ideas") and also facilitates more peer-to-peer interactions ("classmates can explain things better than teachers sometimes," girl, fifth grade; "it is certainly more fun with classmates," boy, fifth grade). Benefits reported about this specific element of the learning expeditions might be related more to the fact that children engage in outdoor physical movement than to arts or sciences (Juelskjær et al. 2008).

However, by engaging in an outdoor expedition that integrates multiple dimensions (artistic expression, scientific method, outdoor experience, physical movement), the students often challenge their physical, cognitive, and emotional limits. Indeed they report that the physical part of these expeditions can be challenging, but they motivate each other through positive peer pressure synergy. When students read the program of the expeditions, they are not always excited about it. They report that they try to motivate themselves by formulating positive expectations ("when I'm out to experience it, it will be good," boy, fifth grade) and being open to surprises, once it improves. They admit, though, that it can be hard "if you are not interested in it" (boy, fifth grade). This poses a complex dilemma to teachers. Some children do not have access to cultural experiences at home and might be skeptical toward educational programs that imply direct and active engagement with the arts and outdoor nature experiences. Preparing the expeditions might require a fine-tuned introduction to these experiences and to the meeting of different cultures: the languages of art and nature, experts from the real world, learning in the making.

11.5.3 Learning Outputs

It is important to emphasize that the purpose of VP is learning at any stage of the process. VP sessions are designed to achieve academic or affective learning outcomes. Students are expected to increase their knowledge of school subjects (such as natural history, science, art) and to increase their numeracy, literacy, and digital skills (for instance, by means of a short film), but they are also expected to work in groups, making use of their collaborative and relational skills. However, learning is not only intended as accountable test results for a few school subjects. Expeditions are conceived as interest based, and the teachers present the activities to the students as stimulating "we have chosen the places that we believe can be of some interest to you" (female teacher). Tasks are defined by children's choices of what they perceive as more interesting. Interest is commonly mentioned in the interviews by students, who are extremely eager to share their impressions and well at ease with reflecting on their experience.

Participation is clearly observed in the involvement of students in the diverse activities, which offer a high degree of variation (lecture, hands-on, eyes-on, peer-learning, documentation by different media) and of engagement. Activities are experiential and expressive, as Dewey would define them (2005). The students in the Moclay project had to listen (lecture and storytelling), observe, report, interview, paint, dig fossils in clay, and make a film. In this way, students could find personal or alternative routes to their learning. As Gardner explains, learning a specific school subject might require a "*secondary route* to the solution to the problem, perhaps through the medium of an intelligence that is relatively strong for that individual" (1993, p. 33, italics in text). In other words, the presence of a variety of experiences, media, and materials allows for these alternative learning strategies, when the main route to learning is not accessible. As the Gardner quote above suggests, multiple approaches to learning allow for alternative learning strategies to emerge. The depth of students' awareness of these learning possibilities remains unclear. The children interviewed showed either a generic awareness of different learning approaches that allowed for several attitudes (there is something for those who like to sit in school and for those who want to be active and move around), or a well-described awareness of the expeditions' interdisciplinarity, where the whole curriculum is connected, history, Danish, nature, and technology. However, children's school evaluations do not show a deeply nuanced approach to the material studied, and some of their responses to the interviews sound like a reproduction of what they might hear from teachers.

11.6 The Role of Emotions When Arts and Sciences Merge

The relationship between art and emotions is often taken for granted but not fully addressed. The uncritical stereotype of the art/emotion connection has its background in the Romantic view of artists as driven by emotions as discussed at the beginning of this chapter: mad geniuses who are prey to irrational feelings and on the verge of mental illness (Chemi et al. 2015). However, more nuanced views are voiced in Dewey (2005) and Langer (1953), who define artistic processes as a symbolic act that mediates not only emotions, but also opinions, values, ideas, ideals, and archetypes. Emotions can be defined as the bodily response to events, sensations, or thoughts, with the role of focusing attention to a given situation or need for reaction, whereas feelings can be defined as the individual's awareness of these bodily responses (Damasio 1999). Emotions permeate the whole artistic experience, as compositional materials, as part of artistic creation – the making – or as appreciation of artworks. In different ways artists feel a variety of emotions while they struggle to shape their vision of a particular material, using a specific medium. Within the field of aesthetics, some scholars have variously conceptualized emotions and art or seen art as the education of emotion (Hjort and Laver 1997). The role of emotions in scientific thinking and experimentation or in learning science is still regarded as being secondary to the role of logic, rationality, and cognition. According to Glaveanu (2014), science is an interesting example of how individuality and eminence tend to obscure the fact that scientific creativity is actually deeply collaborative. Scientists, like artists, often think and create holistically; they make use of emotions, intuition, and heuristics (Kahneman 2011) as an integrated part of their thinking even when they are rooted in their domain.

In the VP expeditions, emotions are fully part of the learning process. Emotional experiences and learning are closely linked to each other (Järvelä 2011). As students report, expeditions stimulate feelings of coziness and ease ("we are a big family," boy, fifth grade) and engagement due to the many opportunities of active participation and collaboration. More importantly for learning outputs, expeditions allow for student ownership of the learning process. In one expedition observed, students had to write, set up, and perform an original performance inspired by the life and works of Danish astronomer Tycho Brahe (1546–1601). Students contributed with technical or dramaturgical ideas, solutions, and new perspectives on the subject matter, which are all taken seriously and discussed collectively. Students might have experienced the pleasure that individuals feel when they recognize perceptive inputs that they have been exposed to in the past. These occasions of cultural recognition seemed to strengthen a cultural identity and a feeling of belonging to the geographic area and its cultural heritage. The teachers interviewed mentioned a sort of cultural pride as being part of their intended outputs for the expeditions. Their explicit ambition was to provide their students with the ability to feel at ease in the territory, in order to be able in the future to engage themselves in private expeditions with their own families.

These children were given broad opportunities for stimulating a variety of approaches to thinking, tinkering, experiencing, and solving problems. Optimal learning happens in multidisciplinary environments, as both artistic and scientific collaborations prove (Glaveanu 2014, p. 37).

Experiences in natural sciences and arts imply aesthetic elements; both domains are based on sensory recognitions and (in some aesthetic theories) on the recognition of beauty. In both domains sensory perceptions and observations by means of senses are the basic steps of knowledge, judgment, and action. For the students of the VP expeditions, this means, concretely, that walking through nature, artistic creations, and cultural experiences implies a deep dive into a complex network of cognitive, volitional, and affective activities. Even though this is common to a number of human experiences, the VP approach stands out with its creative pedagogy – an approach that allows for experimentation and acceptance of complexity. Children meet complex learning environments and deal with them not by strategies of complexity reduction, but rather by navigating through complexity. They learn to understand connections (e.g., how Moclay can be used to find fossils or to paint); to collaborate with peers, experts, and teachers; to express learning in multiple ways (iPads, written texts, dramatizations, paintings, films); to deal with a number of different tasks (from trivial accommodation chores to more cognitively demanding school assignments); and to activate multiple learning strategies (sensory, bodily, cognitive, affective, relational). According to Prinz (2011), aesthetic aspects of experience urge individuals to engage in a complex evaluation and assessment of the object contemplated (artwork or natural site). Appreciation and its negative antonym, depreciation, seem to happen against a background of emotions, to such an extent that it is possible to claim, with Prinz, that "when we appreciate a work, the appreciation consists in an emotional response" (2011, p. 71). From this perspective, emotions are not only the response to art (Hjort and Laver 1997) or the content that art communicates and expresses but also the background for every aesthetic recognition, judgment, assessment, and appraisal. Evidence for this perspective can be collected from several scientific fields (Schellekens and Goldie 2011): philosophy, psychology, neuroscience, biology, and anthropology. Neuroscience, in particular, offers an interesting interpretation of how emotions and learning are intertwined in aesthetic experiences. Put briefly, MRI studies of the brain's activity when assessing works of art show activations in brain areas associated with emotions or with motor response. Motor response is the basic emotional drive toward action; through engagement, it pushes individuals to action. This brings emotions close to learning and development processes because emotions that arise during aesthetic appreciation influence aesthetic preference, judgment, and actions (Prinz 2011). Prinz (2011) even suggests that aesthetic appreciation might be identified with a "kind of wonder" (p. 83), which is the reaction to surprise, to the unknown, and to the unfamiliar that astonishes and amazes us. Several learning theories conceptualize learning itself as a drive toward knowledge and change or as a generic sense of wonder (Illeris 2009). I wish to launch the hypothesis that learning, emotions, and aesthetic experiences might be interconnected through trajectories of art appreciation. In other words, the source of the judgments that individuals (learn to) formulate about artworks and art experiences is the special interconnection between emotions, learning, and aesthetics.

Emotions are, from this perspective, the background for learning, insofar as they are essential to cognitive, neural, and physiological processes, to memory and judgment, to assessment and appraisal, and to action and expression (Prinz 2011). In other words, emotions are what trigger the brain into assessing and judging aesthetic experiences, but they are also the drive behind human agency (motivation to action). Thus, emotions activate complex cognitive processes that are the prerequisites for individuals to learn by means of remembering, finding and solving problems, decoding, constructing and reconstructing, expressing, and communicating. Aesthetic experiences require emotional processes to activate and facilitate the learning described above. Learning environments that are rich in a variety of aesthetic experiences, integrating the arts with sciences, like the VP expeditions, induce learners to deep, unconventional, and engaged learning. My conclusion is that these learning environments ought to be closely and systematically examined and thoroughly encouraged in school systems.

References

BMMK – Børns Møde Med Kunst. (2013). *Vandringer på Fortællingens Hovedvej i Sallingsundlandet: en Praksisfortælling vedr. Læring for Fremtiden*. Region Nordjylland.
Chemi, T. (2014). *The art of arts integration*. Aalborg: Aalborg University Press.
Chemi, T., & Kastberg, P. (2015). Education through theatre: Typologies of science theatre. *Applied Theatre Research, 3*(1), 53–65.
Chemi, T., Jensen, J. B., & Hersted, L. (2015). *Behind the scenes of artistic creativity: Processes of learning, creating and organising*. Frankfurt am Main: Peter Lang.
Connery, M. C., John-Steiner, V., & Marjanovic-Shane, A. (2010). *Vygotsky and creativity: A cultural-historical approach to play, meaning making and the art*. New York: Peter Lang.
Csikszentmihalyi, M. (1999). Implications of a systems perspective for the study of creativity. In R. J. Sternberg (Ed.), *Handbook of creativity* (pp. 313–338). New York: Cambridge University Press.
Damasio, A. (1999). *The feeling of what happens: Body and emotion in the making of consciousness*. Orlando: Harcourt.
Deasy, R. J. (Ed.). (2002). *Critical links: Learning in the arts and student academic and social development*. Washington, DC: Arts Education Partnership.
Denzin, N. K., & Lincoln, Y. S. (Eds.). (2005). *The SAGE handbook of qualitative research*. Thousand Oaks: Sage.
Dewey, J. (2005). *Art as experience*. New York: Berkeley Publishing Group.
Eisner, E. W. (1981). On the differences between scientific and artistic approaches to qualitative research. *Review of Research in Visual Arts Education, 7*(1), 1–8.
Feist, G. J. (2010). The function of personality in creativity: The nature of the creative personality. In J. C. Kaufman & R. J. Sternberg (Eds.), *The Cambridge handbook of creativity* (pp. 113–131). New York: Cambridge University Press.
Fiske, E. B. (Ed.). (1999). *Champions of change: The impact of the arts on learning*. Retrieved from http://artsedge.kennedy-center.org/champions/pdfs/ChampsReport.pdf.
Fredens, K., & Fredens, M. (2014). Vandringspædagogik frekvenstabeller (Working paper).
Gardner, H. (1993). *Multiple intelligences: The theory in practice*. New York: Basic Books.

Glaveanu, V. (2014). *Distributed creativity: Thinking outside the box of the creative individual*. Cham: Springer.

Goldwater, R., & Treves, M. (Eds.). (1976). *Artists on art: From the 14th to the 20th century*. London: John Murray.

Hjort, M., & Laver, S. (Eds.). (1997). *Emotions and the arts*. New York: Oxford University Press.

Horvath, L., & Marshall, J. (2014). The natural sciences: Understanding the natural world. In J. Marshall & D. M. Donahue (Eds.), *Art-centered learning across the curriculum: Integrating contemporary art in the secondary school classroom* (pp. 35–57). New York/London: Teachers College, Columbia University.

Illeris, K. (2009). *Contemporary theories of learning: Lerning theorists… in their own words*. London: Routledge.

Immordino-Yang, M. H., & Damasio, A. (2007). We feel, therefore we learn: The relevance of affective and social neuroscience to education. *Journal Compilation*. International Mind, Brain, and Education Society and Blackwell Publishing, *1*(1), 3–10.

Järvelä, S. (2011). *Social and emotional aspects of learning*. Oxford: Elsevier.

Johnson, M. (2007). *The meaning of the body: Aesthetics of human understanding*. Chicago: Chicago University Press.

Juelskjær, M., Moser, T., & Schilhab, T. (Eds.). (2008). *Learning bodies*. Århus: Aarhus University Press. Retrieved from http://www.ebrary.com

Kahneman, D. (2011). *Thinking, fast and slow*. London: Penguin.

Koppman, D. (2010). Arts integration. One school, one step at a time. In D. M. Donahue & J. Stuart (Eds.), *Artful teaching: Integrating the arts for understanding across the curriculum, K-8* (pp. 79–88). New York/London: Teachers College, Columbia University.

Kukla, A. (2000). *Social constructivism and the philosophy of science*. London/New York: Routledge.

Langer, S. K. (1953). *Feeling and form*. London: Routledge & Kegan.

Lund, B., & Chemi, T. (Eds.). (2015). *Dealing with emotions – A pedagogical challenge to innovative learning*. Rotterdam: Sense Publishers.

Marshall, J., & Donahue, D. M. (2014). *Arts-centered learning across the curriculum: Integrating contemporary art in the secondary school classroom*. New York/London: Teachers College Press.

McLellan, R., et al. (2012). *The impact of creative initiatives on wellbeing: A literature review*. Newcastle upon Tyne: Creativity, Culture and Education.

Morwood, J. (Ed.). (2005). *Pocket Oxford latin dictionary: Latin-english*. Oxford University Press. Retrieved 2 Jul. 2015, from www.oxfordreference.com/view/10.1093/acref/9780191739583.001.0001/acref-9780191739583.

Prinz, J. (2011). Emotion and aesthetic value. In E. Schellekens & P. Goldie (Eds.), *The aesthetic mind: Philosophy and psychology* (pp. 71–88). Oxford: Oxford University Press.

Runco, A., & Albert, R. S. (2010). Creativity research. A historical view. In J. C. Kaufman & R. J. Sternberg (Eds.), *The Cambridge handbook of creativity* (pp. 3–19). Cambridge/New York: Cambridge University Press.

Sawyer, R. K., John-Steiner, V., Moran, S., Sternberg, R. J., Feldman, D. H., Nakamura, J., & Csikszentmihalyi, M. (Eds.). (2003). *Creativity and development*. New York: Oxford University Press.

Schellekens, E., & Goldie, P. (Eds.). (2011). *The aesthetic mind: Philosophy and psychology*. Oxford: Oxford University Press.

Sternberg, R. J. (2003). *Wisdom, intelligence and creativity synthesized*. Cambridge, MA: Cambridge University Press.

Vickery, J., & King, I. (2012). *Experiencing organisations: Exploring new aesthetic perspectives*. Farington: Libri Publishing.

Weisberg, R. W. (1993). *Creativity: Beyond the myth of genius*. New York: Freeman.

Willner, S. (2010). Musical people, a musical school. In D. M. Donahue & J. Stuart (Eds.), *Artful teaching: Integrating the arts for understanding across the curriculum, K-8* (pp. 89–100). New York/London: Teachers College, Columbia University.
Wilson, S. (2010). *Art + science now*. London: Thames & Hudson Ltd.
Winner, E., & Cooper, M. (2000). Mute those claims: No evidence (Yet) for a causal link between arts study and academic achievement. *Journal of Aesthetic Education*. Special Issue: The Arts and Academic Achievement: What the Evidence Shows, *34*(3/4), 11–75.
Winner, E., Goldstein, T. R., & Vincent-Lancrin, S. (2013). *Art for art's sake? The impact of arts education*. Paris: OECD Publishing.

Tatiana Chemi is an associate professor at Aalborg University. She is the chair of Educational Innovation, where she works in the field of artistic learning and creative processes. She is currently involved in research projects examining artistic creativity, arts-integrated educational designs in schools and the role of emotions in learning. In 2013, Aalborg University Press named her Author of the Year.

Part III
Wellness and Wellbeing in Science Education

Chapter 12
Jin Shin Jyutsu and Ameliorating Emotion, Enhancing Mindfulness, and Sustaining Productive Learning Environments

Kenneth Tobin, Konstantinos Alexakos, Anna Malyukova, and Al-Karim H. Gangji

> Ken: "Oh my goodness, that was a nightmare! How could he stand to come to school and experience that, 5 days a week?" It was 1984 and I had just finished being a participant observer in a high school science class in suburban Perth, Western Australia. I was debriefing with a senior colleague from the United States following a class in which the students were extremely disruptive. The totally dysfunctional learning environment was a shock to both of us. Students were verbally disrespectful to the teacher and openly violated etiquette by moving around the classroom, often engaging in horseplay as they pushed one another, laughed boisterously, and loudly called out to peers. In fact, my colleague and I were both highly emotional as we fought back tears. The teacher was a personable, self-assured, well-qualified science teacher, who was also a prominent, highly accomplished martial artist (a status that did not deter students from being disrespectful and interacting with one another physically). The teacher never had a chance to be effective with his students. As science teacher educators, we were very sad about what we had experienced. First, we were sad for the teacher who never gave up despite repeated failures to teach effectively. How long would this teacher continue in such unrewarding conditions? Second, we were sad for the students who had no chance to succeed in such an unpleasant and unproductive learning environment.

(continued)

K. Tobin (✉) • K. Alexakos • A. Malyukova • A.-K.H. Gangji
Learning Sciences, Urban Education, The Graduate Center of CUNY, New York, NY, USA
e-mail: ktobin0@gmail.com

A. Bellocchi et al. (eds.), *Exploring Emotions, Aesthetics and Wellbeing in Science Education Research*, Cultural Studies of Science Education 13,
DOI 10.1007/978-3-319-43353-0_12

As we struggled to make sense, we called on our extensive experiences as science teachers, science teacher educators, and classroom researchers. We prided ourselves on being up-to-date with the latest theories and empirical research and had recently embraced interpretive research methods in an endeavor to address what we considered to be the big questions that faced educators – including macrostructures such as the assessment system and tracking that tended to structure schools throughout an entire state of Australia. As rich as our research proved to be, it was silent on the role of emotions and the inability of the teacher and students to ameliorate an excess of emotion when and as necessary (Tobin and Gallagher 1987). Our silence about the role of emotions and our failure to intervene in what happened in that study lead us to two important points – first, theoretical frameworks illuminate experiences in particular ways that simultaneously obscure other important patterns and contradictions that simply remain unnoticed. Probably it is a good idea for researchers to ask sincerely – what else is happening? What more is there? In asking such questions, it is essential to listen and learn from different voices that reflect different life experiences. Second, there is little doubt that our sorrow after the participant observation at the school was genuine and deep. We were shaken up. However, we also benefited a great deal from the research – in ways that did not extend to the teacher and his students. Sure, we could argue that generations of teachers and students would benefit from our hard work in producing good research and publishing what we learned in high impact journals (Gallagher and Tobin 1987). However, just as we now have a plethora of theories to illuminate classroom experiences, we also have new methodologies and ways of generalizing from research to ensure that participants benefit from their involvement in the research. Although it took two decades to happen, we now adopt methodologies that are designed to benefit all participants equitably (Tobin 2015a). In the next section of this chapter, we describe key features of authentic inquiry, a methodology that addresses equity of individuals and collectives.

12.1 Authentic Inquiry

12.1.1 Ontological Authenticity

Over the years, we have developed authentic inquiry from Ebon Guba and Yvonna Lincoln's (1989) fourth-generation evaluation (Alexakos 2015; Tobin 2015b). Basically we adopt four authenticity criteria (i.e., ontological, educative, catalytic, and tactical) identified by Guba and Lincoln, adapting them to fit in a multilogical approach to social inquiry (i.e., multiple methodologies that complement one another). Ontological authenticity addresses the perspectives different stakeholder groups have about what is happening and why it is happening. As a result of doing research, we seek to use procedures to afford all stakeholder groups changing their perspectives about questions such as the following: What is happening? Why is it happening? What is normal? What is preferred? What should happen next? Ontological authenticity recognizes that theory illuminates some things that happen and obscures others. Hence, by heightening awareness and understanding of new theories and their applicability to the classroom, all participants have opportunities to see differently, discern changes, and vary their expectations for what ought to happen.

12.1.2 Educative Authenticity

Educative authenticity provides opportunities for all participants to learn about others' perspectives – to understand them, see their value, and show respect for difference. As a result of being involved in research, all participants should be educated about others and expand their knowledge base to include different perspectives and the values they show for social differences/categories and knowledge systems that differ from what they consider to be true and of value.

12.1.3 Catalytic Authenticity

Catalytic authenticity addresses the importance of institutional changes occurring as a result of being involved in research. Collective improvements should occur during the conduct of research, not just higher attainments of individuals. In the case of the research reported in this chapter, catalytic authenticity would entail changes in the teaching and learning practices associated with a number of collectives involved in the research – including classes in which the four coauthors are currently involved as teachers and/or learners and the enactment of a monthly seminar program. What is learned from the conduct of the research reported here should lead to changes in conduct that goes beyond individuals changing what they do, believe, and value. Specifically, changes in ontology and being educated about others' perspectives will change the nature and quality of interactions – thereby changing the nature and quality of networks.

12.1.4 Tactical Authenticity

Tactical authenticity involves individuals benefiting equitably, in ways that are of value to them, as a result of their participation in a study. For example, researchers might benefit from publication of what they learned, and teachers might learn new ways to be effective and earn promotion to leadership positions. Some students might learn how to be more effective learners, and others might learn how to focus and sustain their participation and engagement. Care must be taken to help people who cannot readily help themselves succeed. That is, through the design and enactment of a study, all instances of inequity should be addressed proactively. The responsibility extends beyond becoming aware to include designing interventions and ensuring that positive changes occur.

In the following sections, we describe the central tenets of a methodology for the research described in this chapter, what we learned from the research and implications for professional changes and for literate citizenry. First, we lay out some central sociocultural perspectives.

12.2 Sociocultural Perspectives

William Sewell Jr.'s paper on agency and structure (Sewell 1992) provided us with a connection between agency, structure, and culture – all central tenets of our research in urban science classes, situated in a comprehensive high school in West Philadelphia. The paper served as a catalyst that led to our equating teaching as enactment and cultural production. Having located Sewell's paper, we then searched, found, and read other papers he had published as chapters in books in 1999 (Sewell 1999a, b). Taken as a set, the three papers authored by Sewell extended our sociocultural framework, which was grounded in Pierre Bourdieu's reflexive sociology (Bourdieu 1992). Key to a reflexive approach is that we endeavor to heighten participants' awareness about their own and others' practices (i.e., what is happening from different perspectives). Use of reflexivity is of crucial importance in our research and to all four of the authenticity criteria described in the previous section.

Culture is enacted (i.e., produced) by participants in social fields, which are unbounded (like magnetic fields are unbounded) and are constituted by structures, which are resources for the actions of participants. Action, in all of its forms, is synchronous with enactment, or cultural production, that always is both reproductive and transformative. Social agents, referred to here as actors or participants, are both aware and unaware of their enactments, which are considered to be partly agentic and partly passive. The structured flux associated with a field provides resources for action and is simultaneously reproduced and transformed by action. Accordingly, the structural flux associated with a particular field is similar over space and time (i.e., it is reproduced), and it also is different (i.e., it is transformed). Often, we refer to the structures of a field as bearing a family resemblance to the structures associated with the field at another time and/or another place. As culture is enacted in a field, individuals experience it as a structural flux (i.e., experiential), and as they experience the structures, through interaction, they reproduce and change them. Another way to say this is that actions are structured and structuring. Accordingly, structures do not determine action, and neither are they determined by human action.

To cope with the complex maze of structures, participants act according to their familiarity with a particular field. Bourdieu (1992) referred to this as *sens pratique*, a sense of the game, and used the term habitus for situations in which structures of a field support particular, familiar, forms of practice rather than others. When culture is being enacted fluently, that is in settled times, actors think about some aspects of what they are doing and are not consciously monitoring other practices. In such circumstances, actions are appropriate, anticipatory, and timely. That is, enactment is fluent. A person's actions contribute to the structural flux of a field and are thereby resources for others' actions and agency. One way to think about this is in terms of participants acting to expand others' agency.

If something unanticipated happens, such as when something inappropriate is enacted or if actions are out of sync (i.e., not timely), then fluency can be disrupted and habitus can break down. That is, habitus breakdown reveals itself to actors and other participants, in its breakdown. Settled times become unsettled, and flow (i.e., fluent enactment of culture) needs to be restored/repaired (Swidler 1986). When habitus breaks down, an actor might quickly resume fluent interactions in some cases, whereas in other cases, additional learning (i.e., new productions) may be needed. In such cases, the collective can act to support a resumption of productive cultural enactment. On the other hand, a breakdown in fluency is an opportunity for those who would like to prevent particular forms of culture from being enacted, to show resistance through their subsequent actions. This is a very familiar scenario in many urban classrooms.

From Sewell's standpoint, culture can be experienced sensually as patterns that have thin coherence and contradictions to those patterns. The contradictions are reflective of the opportunities individual actors have to appropriate resources/structures for their own purposes and associated forms of agency.

The above sociocultural framework evolved in our research in the period from 1998 through 2006. The most central components were dialectical understandings of sociological constructs, represented here with a vertical bar, |, to depict that each related construct is a constituent of a whole, that dialectically related constructs presuppose the existence of one another, and that changes in one constituent will be associated with changes in others. Of course, more than two entities can be related in this way and we use the term multilectical to represent "dialectical" relationships among two or more constructs (Fellner 2014). Some of the more salient multilectical relationships in our ongoing research are agency | structure, passivity | agency, passivity | structure, individuals | collective, coherence | contradiction, aware | unaware, part | whole, micro | meso | macro, and reproduction | transformation. Central to our research is the epistemological position that knowledge is cultural production. As such it can be regarded/studied through the lenses outlined above; it is field dependent, and it can be viewed across multiple levels (i.e., micro, meso, macro). Participants interacting in fields enact culture, here considered as socially produced knowledge (reproduced | transformed). Like all forms of culture, knowledge is not contained in the fields of production, but is transmitted to other fields where it serves as a resource for participants' actions in those fields.

We use the construct of field to consider and study cultural enactment. When the focus is on one field, either as a lived experience (i.e., teaching a class) or as a research focus (i.e., doing research in a doctoral class), a field is a social space that can be defined by its primary activities – that is, the individual goals and collective motives of participants (i.e., goals | motives). Consider an example of participants in a research methodology class – which we can regard as a field. When participants do homework, talk with friends about a course, or just think over what happened in class, they are participating in that field no matter what the time is, where they are physically located, and who else is with them. The relationship between participation in a field and the lifeworld more generally is represented as meso | macro.

Because of the dialectical relationships between social fields (i.e., micro | meso | macro), an individual is continuously dealing with the flow of structures from all fields in his/her lifeworld. Thus, individuals experience a unique structural flux that is the foundation for social reality. Persons sharing space and time are not experiencing social life in the same way, even in those social spaces and at the times they are together. Importantly, at all times and in all places, individuals have unique experiences even though the structures produced in a shared field will bear a family resemblance among participants. Importantly, experiences are historically constituted and also reflect structures that flow across and through the fields in each person's lifeworld.

In the next section, we address multilogicality (Kincheloe 2008), a framework that situates authentic inquiry and other theories as complementary methodologies that afford numerous ways of making sense of broad questions such as: What is happening? Why is it happening? What more is there to be seen and learned?

12.3 Multilogicality

Ken: Elsewhere I have argued that science education is dominated by a mainstream that is grounded in conceptual change theories (Tobin 2015b) and crypto positivism (Kincheloe and Tobin 2009). These ways of thinking about research and science education have marginalized other ways of making sense of science and being a science educator. The mainstream has virtually become monosemic (i.e., one way of making sense), is hegemonic to marginalized collectives, and seems impotent to address grand challenges that face humanity, such as climate change, sustainability, and healthy lifestyles (Powietrzynska et al. 2015). Accordingly, we resisted forces to conform to the mainstream and forged new methodologies to improve science education and, more generally, teaching and learning throughout participants' lifeworlds.

A multilogical approach to research affords our use of a variety of frameworks to design and do research. We have heightened awareness of experience being radically polysemic (i.e., accommodating difference by allowing for many meaning systems to make sense of experience). Accordingly, we seek to maximize our learning by searching for appropriate frameworks and using them to illuminate what we learned from research. To accept a theory as part of a multilogical bricolage is not to regard it as a social truth but to accept it as a viable way of seeing social life from different perspectives. A consequence of adopting such an approach is that research in science education will be characterized by difference, claims will be nuanced, and many of the differences and nuances will have novel implications for practices of teachers, learners, researchers, and education for literate citizenry.

For the most part, our research is grounded in poststructural hermeneutic phenomenology (Madison 1988), with ethnomethodology (Garfinkel 1967) providing an invitation to disrupt social equilibria in order to reveal forms of cultural enactment that might otherwise remain invisible. Rather than randomly select "subjects" for study, we prefer to select participants for our research purposefully and serially. We never leave who will be involved in a study to chance, and we never regard participants impersonally, as objects of research. Instead, we select participants from whom we can learn and who can contribute a different set of voices to the research. Selecting participants one at a time affords intensive research; selecting additional participants contingently – if and as necessary – depends on what is being learned and what we would next like to learn. Accordingly, the research is emergent and contingent – always dependent on what happens and the sense we make of what happens, why it happens, and what more there is beyond what we have learned already.

12.4 Researching Emotions

Ken: In the study we report in this chapter, we wanted to learn more about ways in which individuals access and ameliorate emotions as they enact education – which we regard as an important aspect of social life. Our interest in research on emotions began more than a decade ago, in a context of research in urban high school science in inner city Philadelphia. Classes were characterized by intense emotions that cover the entire spectrum of possibilities, including anger, fear, sadness, happiness, pride, surprise, and disgust. We studied emotions as they were expressed in the face, gestures, body orientation and movement, and the voice (e.g., Roth and Tobin 2010).

(continued)

When our research was situated mainly in New York City, we focused increasingly on relationships between excess emotions and poor health (Tobin and Llena 2012). Because teachers and students struggled for control in many of the classrooms we studied, we observed at firsthand ways in which emotional intensities cascaded out of control, often increasing stress levels and threat of physical violence. Adherence to authenticity criteria (catalytic and tactical) focused our attention on ways to reduce emotional intensity and to better understand relationships between emotions and well-being. We began to study emotions and their physiological expressions and sought to discover different ways in which emotional intensity could be ameliorated if and when necessary. Although we were concerned about the health of teachers, especially those involved in our research, we were also very much concerned with students and their well-being. It became a priority for us to design interventions to ensure that teaching and learning were not hazardous to the participants' health. We have already taken steps to learn about different knowledge systems that address wellness and to explore the potential applicability in different facets of education – including research and teaching and learning at the college level (i.e., Tobin 2016, 2015a, b; Tobin et al. 2015). All four of the authors of this chapter have had experience as learners of Jin Shin Jyutsu (hereafter JSJ) and have been involved in independent and collaborative studies in which it has been used as part of an interpretive framework in ongoing studies of mindfulness and wellness. A number of introductory resources are provided after the references to provide readers with additional information about JSJ.

12.5 Jin Shin Jyutsu

JSJ is a knowledge system that is inherently multilogical, grounded in traditional medicine practices from across numerous Asian countries and dating back thousands of years. It was a healing art that was passed on orally and gradually was lost in large part because of the global dominance of Western medicine. Its reemergence in the twentieth century was initiated by the intensive and extensive studies of Jiro Murai (1886–1960), a Japanese philosopher, who studied ancient texts such as the Kojiki (Phillipi 1968) and semisecret knowledge systems such as Tantric Buddhism that originated in India. Because JSJ is multilogical and involves forms of knowledge as practice that predated widespread use of writing, the retrieval of the "lost knowledge" involves interpretation of art objects such as paintings, sculptures, and jewelry and documents written in ways that were frequently intended to shroud understanding (e.g., the Kojiki, the Record of Ancient Things, is a story of creation that uses allegories). Murai did the hard work of extracting JSJ from these sources and then developed it empirically. In addition, Murai developed deep knowledge of acupuncture and Chinese medicine. In parallel, he undertook detailed, systematic studies of what came to be called JSJ and its applications to myriad health/wellness projects that he and others presented over several decades of the twentieth century. Notably, Murai catalyzed this impressive lifetime of research by self-help – overcoming a personal diagnosis of terminal illness at the age of 26, resumed good health, and dedicated himself to a lifetime of clinical research and practices. Because his earlier practices involved documents from the Royal Archives of the Imperial Palace, he was restricted from sharing what he learned until after World War II. Murai met Mary Iino (1918–2008), an English-speaking Japanese-American in 1946, and she agreed to participate in his lecture series with the designated purpose of disseminating JSJ to the English-speaking world. With a very small handful of others, Iino studied with Murai for 5 years before returning to the United States, where she continued her collaboration with Murai until his death in 1960. On her return to the United States, Iino married, becoming Mary Burmeister, and gradually began to add to what she learned in Japan and subsequently codeveloped during correspondence with Murai.

When Iino left Japan in 1953, the group that continued to work with Murai included Iino's father, Uhachi Iino (1886–1961), and Haruki Kato (1928–2014), recommended to Murai by Uhachi. Not surprisingly, Burmeister's approach to learning and disseminating paralleled her master's careful and somewhat conservative approach – developing the knowledge base through careful study and analyses and gradually developing a clinical practice grounded in many years of systematic study. Within an overarching framework of concern for misuse of JSJ, Burmeister also seemed reluctant to share what she knew until the knowledge base was clear in her own mind, and she had developed a network of trusted practitioners. Accordingly, it was not until 1965 that Burmeister began to teach a small number of students; thereby initiating what eventually became global dissemination.

The central tenets of JSJ involve the flow of universal (or revitalizing) energy through the body in circuits that pass through 26 pairs of safety energy locks (SELs). Unlike acupressure points, SELs are spheres with a diameter approximately the size of the palm of the hand. The flows run vertically, through the SELS, on the center, left, and right of body, diagonally, and through 12 organ flows (e.g., the stomach, the spleen). The SELs can stop or facilitate the flow of universal energy along these pathways. Figure 12.1 contains a schematic representation of the location of the 26 pairs of SELs on the body.

The multilogical approach employed in this study employs a number of perspectives in a dynamic frame that focuses on body movements and interactions and their possible meanings and purposes. In an interview with Melissa Higgins (1988), Burmeister, perhaps the most influential contributor to the recovery of JSJ, commented: "Plato said, 'learning is remembering.' There is nothing we have to learn. We are always utilizing part of Jin Shin Jyutsu naturally …." We adopt this perspective as a point of departure for observing interactions involving body movements, orientations, touches, holds, and flows (i.e., sequences of holds and/or touches) during research activities. Similarly, Burmeister connected emotions, which she referred to as attitudes, with wellness, noting that emotions are the root of many health problems (she uses the term project to reflect the temporary nature

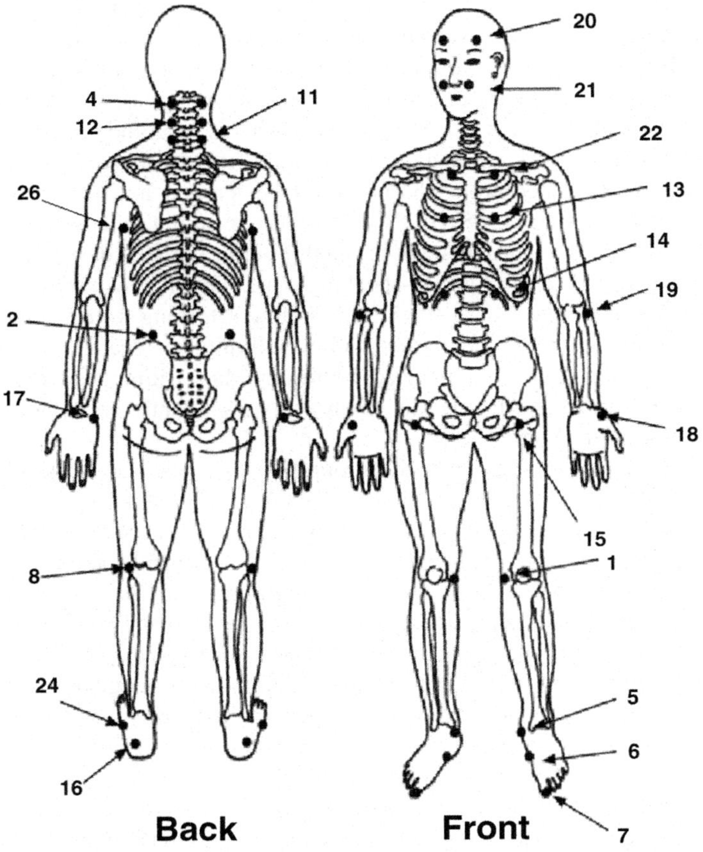

Fig. 12.1 Selected SELs that have salience to this research

of health issues). To paraphrase Burmeister (Higgins 1988), revitalizing energy can be blocked at any of the 26 SELs when we abuse our bodies in everyday life – mentally, emotionally, digestively, and physically. When blockages occur, the body can respond emotionally and physically with symptoms that reflect illness. Using the analogy of jumper cables, Burmeister explained how a person can use his/her own hands, or someone else's, to get the blocked or diverted energy flowing along the appropriate pathways.

Burmeister's standpoint, as articulated above, provides a premise for this paper, which explores the extent to which touches, holds, and flows consistent with JSJ occur naturally in activities associated with the teaching and learning occurring in a graduate course in urban education at the Graduate Center of CUNY, in New York City. Kenneth Tobin, first author of this chapter, teaches the course. Anna Malyukova, a coauthor of this chapter, is a student in the course. An additional site for the research reported in this chapter is a monthly seminar at the Graduate Center of CUNY, USER-S, which is a forum for faculty, graduate and undergraduate students, and other coresearchers to disseminate what they learn from their research. As well as providing a means for keeping up with the research of a number of research squads within CUNY and other universities in the New York City region, it is a safe haven for graduate students and faculty to present ongoing research to peers. Although the number of participants varies from month to month, there are typically around about 30–40 at each seminar. In this study, we undertook research with Anna and Al-Karim Gangji as they enacted roles as critic for several presentations presented at USER-S.

12.6 Event-Oriented Inquiry

Based on Sewell's (2005) uses of event analysis in research on history, we developed event-oriented inquiry (Tobin 2015a) for uses in multilevel research that involves video analysis. An event is conceptualized as a "spike in the curve." That is, an event consists of phenomena that stand out from what has been happening. It is something that is of interest and has the potential to be transformative after it has been researched.

Since this study is part of several years of inquiry in which we have researched intuitive uses of JSJ in different educational contexts, we have a strong desire to use what we learn from this study to design toolkits to afford all citizens to use self-help techniques to resolve health projects and ameliorate emotions when and as necessary (e.g., Tobin 2015b). Accordingly, we looked for relevant events that stood out from what else was happening. We then identified a start and finish of each event and analyzed them in terms of a multilogical framework that we described in the earlier parts of this paper. As it happened, there were many more events to analyze than could be included in this chapter. These events will be addressed in subsequent publications. The following sections address what we learned from this research and an exploration of potential implications.

12.7 Enacting the Role of Critic

One format we have often used at the USER-S for doctoral students is to schedule three of them to present in a 1-h time slot. We divide the number of participants into thirds and each third is assigned to one presenter. After 12 min, the groups rotate until each presenter has presented to each group (we refer to this activity format as a round-robin). The round-robin format provides new researchers with a comfortable arrangement of space, time, and audience size, allowing them to repeat their presentations three times.

To expand the academic conversation, the presentations conclude with a panel discussion composed of the presenters and two graduate students as discussants/critics so that they can learn how to provide productive critique. This allows the discussants/critics to provide their remarks to a plenary session and for the presenters and critics to respond to questions and comments from participants/audience.

At a recent seminar, September 2015, the three presentations from the doctoral students focused on research on mindfulness and wellness. The critics were both graduate students – Anna, finishing up a master's degree in liberal studies, and Karim, who is writing his doctoral dissertation. We analyze their presentations during the panel activity in terms of the occurrence of JSJ-like touches and holds. Karim provided the first critique in 1 min 36 s. Anna followed, remaining seated for the entire delivery, which took 2 min 25 s.

12.7.1 JSJ-Like Touches and Holds

Karim is to the left of the other panelists, facing them as he leans back on a lectern (see Panel 12.1). His critique provides a critical review of the three presentations and connects to his own research on mindfulness and science teacher education.

The other panelists listen to Karim, who is leaning on the lectern with his right leg over his left leg, touching lightly at SEL 1 on each leg. On an emotional level, harmonizing SEL 1 can strengthen self-confidence, calm nerves, and release stress. Accordingly, Karim's body orientation allows him to touch both right (R) and left (L) SEL 1s together and perhaps synchronize the downward energy flow. Keeping the 1s open keeps things moving, including the breath. By holding the 1s, the chest is opened up, and the lungs can function to exhale and receive new breath. As Karim leans on the lectern, his right arm engages with the top. Although it is not visible,

Panel 12.1 Karim as critic

the contact of his wrist and elbow would likely engage R SEL 17 and R SEL 19. In addition, it appears as though R SEL 25, R SEL 2, and R SEL 23 are touching the lectern. If so, the pressure of the body on the lectern is a way to harmonize a number of SELs.

Holding SEL 17 can minimize nervousness and foster intuition. SEL 19 can increase self-confidence and enhance authority. SEL 25 can promote inner stillness by reducing excitement or, at the other end of the energy spectrum, overcome exhaustion. SEL 23 can overcome fear, reduce stress, increase patience, and avoid self-judgments. SEL 2 can overcome doubt and promote self-knowledge while clarifying perceptions. If a person is sad, harmonizing SEL 2 can ameliorate the intensity of the sadness.

Karim: While I was presenting, I was fairly nervous and at the same time was extremely aware of what I was trying to express with my selected words and keeping in mind to never come across arrogant or too sure of myself. I tried to relate the presenter's work to my work on mindfulness in education. I like humor to break the tension and maintain focus and also feel the emotions of my audience at the same time. In addition, in communicating with a larger audience, I prefer to stand. Standing helps me relax and allows my hands and legs to be a bit freer to express my gestures clearly. And if there is a support as in this situation luckily, I had a lectern behind me, which allowed me to lean back slightly and cross my legs to calm myself and indicate to the audience that I was comfortable talking.

During Karim's presentation, Anna maintains a reasonably constant body positioning and orientation. Her L SEL 8 engages her R SEL 1 as her left leg is crossed over her right leg. Her left hand rests on her left leg, making contact with L SEL high 1. In addition, her right hand is holding her left wrist at L SEL 17.

When the energy is blocked at SEL 8, it can produce anger. Accordingly, harmonizing SEL 8 can reduce anger and build trust. Blocked 8s can increase rigidity and stubbornness; so, opening 8s can still the mind to receive new ideas. That is, opening the 8s minimizes a tendency to resist while enhancing willpower and determination. As we mentioned above, engaging SEL 1 releases stress and calms the nerves. Therefore, crossing the legs in this way is potentially synergistic. Furthermore, holding SEL 17 reduces nervousness and sets up the mind in ways that are similar to meditation, clearing the mind of the emotions and setting up a state that is ready to receive knowledge. Holding SEL 17 can help to accept new perceptions/ways of seeing, thinking, and acting. By expanding intuition, holding 17 can foster creative thinking and assist Anna to stay in the moment.

Anna: From the analysis of JSJ holds which I did while listening to Karim, I see that I was trying to build trust and be able to receive new ideas by the way my body aligned itself, and that makes me reflect back and remember how I felt at the moment. What comes to my mind first is the overwhelming emotion of anxiety, which I felt at the moment most strongly, and which were characterized in holds on SEL 1 and SEL 17, which releases stress and anxiety, but what I can also see and reflect on are the clues about how intelligent my body is without me even knowing it. Anxiety is all I can remember, but my hands and my legs positioned themselves in such a way that it allowed me to become not only more relaxed but more susceptible to new ideas and new information coming from Karim, which were seen by other holds, like crossing my legs and unlocking the SEL 8s.

Karim sits down, and as he does so, he crosses his arms, resting his left hand on R SEL 14 and his right hand on L SEL 14. Because the right arm is on top of the left arm, R SEL 17 more than likely makes contact with L SEL 18. Karim maintains this posture throughout Anna's critique of the three presentations. For the most part, Karim's holds reflect those that have been mentioned earlier. The exception is SEL 14, which is situated on the front of the body and controls downward energy flows related to all digestive processes associated with organs such as the stomach, spleen, pancreas, liver, and gallbladder. When SEL 14 is not in harmony, worry and stress can build up. Holding SEL 14 in the way that Karim holds it as he listens to Anna's critique affords him staying calm, in a state of equilibrium. Clearing SEL 14 clears the mind.

There is a possibility that the folded arms are reaching back to the edges of the shoulder blade and that Karim is in fact giving himself a hug, by holding each of the SEL 26s. SEL 26 is located at the edge of the shoulder blade where the shoulder meets the back. Disharmony in the 26s can produce stress, diminish energy, and unsettle the mind. Accordingly, harmonizing the 26s can increase mindfulness by harmonizing all emotions and thereby diminishing stuckness. Harmonizing of the 26s also can involve holding the center of the palms of the hands or the soles of the feet.

Karim: As I finished and introduced Anna, I sat down with a sigh of relief but at the same time was wondering if my critique of the presenters was in sync with my work on mindfulness and the well-being of my students. However, I immediately focused on Anna's response, and as she spoke about each presenter and their work, I kept thinking and feeling how much it was related to my personal life regarding my parents and my children. The habit of crossing my hands, I have realized, allows me to listen with great care and at the same time helps me calm down. In a very short period of time (less than half a minute), I could feel my heartbeat return to a normal relaxed rate and could also feel my breathing pattern to be more complete (full breaths) and began feeling

(continued)

calm. Yet at the same time, I felt Anna actually tied together the presenters' research work as she individually commented on their talk and connected her research work. While listening to Anna, I reflected that I might have failed in that respect and focused on the larger picture and wasn't too specific as a critic. I was aware of what I think was my failure, and it kept me rather uneasy while I was listening to Anna. I managed to listen, but I think I was consumed with the thoughts of not completing my responsibilities to the best of my abilities. At the end after the presentation, I had a difficult time recalling Anna's remarks. After watching the video, I was relieved that I was not off topic and I actually acknowledged each presenter and then talked about my related research.

Frame-by-frame analyses of the video file raise the possibility that Karim is short of breath, possibly related to the emotion he describes in the above narrative. He seems to be breathing through his mouth at various times, and his patterns of breathing appear to be mainly thoracic with an in-out period of about 4 s. Philippot et al. (2002) reported that primary emotions of happiness, anger, fear, and sadness were associated with characteristic breathing patterns. Use of the characteristic breathing patterns associated with an emotion led to the production of that emotion in the case of happiness, anger, and sadness. Ambiguous results were obtained for fostering fear through the use of fast, irregular, shallow breathing. In contrast, happiness/joy was produced when participants in the study used slow, deep, regular breathing through the nose. Similarly, anger was produced with fast, deep, irregular nasal breathing, and sadness occurred when participants used nasal breathing with average amplitude and frequency. Although the following suggestion is highly tentative, it is possible that Karim's breathing pattern was associated with low-grade fear about his performance. Salient to this possibility is Karim's body posture (see Panel 12.2). It is possible that sitting in this way helps to harmonize the kidney and bladder organ energy flows. The orientation of Karim's legs presses both little toes onto the floor, his SEL 25s are engaged firmly with the seat, and his SEL 23s are engaged with the backrest part of the chair. We raise the possibility that this posture is used intuitively to ameliorate buildups of fear, stress, and self-criticism.

As Anna begins to speak, she changes the orientation of her legs such that the left and right SEL 16s are in contact. As she did while Karim was presenting, Anna held L SEL 17 with her right hand. Anna held this posture during her critique of the first presenter. As she switched her attention to her critique of the second presenter, she began to gesture with the right hand, and the tendency to touch and hold SEL 17 diminished. Throughout her critique, her left hand remained in contact with her left SEL high 1. Even though the contact was not direct, the jumper cabling between the left hand and the L SEL high 1 is effective, penetrating through the notepad in this case. Finally, as Anna began to review the third speaker, she broke contact between the SEL 16s and placed both feet flat on the floor. During this time, the fingers from her right hand overlapped the backside of her left hand and fingers, and the left hand retained its contact with L SEL high 1.

Panel 12.2 Anna as critic

Table 12.1 Relationships of fingers to organ flows, safety energy locks, and emotions

Finger	SEL	Organ flow	Emotion
Thumb	1, 9, 16, 19, 21	Spleen	Worry
		Stomach	
Index	5, 8, 10, 11, 22	Kidney	Fear
		Bladder	
Middle	3, 6, 12, 13, 25	Liver	Anger
		Gall bladder	
Ring	2, 4, 7, 14, 17	Lung	Sadness
		Large Intestine	
Pinky	15, 18, 20, 23, 24	Heart	Pretense
		Small intestine	

Each of the fingers is associated with particular emotions, safety energy locks, and organ flows (see Table 12.1). Accordingly, touching and holding fingers can make an enormous difference to the emotional and physiological harmony of the body. For example, the thumb is associated with the spleen and stomach organ energy flows, SELs 1-4, and worry. Each of the fingers is interconnected with different SELs, organ flows, and emotions in similar ways. Hence, when Anna periodically touches her fingers together, she is mediating all of the body's energy flows. Interestingly, when Anna mistakes Ferzileta's name, she instinctively places her hand on Ferzileta's SEL 26. As we mentioned above, this would have the effect of minimizing any stress that might be associated with the error and promoting a state of comfort and assurance. In countries like Brazil, Anna's gesture of touching SEL 26 often accompanies an apology.

Anna: Blood was pumping in my ears really loud, as I was trying to absorb what Karim was saying about his experience as a science educator. At the same time, I was trying to collect my thoughts about what I needed to say when my turn came, just a few sentences away from that moment. Meanwhile, I wondered if everyone around me could hear my heart beating just as loud as I could. I realized how nervous, how insecure I felt at that moment. In the moment I was aware of how I was being perceived by others. I took a shallow breath and hoped that it was not as evident and overwhelming to others, as it was to me. Later, when I saw the video of the event, I wanted to watch it anxiously. I wanted to observe what I was not able to see while I was in the focus of the camera. I wanted to see what I sounded like, how nervous I seemed, how much my body gave me away. If not for JSJ and the way my body responded to the changes from the outside to harmonize it, I could hardly believe how well I did. I imagine that only a person who knows me really well would be able to pick up on the nonverbal cues of how I turned my head on the side, or the way I pursed my lips, when I experienced one emotion or another. And how, without any knowledge of it, my body responded to restore the energy within and help itself in the most poignant (or not) moments, by placing a hand at one SEL or another to help lessen anxiety or help stay awake, helps me think or release the anger building inside. Once you learn about JSJ at least a little bit, and I am saying it from a perspective of someone who is only beginning to understand this alternative framework of Eastern medicine, you cannot help but notice it everywhere: among the people around you, your friends and family, on TV or subway, outside on the streets of the city, and finally in yourself.

12.8 Writing Activity

Ken: A 10-min free writing activity is an integral part of the graduate class, following directly after a 5-min breathing meditation that is intended to afford a higher state of mindfulness. On this occasion, the free writing was a little unusual since I asked the students, prior to the breathing meditation, to specifically address possible changes to the assessment tasks they would undertake for the remainder of the course. Most of the students did not anticipate that I would provide a focus for writing and nor did they expect I would recommend more autonomy for them in terms of how they would be assessed in the course. Anna's writing reflected the unexpected nature of the structures I provided and the difficulty she had in the meditation of letting go of the focus of the upcoming writing task – it proved to be a distraction for her. During the writing activity, Anna wrote 192 words.

Here are two excerpts from Anna's writing:

> Anna: The provision of a focus for the writing activity was a bit unexpected. Wow! I was totally not trying to think about the question and, that of course, I could not help myself but think of it. And yet, I was able to catch myself on the thought and let it go. I finally imagined the waves of the ocean, and all of a sudden a thought came to me that looked like the answer to the question for the free write. I acknowledged that thought too and tried to let it go as best as I can ….
>
> Anna: I have many interests involving mindfulness and wellness, and, I think, what I would like to do is to write short essays about different projects (of a more personal nature), mindfully surfing, mindfully pregnant, mindfully learning, mindfully critiquing, and then analyze them from the theoretical frameworks/perspectives that we have been talking about in class.

Eight minutes of the video recording of the writing task showed Anna in the frame. Her upper body was clearly visible for the entire 8 min, with a front view. Table 12.2 contains a brief summary of the touches and holds that occurred during the writing activity. As Anna writes during the writing activity, we examined the ways in which the fingers touched one another as she wrote. Anna's thumb touched her index finger, and other students grasped their pens/pencils in different ways that led to the thumb being over the index finger, and in some cases the thumb over both the index and middle fingers. This opens up the salience of mudras being used intuitively for the purposes of ameliorating emotions and fostering wellness (Hirschi 2000).

Three of the SELs accessed during the free writing (21, 20, 4) involved clearing the mind and improving the quality of thinking. SEL 21 is concerned with worry, mental tension, and unproductive thinking. Clearing SEL 21 can provide a boost of energy and an increase in understanding. Placing the hand on SEL 21 can afford improved thinking. Touching and holding SEL 20 promotes clarity of thought and increases the use of common sense, promoting awareness and intuition. SEL 4, when it is cleared, can promote peace and clarity, calming the thoughts and affording logic and common sense. Holding SEL 4 is relaxing.

Touching SEL 12 can facilitate mindfulness in that clearing SEL 12 can release all emotions – worry, fear, anger, sadness, and pretense. Accordingly, touching SEL 12 is a good way to become "unstuck," affording emotional balance, personal humility, and increased tolerance. Harmonizing the 12s can minimize personal desires in favor of "good for all."

Table 12.2 JSJ-like touches and holds during a writing activity

Time	SEL touch or hold
0:00	Left on R14 until
0:16–1:12	Left on L11/12 until
1:15	Left flat on table
3:40	Left brief touch on L11/12
03:51–3:54	L on L 20 (extending to R 20)
5:25	Interesting mudra with right when writing
05:29	L on L 21
06:34	L on L 20
06:39	R on R hi 1
06:45	L on L 20 (clears hair). Hand moves around to SEL 4-SEL hi 4
06:51	Head leans on L 4/12 region on the side/front of the neck. This hold continues until about 7:23, when Anna begins to write again. At 6:54 she is resting on 21
07:59	L on L hi 1 (probably – no clear view)
08:16	Anna hands her paper to the professor

12.9 Anna as a Student in Class

At the conclusion of the writing activity, two graduate students cotaught a lesson centered on critical pedagogy. The focus of the lesson was on the methodology and associated methods used in a study undertaken by Tricia Kress (2015). We reviewed 47 min of the class, and all students in the captured video images continuously used touches and holds that engaged the SELs on the face, neck, and shoulders. For every student, including Anna, the hands were continuously engaged in touches of various SELs. Because of the camera location and the positioning of bodies in relation to the rectangular tables at which the students were seated, only one student could be viewed holding/touching SEL hi 1s. However, the orientation of the arms in relation to the body suggest that SEL hi 1 may have been the most common hold during the entire lesson. Similarly, the orientations of legs and ankles were only visible for Anna and a student seated adjacent to her. For long periods of the lesson, Anna crossed her left leg over her right – probably engaging L8 with R1, as she did in the panel activity. Similarly, when her legs were uncrossed, there were times when she crossed her ankles, engaging L SEL 16/24 with R SEL 24.

The video record provided clear images of the SELs from the waist upwards for six students. Each of these students was rotating her head to have eye contact with the coteachers, referring to her textbook, and writing notes. As well, each student used both hands to continuously touch, hold, and trace over SELs on the face and neck regions. The pattern of touching and holding did not seem to be synchronized, although we might find synchrony in the touches and holds of coteachers.

Anna: In the 47 min of class, where I am a student, I engage in many JSJ holds, some are very frequent and some rare but just as significant. As I am staying quiet throughout the entire class, there are moments when I felt the urge to speak up, and there is some definite JSJ activity going on at those moments, which is more noticeable, at least to me, as I watch the video. I am mostly trying to stay connected to the speaker and understand their line of thought, and when I am struggling, my hand is always around SEL 21, which boosts energy and understanding, as well as alleviates mental tension. For the most time, the class is an intense exercise in mental thought, and as I work hard on thinking, comprehending the concepts and ideas expressed, I notice my hand on my forehead a lot, which is SEL 20. SEL 20, when unlocked, is associated with clearing the thoughts and promoting awareness and intuition.

Since I am familiar with JSJ, I often notice the way my body acts not just on the video but in the moment as well. But when I observe it from a side, I see it as a kind of art, which most of us are unaware of. It is an incredible dance of hands and legs guided by the universal energy flowing in a body. And our body, intelligent beyond our understanding, performs these natural flows and holds in a way, which not only reveals to us as observers, of what goes on in a person's body, but also assists itself in restoring balance and harmony.

Above the waist, Anna consistently engaged all SEL regions on the face, front of the neck, and the SEL 22 region. One pattern that warrants comment involved a sweeping movement of her right finger(s) in an anticlockwise direction that began with R SEL 21 and swept upward toward the outside corner of the right eye before descending down the jaw line to the neck and SEL 22. SELs 21, 12/11, and 22 were involved in the sweep, which was repeated frequently (as many as 10 times on one occasion). Commonly, Anna grasped her hair as she moved it away from her face, completing the sweep as described above. From a JSJ perspective, the path resembles the reverse direction of part of the right stomach flow. Some JSJ practitioners instruct clients to use their index finger to trace over the stomach flow, first in the reverse direction and then in the direction the flow takes when it is in harmony – hereafter referred to as the forward direction. The stomach energy flow is associated with many physical ailments, such as gastritis and myriad emotional imbalances, such as worry, that can affect concentration and fray the nerves. It is worth considering whether worry and frayed nerves could be reduced through the conscious use of reverse and forward tracing of the stomach flow. In this particular case, the flow likely extended to SEL high 1 and may have also involved the ankles and feet. This is a possibility that might be a focus for follow-up research, or it might be anticipated in the design and use of a flow that is taught to people seeking to ameliorate worry and frayed nerves. In the context of designing self-help packages, the tracing technique might complement techniques such as holding the thumbs, holding the palms of the hands, and using specified mudras.

Another potential flow that could possibly be involved in Anna's intuitive uses of JSJ is the gallbladder function energy flow, which involves SELs 12, 20, and 22. The forward flow circles around the cheekbone, past the outer edge of the eyebrow, turning around the back of the ear before circling onto the back of the head and up to the forehead. Of course the gallbladder flow also involves SELs below the waist, which was not visible in the video record. The emotions associated with the gallbladder flow can involve high- and low-grade anger, commonly observed as frustration in many classes. As was the case with the stomach energy flow, tracing the fingers and/or hand over the reverse and forward flows of the gallbladder energy flow might produce conditions of harmony in that flow and ameliorate emotions and physical ailments such as headaches, pains in the side of the body, and stiffness in the neck. If a self-help package were to be designed, it might include suggestions about use of the gallbladder flow, finger holds, and mudras together with touching the center of the palm and hugging the SEL 26s.

12.10 What Did We Learn and How Did We Change as a Result of This Study?

Authentic inquiry is a central component of the multilogical bricolage that underpins this study and other recent research we have undertaken (e.g., Alexakos 2015). Our ongoing research on mindfulness, emotion, and wellness has systematically disseminated what we've learned in order to change our approaches to teaching and learning in our universities and high schools and to modify the designs of ongoing studies. In this regard, interventions designed to be used in research, including breathing meditation, mindfulness heuristics, and physiological expressions of emotion (e.g., heightening awareness of oxygen dissolved in the blood), have been systematically infused into our teaching, research, and international dissemination activities (e.g., conference presentation, publications). Accordingly, what we know from our research makes a discernible difference to our lifeworlds – hopefully benefiting those with whom we interact.

In this study, we set out to look at the incidence of touches, holds, and flows in a variety of activities in which graduate students are involved. Because the teaching and learning environments we build in our classes have become increasingly informed by sociocultural theory, we wondered whether the changing roles of teacher and students would be associated with an increase in emotions and perhaps excesses of emotion. If this were to be the case, we wondered whether JSJ-like touches and holds would occur naturally as participants in classroom activities interacted with one another.

As we considered what we might study and report in this chapter, we considered the new roles for students in relation to a number of activities that included coteaching involving 2 or 3 graduate students from the class, participating in a writing activity following a short breathing meditation, being a student in a class that is cotaught by peer coteachers (in this case two graduate students), and participating in an exit cogenerative dialogue in which one graduate student participated with two professors, who were coteachers responsible for the course (in this case the first two authors of this paper). The final activities concerned graduate students participating in research seminars, as presenters and critics.

Since the design of the study was emergent and contingent, we began with Anna and Karim, enacting roles as critic for three peers, graduate students making presentations based on research undertaken in their doctoral programs. In this study, we focus first on Karim and then on Anna. We were not quite sure what to expect, but benefited from Karim enacting his role while standing up and leaning on a lectern. In contrast, Anna remained seated and crossed her legs and ankles as she delivered her critique. Through the study of the two participants, we began to see different emergent patterns and contradictions.

Once we noticed that Karim appeared to be using the lectern to make contact with several SELs that could ameliorate a buildup of emotions, we could see that this was an extension of doing much the same thing by crossing the legs and ankles. This expanded our way of analyzing what was happening. The way Karim was sitting in his chair seemed odd, but through the lenses of using objects to engage SELs, we were able to see how the chair engaged, in this case, SELs 25 and 23. Similarly, the orientation and movement of his feet may have engaged the little toe on each foot.

When multiple SELs are engaged, this can point to a flow, such as one of the organ flows (e.g., the bladder, kidney) or a special flow that is designed to address a particular health project (e.g., fatigue flow). In this particular case, we felt that Karim might be accessing intuitively the bladder and/or the kidney flows. There were other signs that this might be the case. He seemed short of breath. Evidence of this was that from time to time he appeared to be breathing through his mouth and his breathing pattern was mainly thoracic, shallow, and high frequency. Using frame-by-frame analysis, we could clearly see the in and out breaths in the rise and fall of his elbows. It was possible that concerns about his presentation together with issues from his lifeworld may have contributed to a buildup of fear and associated stress. The buildup could have been enhanced further by his breathing pattern. In an effort to offset his attachment to emotions, he may have intuitively engaged SELs to harmonize the bladder and kidney flows and to clear emotions.

Based on our analysis of Karim's practices during his own critique and then as he listened to Anna's critique, we identified touches, holds, and flows being enacted intuitively as Karim and Anna enacted roles of critic. After the critic roles had been studied, we decided to research Anna's intuitive uses of JSJ in a writing activity. The additional understanding we gleaned from this analysis is highly salient to JSJ in part because Murai used mudras when he successfully treated his terminal illness in a process of commencing his rediscovery of JSJ. As the students wrote, including Anna, the fingers engaged with one another and the pen, in many different configurations. The students were intuitively using mudras, which would likely have implications for emotional expression and wellness. We signal this as an important area for further research. To view an example of three different students, including Anna (Panel 12.5), refer to Panels 12.3, 12.4, and 12.5. It seems as if the students show their uses of two types of mudra.

In the segment of the class in which Anna was a student, immediately prior to her coteaching with a peer for an hour, Anna employed JSJ-like maneuvers continuously. Consistent with Karim possibly deploying part of two organ flows, Anna utilized light movements on both sides of her face that were consistent with the stomach flow. A sequence of five moments is captured in Panels 12.6, 12.7, 12.8, 12.9, and 12.10 as an illustration of where her fingers traced – either lightly touching the skin or just above it. The interpretation that her actions were an intuitive enactment of part of the stomach flow is highly speculative, but warrants closer microanalysis.

Panel 12.3 Mudra (a)

Panel 12.4 Mudra (a)

Panel 12.5 Mudra (b)

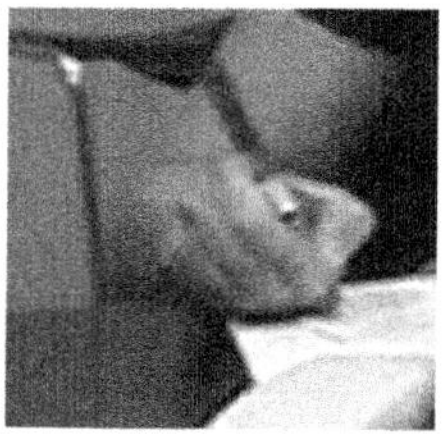

Finally, throughout the time in which Anna was a student, she momentarily held one finger just above the top lip and beneath the center of her nostrils. A variant was to sometimes trace the finger along the top lip and at other times place one finger above the top lip and one or more fingers below the bottom lip. The possibility we raise, based on some of the experiences described in lessons conducted by approved JSJ teachers, is that language functioning that may be lost during a stroke can be partially (at least) restored with a similar hold. Furthermore, if people who have previously lost language functionality begin to struggle in remembering or speaking words, a hold involving the fingers above and below the lips can restore/improve functionality. A speculation that is worth researching is that nonnative speakers might benefit from similar holds and might use similar holds intuitively when they read, speak, and listen to a nonnative language. Since this lesson was in English and Anna's native language is Russian, her uses of holds like those in Panels 12.11 and 12.12 are consistent with our speculation.

By the time we finished our analyses of the writing activity, we realized there would be insufficient space in this paper to include analyses of the use of JSJ-like touches, holds, and flows during coteaching and cogenerative dialogue. Accordingly, we truncated the empirical part of our analysis after we had descriptively analyzed the intuitive uses of JSJ-like touches, holds, and flows. The analyses of JSJ -like maneuvers while coteaching and participating in cogenerative dialogue are studies that are bound to be productive.

Panel 12.6

Panel 12.7

Panel 12.8

Panel 12.9

Panel 12.10

Panel 12.11

Panel 12.12

12.11 Next Steps

Our next step is to disseminate what we have learned from this study to the graduate students in the classes involved in the research. Already many of them have read this chapter and participated in conversations about what we have learned. Importantly, we will disseminate what we have learned to colleagues who are teachers and teacher educators, not just in schools and universities but in any institutions involved in teaching and learning. We are not arguing that every touch represents an intuitive JSJ-like maneuver. However, we are struck by an almost continuous flow of touches and holds that do appear to be used to address disharmonies in the flow of universal energy. It might be helpful to heighten awareness about the occurrence and possible purposes of JSJ-maneuvers so that, when and as necessary, interventions can be used consciously for self-help. We will create intervention packages based on our growing knowledge of JSJ to ameliorate emotions, increase mindfulness, and enhance wellness. It seems imperative that all participants in social life should have the tools to decrease the intensity of emotions that are building to excess, detach conduct from such emotions as they are produced, and attend to minor health projects that arise in the moment-to-moment flow of life.

The dysfunctional classroom from suburban Western Australia, described in the introduction to this chapter, has been a staple feature of our research in high school science from the mid-1980s to the present. That is, for more than three decades, teachers and students have displayed excesses of emotion as they participated in science classes. The emotions stuck to and saturated their conduct, making learning virtually impossible. The toll on learning seems to be potentially enormous. But what have been the tolls on staying involved and maintaining good health? If Burmeister is right in her assertion that emotions underlie many health projects, then it makes sense to ask whether teachers and students have created and sustained poor health because of deleterious emotional climates that have minimized mindful participation in learning tasks, produced excess emotions, and disharmonized bodies for sustained periods of time. We offer the possibility that toolkits like those we propose to design and enact could pave the way for all participants in science education to ameliorate emotions when and as necessary. The incentive to do this is to stay healthy and learn more science.

References

Alexakos, K. (2015). *Being a teacher | researcher: A primer on doing authentic inquiry research on teaching and learning*. Rotterdam: Sense Publishing.

Bourdieu, P. (1992). The practice of reflexive sociology (The Paris workshop). In P. Bourdieu & L. J. D. Wacquant (Eds.), *An invitation to reflexive sociology* (pp. 216–260). Chicago: The University of Chicago Press.

Fellner, G. (2014). Broadening our lenses of perception to advance learning: An introduction to multilectics. *Teaching and Teacher Education, 37*, 169–182. doi:10.1016/j.tate.2013.04.015.

Gallagher, J., & Tobin, K. (1987). Teacher management and student engagement in high school science. *Science Education, 71*, 535–555.

Garfinkel, H. (1967). *Studies in ethnomethodology*. Englewood Cliffs: Prentice Hall.

Guba, E. G., & Lincoln, Y. (1989). *Fourth generation evaluation*. Thousand Oaks: Sage Publications.

Higgins, M. (1988). An interview with Mary Burmeister, Master of Jin Shin Jyutsu. *Yoga Journal, 79*, 24–29.

Hirschi, G. (2000). *Mudras: Yoga in your hands*. San Francisco: Weiser Books.

Kincheloe, J. L. (2008). *Knowledge and critical pedagogy: An introduction*. Dordrecht: Springer.

Kincheloe, J. L., & Tobin, K. (2009). The much exaggerated death of positivism. *Cultural Studies of Science Education, 4*, 513–528. doi:10.1007/s11422-009-9178-5.

Kress, T. M. (2015). Can't you just know?: Critical research as praxis. In K. Tobin & S. R. Steinberg (Eds.), *Doing educational research* (2nd ed., pp. 167–179). Rotterdam: Sense Publishing.

Madison, G. (1988). *The hermeneutics of postmodernity: Figures and themes*. Bloomington: Indiana University Press.

Philippot, P., Chapelle, G., & Blairy, S. (2002). Respiratory feedback in the generation of emotion. *Cognition & Emotion, 16*, 605–627. doi:10.1080/02699930143000392.

Phillipi, D. L. (1968). *Kojiki* (translation). Tokyo: University of Tokyo Press.

Powietrzynska, M., Tobin, K., & Alexakos, K. (2015). Facing the grand challenges through heuristics and mindfulness. *Cultural Studies of Science Education, 10*, 65–81. doi:10.1007/s11422-014-9588-x.

Roth, W.-M., & Tobin, K. (2010). Solidarity and conflict: Prosody as a transactional resource in intra- and intercultural communication involving power differences. *Cultural Studies of Science Education, 5*, 807–847. doi:10.1007/s11422-009-9203-8.

Sewell, W. H., Jr. (1992). A theory of structure: Duality, agency and transformation. *American Journal of Sociology, 98*, 1–29.

Sewell, W. H., Jr. (1999a). The concept(s) of culture. In V. E. Bonell & L. Hunt (Eds.), *Beyond the cultural turn* (pp. 35–61). Berkeley: University of California Press.

Sewell, W. H., Jr. (1999b). Geertz, cultural systems, and history: From synchrony to transformation. In S. B. Ortner (Ed.), *The fate of culture: Geertz and beyond*. Berkeley/Los Angeles: University of California Press.

Sewell, W. H., Jr. (2005). *Logics of history: Social theory and social transformation*. Chicago: University of Chicago Press.

Swidler, A. (1986). Culture in action: Symbols and strategies. *American Sociological Review, 51*, 273–286.

Tobin, K. (2015a). The sociocultural turn: Beyond theoretical imperialism and the imperative of learning from difference. In C. Milne, K. Tobin, & D. deGennaro (Eds.), *Sociocultural studies and implications for science education* (pp. 3–31). Dordrecht: Springer. doi:10.1007/978-94-007-4240-6_1.

Tobin, K. (2015b). Connecting science education to a world in crisis. *Asia-Pacific Science Education, 1*, 1–21. doi:10.1186/s41029-015-0003-z.

Tobin, K. (2016). Collaborating on global priorities: Science education for everyone – Any time and everywhere. *Cultural Studies of Science Education, 11*, 27–40. doi:10.1007/s11422-015-9708-2.

Tobin, K., & Gallagher, J. J. (1987). What happens in high school science classrooms? *Journal of Curriculum Studies, 19*, 549–560.

Tobin, K., & Llena, R. (2012). Colliding identities, emotional roller coasters, and contradictions of urban science education. In M. Varelas (Ed.), *Identity construction and science education research: Learning, teaching, and being in multiple contexts* (pp. 141–156). Dordrecht: SensePublishers.

Tobin, K., Powietrzynska, M., & Alexakos, K. (2015). Mindfulness and wellness: Central components of a science of learning. *Innovación Educativa, 15*(67), 61–87.

JSJ Resources

Burmeister, M. (1997a). *Text 1*. Scottsdale: Jin Shin Jyutsu Inc.
Burmeister, M. (1997b). *Text 2*. Scottsdale: Jin Shin Jyutsu Inc.
Quan, J. (Ed.). (2013). *Zero to twenty-six: Excerpts from the main central*. Scottsdale: Jin Shin Jyutsu Inc.
Riegger-Krause, W. (2014). *Health is in your hands*. New York: Upper West Side Philosophers, Inc.
Waldeck, F. (2012). *Jin Shin Fee: Healing method by Master Jiro Murai & Mary Burmeister*. Munich: Creative-Story.
Watkins, B. (2014). *26 keys to unlock my inner treasure*. Raleigh: Lulu Enterprises Inc.

Kenneth Tobin is Presidential Professor of Urban Education at the Graduate Center of CUNY. In 1973, Tobin began a programme of research on teaching and learning that continues to the present day. The current emphasis of his work involves mindfulness, wellness, environmental harmony and the transformative potential of social research. Tobin has published more than 20 books, 200 refereed journal articles and 125 book chapters. He is recipient of numerous awards, including Distinguished Contributions to Science Education through Research Award (2007, National Association for Research in Science Teaching), Mentoring Award as an exemplary scholar and mentor (2008, Division G, American Educational Research Association) and the National Science Foundation Director's Award for Distinguished Teaching Scholars (2004).

Konstantinos Alexakos is a professor in the School of Education at Brooklyn College (CUNY) and at the Graduate Center of CUNY. His interests include teaching and learning with a focus on emotions, mindfulness and wellness, including emotional bonds and fictive kinships (close personal friendships), coteaching, radical listening, cogenerative dialogues, emotional climate, breathing and mindful practices, with the goals of improving learning and teaching, personal wellness and the emotional climate in the classroom, as well as creating safe spaces for discussing challenging topics, valuing difference and learning from each other.

Anna Malyukova is a doctoral student at the Graduate Center. Anna is very passionate about mindfulness and emotions in the process of learning on the birth-death continuum. She began her education in Russia in 1997 pursuing her degree in engineering, but has since become interested in education and received an AS in early childhood education and a BA in liberal arts from CUNY. In her collaboration with Kenneth Tobin and Konstantinos Alexakos, she became very interested in the role of emotions in the process of learning. She has recently had a publication with Carol Korn-Bursztyn in *Immigrant Children and Youth: Psychological Challenges* (2015).

Al-Karim H. Gangji is a high school physics teacher and an adjunct lecturer in the Physics Department at Queens College of CUNY. He is currently a doctoral student in the Urban Education Program at the CUNY Graduate Center. His research interest involves mindfulness in preparing elementary teachers to teach science and mathematics. He is the recipient of the Queens College President's Award for Excellence in Teaching by Adjunct Faculty (2013), the MIT Inspirational Teacher Award (2009), Educator of Distinction from the Coca-Cola Scholars Foundation (2007) and the Hofstra University/New 12 Long Island Educator of the Month (2003).

Chapter 13
The Role of Care in Environmental Education

Cassie F. Quigley and Renée Lyons

Environmental education and environmental advocacy have a complex relationship (Fien 2003). Their association is tenuous for many educators, including us, because advocacy has the potential to lead to indoctrination. As environmental educators and activists, we care deeply about the environment including the well-being of all people who inhabit it. How this notion of care is translated to our students presents a concern as a fine line exists between environmental education, advocacy, and indoctrination.

Science has long been held as a neutral and objective subject, free of any bias or political agenda (Namenwirth 1986). This claimed neutrality has led some to believe that science education is not the proper vehicle for the value-laden issues of environmental education (Lucas 1980). Learning about science is not enough to decide among competing environmental policies, and hence values are involved in any decision about what constitutes an environmentally friendly behavior. Preservice science teachers are taught that a variety of techniques (e.g., open dialogue, provide opportunities to engage in multiple viewpoints) can ensure a balanced approach and neutrality in science education (Cook-Sather 2002). However, the notion that science education is or can ever be neutral is not accurate. Curriculum design, learning objectives, and pedagogical techniques are value laden. John Fien (1997) insists that all educational objectives, curriculum, and instructional approaches are ultimately a selection of culture, and as such curriculum planners and teachers enacting these practices are making decisions and selections within certain values to the estab-

C.F. Quigley (✉)
Department of Teaching and Learning, College of Education, Clemson University, Clemson, SC, USA
e-mail: cassieq@g.clemson.edu

R. Lyons
Department of Curriculum and Instruction, Clemson University, Clemson, SC, USA
e-mail: reneel@g.clemson.edu

A. Bellocchi et al. (eds.), *Exploring Emotions, Aesthetics and Wellbeing in Science Education Research*, Cultural Studies of Science Education 13,
DOI 10.1007/978-3-319-43353-0_13

lished priorities of that culture. Social and cultural values influence the structuring of knowledge; thus, science education has the ability to perpetuate certain values over others—typically those of the dominant cultural and political beliefs (Longino 1990).

Rather than claim neutrality, we take a committed value stance, consciously and deliberately teaching students to care for the natural world because we believe a socially critical approach is necessary for creating an ecologically just world, which is achieved through teaching about care (Fien 1997). In this sense, we support advocacy; our education is for the cause of creating an ecologically just world. We recognize that not all environmental educators agree with our stance. Bob Jickling (2003) states that education for *any* cause is not education. Edward Johnson and Michael Mappin (2005) caution environmental educators to consider carefully the values that must be drawn upon when establishing the connection between how the world is (ecological science) and normative conclusions about what ought to be done. Despite the call by researchers such as Jickling (2003) for the separation of care and EE, we wonder if environmental education can be taught *without* the notion of care. Ultimately, if you want people to care about the environment, what role does education play? In this chapter, we explore this question by conceptualizing care in environmental education and how this care translates in the teaching EE.

The relationship between education, advocacy, and indoctrination is important to foreground in this paper as we explore the role of teaching about care in environmental education. This chapter is based upon the premise that education is value laden and linked to economic, political, and societal influences (Kelly et al. 1993). This premise is relevant for educators concerned with ethics and values in education; however, the concern should not be with whether a teaching approach is value laden but rather what values are used and how the values are presented to students. This issue is particularly relevant for the topic of environmental education (EE), as EE is not only tasked with teaching individuals and communities about the environment through multidisciplinary (biology, physics, mathematics, earth sciences, economics, sociology) approaches but also exploring environmental problems and how to solve them. Because the ways in which these environmental problems are solved are value laden, we adopt a critical approach to teaching EE.

A critical approach to teaching EE does not mean we support indoctrination. Indoctrination through science teaching is a concern for students, parents, and teachers because it can narrow views and impede critical thinking (Siegel 2013). Environmental education becomes indoctrination when educators teach students to adhere to certain attitudes, assumptions, or values over others (Jickling and Spork 1998). To avoid indoctrination, environmental educators need to be cautious and distinguish between education and advocacy (Johnson and Mappin 2005). Environmental educators need to be conscious of the values, biases, and agendas inherent in environmental issues and policies, rather than unconscious (Gough 2002). Avoiding indoctrination implies decisions about the rightness or wrongness of actions are reached through discussion and sensitivity to differing beliefs and value systems (Tilbury 1995).

In a critical approach to EE, teachers ask students to understand the reasons for environmental problems, encourage students to deconstruct normative rhetoric around environmental issues, and then provide opportunities for students to participate in the reconceptualizing of these issues with the values of social justice in mind (Fien 2003). It is with this critical approach that we reconcile the involvement of advocacy in EE. To do so, we utilize John Newfield and Virginia McEylea's (1984) distinction of promoting values in an ethic of care without teaching the attitudes connected to the issues. By making this distinction, they argue that indoctrination, which occurs when teachers encourage students to form ideas on issues irrespective of any available evidence, can be avoided. Teachers can avoid teaching attitudes instead of values by providing opportunities for evidence to be appraised and disputed according to values presented (Fien 2003).

13.1 Exploring an Ethic of Care in Environmental Education

Fien (2003) argued for environmental education that extends beyond sociological and educational theory and integrates arts, philosophy, and ethics to learn how to care for others and the Earth. To develop his position, Fien utilized the work of Nel Noddings (1992) on caring to consider the moral and ethical obligations of humans to care for the environment. More specifically, Noddings' stated that teachers who employ care as pedagogy have a moral approach to education. She called this moral approach, *the ethic of care*. Noddings uses Milton Mayeroff's (1971) definition of caring as follows, "To care for another person, in the most significant sense, is to help him grow and actualize himself" (p. 1). In education, caring often exists between the student and teacher or the students and each other. In this chapter, we extend this work by exploring what this means for the ethic of care in education when the one being cared for is the environment.

Fien was critical of approaches to caring for the environment that result in a dichotomy between humans and nonhuman elements of the environment (i.e., "nature"). Often, environment is classified as the combination of nature and humans, although this is contested. Tim Ingold (2000), an anthropologist, takes a different view on the human relationship with the environment. He contends that "nature" is the world without humans inhabiting it, while "environment" is actually in relation to humans and therefore is always viewed from the human perspectives. Ingold's work is helpful to our understanding of how we view and define "environment" for several reasons. First, it helps us to understand that "the environment" is fundamentally historical and cultural. Second, by delineating nature and the environment, we avoid seeing humans as beyond the world, and thus in charge of its processes. Last, Ingold's work provides context that the environment is continually changing through the activities of human beings, and thus the environment itself is a process rather than an object.

This understanding of "environment as a process" is important for environmental education as understanding the ways students view themselves in and with the environment shapes the ways in which they connect with the environment and understand the changes within the environment (Li and Ernst 2014). Fien (2003), while connecting Noddings' notion of care to environmental education, recognizes that values direct what is cared for and calls upon environmental educators to promote values and critical thinking of environmental issues according to these values. Peter Martin (2007) extends Fien's research by giving practical ways environmental educators can nurture an ethic of care within their classrooms. Martin's conceptual framework, developed to guide education *for* the environment, presents care as being generated by relatedness, reciprocity, and ways of knowing. Despite providing well-theorized implications for the structure of care in environmental education, Martin does not provide empirical evidence using an ethic of care in the classrooms.

In this chapter, we position this chapter in line with the well-being aspects of the book but also lean on emotions literature to guide the work on care. We do this because the motivation to care involves reasoning but also emotions as individuals make decisions about what to care for. Emotions provide the learner with purpose for engaging in a learning experience and guide the motivation to learn (Dirkx 2001). We strive to continue Fien and Martin's conversation of the ethic of care in environmental education by further theorizing Noddings' notions of care in the context of environmental education and finally highlight an experience of care-based environmental education pedagogy. The experience describes nurturing an ethic of care in environmental science as a middle school teacher at an international school in Phnom Penh, Cambodia. The chapter concludes with a section on further directions for an ethic of care in environmental science, as well as the implications of and complications with teaching students to care for the environment.

13.2 Conceptualizing an Ethic of Care in Environmental Education

As Noddings (1984) argues in *Caring: A Feminist Approach to Ethics and Moral Education*, the goal of educators should be to develop competent and caring people. Noddings believes the priorities of our educational system have been misplaced when we overemphasize content and all the possible topics a child can learn at the expense of nurturing caring individuals. This is not to say that intellectual development is not important. On the contrary, she argues that the intellect is better developed when caring is present. Caring involves an emotional understanding of an issue. Caring implies competence and provides the emotional commitment to work toward enhancing the well-being of what is cared for (Noddings 1995). Deep learning of issues needs to be both an emotional and intellectual experience because an emotional engagement with a learning experience indicates deep involvement of the learners' psyche (Dirkx 2001). An emotional understanding of issues deepens involvement by developing intrinsic motivators for engagement (Goralnik et al. 2012). Emotional engagement of students will guide what a student attends to and thus what he or she will learn (Weiss 2000). The cognitive development of students is important, but of equal importance is the emotional development (Venter and Ferreira, 2014). When educators ask learners to view the world differently, or consider new values and how these values should shape their actions, an emotional transformation is needed (Shuck et al. 2013). Emotions determine how we act. As Dirkx (2006) explains "through the experience of emotions [we] come to recognize what is cognitively and affectively of value (p. 31)."

Noddings (2002) explains that caring is at the heart of morality, which she defines as a universal active virtue. According to Noddings, there is no external standard or principle against which a person must justify behaviors toward other individuals. Instead, any standard for how one ought to act is generated from the remembrance of what it feels like to be cared for. As a person considers how to act toward another, they do so with a reflection upon their own experiences of the warmest and best of human relations (Noddings 2013). A person behaves ethically toward others when he or she treats others with respect and acts with care. The action in caring involves working to preserve the caring relation and promoting the growth of the cared for.

Caring exists on a continuum from natural caring to ethical caring. Natural caring occurs in situations when "I want" and "I must" are born out of desire, and it may be hard to distinguish whether the motive is personal enjoyment or obligation. There are certain things we naturally care for. Noddings relates a natural care to the care a mother exhibits for her child. We may care and attach great importance to doing kind things for a loved one or to spending time with our families. No one taught us to care for these things. We care because we feel enjoyment and love for these things. Ethical care involves an obligation and a commitment to act on behalf of the one cared for even if we don't naturally want to act. When the "I want" is not naturally present, a feeling of "I must" needs to guide the caring action. An example of an ethical care would be a pet owner taking their dog out for a walk even when they are tired after a long day of work and would rather rest at home. As a pet owner, there are times when I do not naturally want to take care of my pet because of other pressing needs. Yet, I feel obligated to care and act. The source of this felt obligation, the "I must," is the remembrance and reflection of being cared for yourself and on watching others exhibit care (Noddings 2013). In this sense, an ethical care is a fulfillment of an ethical ideal; a person acts based on an ideal version of themselves.

An ethic of care includes responsibility and respect for the feelings and worth of the other but also the motivation and willingness to act to protect the other (Noddings 2012). Fien (2003) describes an ethic of care as having compassion. Compassion moves one from the emotional, or natural caring, to the ethical realm and involves not only feeling empathy for another but acting on that empathy to look out for the well-being of another. Ethical caring is certainly important but not superior to natural caring. Natural caring is both the source and the end result of ethical caring. As a person decides to care ethically, emotional barriers are broken down, and relatedness is established that may result in natural caring (Noddings 2013). Returning to our example of caring for a pet, as I continue to ethically care for my pet, I begin to form a closer relationship with my pet that results in an increased natural care. As my relationship grows, I find myself more and more wanting to do selfless things for my pet, and our relationship results in me finding enjoyment and love in caring for my pet.

13.3 Cultivating Care in an Environmental Science Classroom: Investigation of Aedes Mosquito

The tendency in environmental education can be to cover as much information as possible at the expense of covering the most significant information well. Teachers may feel overwhelmed with the amount of content to be taught. There are many big issues students need to know such as extinction and sustainability. A sense of urgency to provide students with as much knowledge on environmental issues as possible may lead teachers to sacrifice the depth of knowledge for the breadth. Within the Western perspective of knowledge, value is placed on efficiency and the amount of information to be covered. As Noddings points out, the problem with this view of education is our students learn much at the expense of learning well (Noddings 1995). Perhaps this model of efficiency is prohibitive to fostering a deep connection and care for nature.

In the following vignette from an environmental science high school class, we provide an example of how place-based education can provide a platform for caring for the environment. According to David Greunewald and Gregory Smith (2014), place-based education "can be understood as a community-based effort to reconnect the process of education, enculturation, and human development to the wellbeing of life" (p. xvi). In this way, by rooting the educational process in the community, place-based education helps to counter the notion that the environment is "something out there"—a notion that has been documented by many researchers including Loughland et al. (2003). This vignette serves to provide an exemplar of how we move toward a pedagogy of care, with first cultivating care in science classrooms.

As a high school environmental science teacher at an international school in Southeast Asia, I (Cassie) had little trouble teaching student activism. For me, I saw environmental science as an active endeavor; it was not enough to teach the content to understand the challenges with our environment but to utilize that knowledge to solve the problems. However, it was not until I examined how activism can sometimes make the students feel forced if the students do not care for the environment. The data in this vignette was collected over a 6-month period, using reflection observation notes, student interviews (the students were in grades 11 and 12), and student artifacts from the unit. We analyzed the data to understand the ways in which care played a role in teaching and to understand the explicit pedagogical techniques that were used during this unit. Although there are many examples in my teaching that I draw upon when examining this relationship between EE and caring, there is one example that exemplifies the ways in which students can care for the environment in ways that are similar to Noddings' ethic of care. In the vignette below, I describe how I nurtured the ethic of care with my students.

On a particularly hot March day, the students voiced a concern about the residue left on their desks from the frequent mosquito spraying that occurred in the school. As one student dragged her finger along the lab bench and then showed her classmates the oily residue, she stated, "this cannot be good for us." Coincidentally, we were beginning individual inquiry projects, and the students asked if they could research about the mosquito spraying and examine if there would be alternative approaches to spraying that would have the desired outcomes (less mosquitos).

As a teacher who readily incorporated place-based/project-based instruction, I seized the opportunity to begin a project that was not only student-driven but also had the potential to impact the health of the school and the local environment. The students invited the pest control company, Pest Destroyers (a pseudonym), to class and asked them to present the types of chemicals used to learn more about their pest control practices. Importantly, the students created a list of pre-questions surrounding content necessary to hold a meaningful discussion with Pest Destroyers employees. Some questions included, "what is the life cycle of a mosquito?" and "what diseases are spread by mosquitos?" and "what is the anatomy of a mosquito?" After the exploration into the current research, we held a discussion with the employees who presented the types of chemicals used (i.e., cypermethrin, thiamethoxam, and chlorpyrifos). The students discovered most of the chemicals were sprayed directly into drains despite warnings not to use them in any water source for potentiality of contamination of water sources. Additionally, almost all of the chemicals warned against use in areas where children were present. During our discussion, Pest Destroyers workers admitted they felt the chemicals were being overused and had warned the school board about the risk of chemical resistance in mosquitos. Despite these concerns, the chemicals were being sprayed, misted, or spread in crystal form, twice a week.

As the discussion came to a close, the students began to conduct investigations to understand the toxicity levels found by the World Health Organization, and alternative uses for chemicals, as a way to rethink the solutions to the health risks of the chemicals. By using various sampling techniques, the students determined the relative densities of mosquitos (both larval and adult) in different habitat types around the school. The focus was on the Aedes mosquito as this is the primary way the dengue fever is spread, which proved to be a major health concern for the community. In this way, a portion of the study was on identification of Aedes and learning the structures of the insect. After obtaining initial data of the relative densities, the students plotted the largest densities on a map to look for patterns.

It was during this conversation about patterns that I realized the students were doing more than learning about scientific content; they were learning to care about the potential risks of the chemical and search for alternatives. It did make me wonder, however, what was the basis for this care. Noddings (1984) notes that proximity is the most powerful determinant of caring behavior.

Would the students have cared if the spraying was occurring in Africa? Or did they only learn to care because it was happening to them and to the people they cared about? As Martin (2007) notes, this form of care can provide a foundation for EE. For example, proximity to the issue they cared about created a space that could be purposefully fostered who found that young people often describe the environment as a place that is essentially separate from themselves. As the students began to care for the environment, most likely due to the proximity of the issue, I was nurturing an ethic of care with my students. I was guiding the students to be critical as they apply values, make decisions, and take action.

Then the students began what we call "care-based scientific work," work that we define as developing of critical environmental consciousness based on the evidence garnered during scientific investigation focused on care. The goal of this work was to reduce the numbers of mosquitos by covering drains and creating systems for drying up pooled water. Then they created a list of high-risk zones, areas that were still problematic, and where intervention was needed. The students recommended landscaping and draining to prevent further pooling of water—a breeding place for Aedes.

After a 3-month investigation, the students analyzed their data and found powerful results. They were able to reduce the amount of mosquitos in a target area by almost 30%, which was above the reduction rate of Pest Destroyers. In addition, because the students were also preventing the breeding of mosquitos instead of reducing the number of live adults, they predicted the numbers to decrease continually. Despite these successes, the exact prediction could not be calculated precisely given our small sample.

The students then presented their findings and alternative solutions at the school board meeting. The school board listened as the students presented and then asked questions about dengue fever and the ability to target the Aedes mosquito. The students responded with evidentiary support about the ability to remove the breeding grounds by reducing the amount of standing water. The school board persisted and stated they need a guarantee the Aedes mosquitos would be killed. The school board president agreed, "In the 8 years we've used Pest Destroyers, we've never had a report case of Dengue on campus." At the end of the conversation, as their science teacher, I was proud. The students effectively and clearly communicated their results and responded to the questions with scientific evidence. Despite this success, I realized I had not prepared my students for one element of caring so deeply—rejection. We were so thrilled by our results that I forgot our audience was a group of businessmen and women who were running a profitable school. As we walked back to the classroom, I could see the disappointment on my students' faces and realized that care-informed practices required something else of students—resilience. As an environmental scientist, I understand the resistance toward change but also understand the need for persistent activists.

13.4 Understanding Care-Based Influences in Science: Unpacking the Investigation of Aedes Mosquito

After reflecting on this teaching experience, I wondered how do we evoke persistence of activism in our science classrooms, and does caring for the environment help to strengthen that persistence? A person's incentive to pursue a task or outcome depends on their efficacy, their belief that they can produce the desired effects (Bandura 1997). Efficacy also affects the resilience and motivation to continue pursuing a desired outcome. We cannot possibly control the outcomes of our students' activism and will not be able to prevent rejection. Instead we can work toward building the efficacy of our students. There are several sources of efficacy teachers can build in order to encourage their students to continue working toward preserving the environment even in the face of the rejection. The first source for increasing the efficacy of a group is social persuasion. Social persuasion occurs when other people either increase or decrease a person's sense of confidence in their ability to succeed (Bandura 1997). As teachers we can serve as this source of encouragement for our students. Stories of our own personal successes at affecting change or of others we know can help build our own students' confidence. A second source for increasing the efficacy of a group is trusting relationships. If your class trusts you and each other, these trusting relationships can boost the group's efficacy to work at affecting change even if faced with setbacks (Bandura 1997). Studies show that activating social ties helps increase a group's efficacy and ability to achieve desired outcomes (Sampson et al. 1999). Close-knit communities are more likely to achieve higher collective efficacy in their group's ability, and a strong sense of camaraderie can greatly impact the success of a group to work together (Bandura 1997). Caring for our students and encouraging them to care for each other can help build the type of community and trust that will build efficacy. Coming together as a class and working toward effecting change can evoke persistence as environmental activists.

During these care-informed practices in my classroom, I noticed student voice, belonging, and agency was prevalent. For example, the student's voice appeared in the choice of the topic, the way in which they conducted the study, and the legitimacy and relevance of situation to their lives. This was a real situation that affected the students. Noddings (2002) notes that the emotional engagement of students can be invoked by basing activities in the real interests of the students. Social justice was present in the form of student agency—the act of influence and effect in a specific circumstance (Larson et al. 2005)—in the opportunity to be agents of change in their school by working to improve the conditions in their school.

Dana Mitra and Stephanie Serriere (2012) found students, particularly middle schoolers, valued their schooling the most when teachers privileged their voices and created a sense of belonging. This concept of belonging includes developing relationships marked by supportive, positive experiences with teachers and peers. This includes opportunities to learn from one another. As feelings of belonging increase, students attach to peers and teachers but also their broader community. This is critical to care-based scientific work as they are situated in local, relevant socio-scientific or "STES" issues (Sadler 2009). This sense of belonging helps to motivate students and provides a support system when faced with challenges in their activism (Mitra and Serriere 2012). Two days later, when I walked into school, I witnessed this persistence. Posted on almost all of the classroom doors were handmade signs that read, "No Pesticide Zone. Children Learning" and "No Spraying in Our Classroom." Without my knowledge, the students had talked to all the classroom teachers and presented the information they discovered about the toxicity of the chemicals used in the spraying and the risks of long-term exposure to high-risk groups such as children. One by one, the teachers agreed and posted signs asking the company not to spray. The school board ignored the resistance, but the company listened. The company stated that it did not feel comfortable spraying in areas where employees were asking them not to do so. As a result, the school board issued a statement, "Teachers have the authority to not have his/her classroom sprayed. The pest company will not spray in any classroom with a sign. The teachers do not have authority beyond their classroom walls. Thus, any shared space, such as offices, cafeterias, etc. will be sprayed according to the guidelines established by the pest company."

When discussing this victory with the students, they felt vindicated but understood the decisions the school board made were not guided by scientific evidence the students presented, but rather their own experience. This weighing of scientific evidence is typical of environmental debates that groups pull out different evidences to support their ideas—and these differences in interpreting scientific evidence are likely to be amplified when the results of inquiry have political, religious, or economic ramifications (Oreskes 2004). The school board was deeply concerned with dengue fever, a life-threatening disease particularly for young children. This frustrated the students but also motivated them. They talked about a sense of belonging to a problem that was situated squarely in their school, and armed with their voice, knowledge, and a sense of obligation to continue, the students persisted despite the school board's decision.

13.5 Translating an Ethic of Care for Environmental Education

In this section, we discuss the implications for nurturing an ethic of care in the context of EE. One strategy for motivating students to care for the environment is to appeal to consequences and fear. According to Noddings, care is not likely to be established by appealing to fears of a faraway and doomed future (Noddings 2013). This fear of ecological problems, known as ecophobia, asks students to deal with problems outside of their control (Sobel 1996). If students see environmental problems as faraway events that are out of their control, this will affect their motivation to care and to act. An ethic of care is best fostered in local or personal spaces where children are able to connect caring for nature to their everyday lives (Noddings 2006). If concepts remain academic, global, and abstract, children are removed emotionally from the issue (Kollmuss and Agyeman 2002). Issues that impact their own backyard will have greater value (Sobel 1996). Focusing on local issues provides opportunities for students to feel the impact of caring for the environment as it affects their own wellness (Willem Postma and Smeyers 2012).

Interestingly, the students during the mosquito program developed this sense of belonging to the problem without my influence. During this investigation, I did not teach them what to care about—they already cared; however, I guided them to see how what they cared about was connected to the environment and their own and others' actions. It is difficult for a student to attach importance to something with which they have little experience. Care is more likely to be generated if students can relate to a problem (Gruenewald 2004). As Noddings explains, in an ethical care, which involves obligation to act, we cannot possibly care for every person or every "other." There are limitations to our obligation to care. The possibility of a completed relation is the primary limitation for a person feeling obligated to care and act. This concept illustrates well the inefficiency of focusing on faraway, global environmental problems.

If we present to our students an issue such as climate change and talk about this issue solely in reference to the polar bears and melting of ice in faraway places, there is very little potential for present relation. The student may feel the initial "I must," but if no relation is established, then the feeling of care is nothing more than a hypothetical caring. "I must" if I am able to, but I see little I can do. The obligation to care grows as the cared-for responds (Noddings 2013). So even if a child genuinely cares and feels obligation for a faraway environmental issue, their obligation to act will wane if they perceive no response or are unable to see any response from the one they are caring for. Focusing on authentic issues in local communities allows students to see the impact of their actions on their own wellness (Willem Postma and Smeyers 2012). Students are often motivated to care and act on issues affecting them, their families, and own spaces than with seemingly faraway issues. These local issues are felt by students as they affect the well-being of the students and the ones they love (Macnaghten 2003).

Teaching students to care is also done best in the present context. Dirick Willem Postma and Paul Smeyers (2012) refer to this concept as developing zenship in environmental education. One of the most distinctive characteristics of zen is the focus on the present moment. In the Western view of time, much effort is dedicated to dissecting actions of the past in an attempt to learn lessons and project what our future will look like based on those actions. Many times we neglect the present or pay little attention to it. Rather than focus our efforts on what the future will look like if our students do not start caring, we should nurture a childlike care and enjoyment of the present moment and experience with nature (Noddings 2013). Zenship involves spending time experiencing nature now. Providing students with time to feel well and happy within nature will allow for a deeper emotional connection to nature; care becomes a natural by-product (Willem Postma and Smeyers 2012).

When students care, they often face barriers. But during these barriers, leadership can provide pathways to persist. Dana Mitra and Stephanie Serriere (2012) explored student voice in an elementary school, highlighting the case of "Salad Girls," three young students who evoked changes in the food choices at their school cafeteria. By conducting an inquiry that involved gathering data school-wide and communicating the results to the decision-makers at the school, the girls were ultimately able to galvanize changes in the school menu. Similar to my students, the Salad Girls faced standoffs. However, unlike our situation, the Salad Girls had the support of the principal, who encouraged them to meet with district-wide officials. So, why did my students in the mosquito project continue to fight without a clear conduit toward change?

As I reflected on what motivated them to continue their quest, I realized the care-informed approach, including opportunities for students to learn about their ecological home—their school. The investigation into pesticide use and mosquito behavior armed them with knowledge and gave them agency to meet with the school board. Even though the school board refused to listen to the student-garnered evidence, the students told me they felt they had a responsibility to improve their community. In one student's reflection, she stated, "Once I understood the dangers of the chemicals, the long term effects, and that there are other ways to reduce the mosquito populations, I felt that I should do something." They were resilient. The content knowledge of environmental education acting alone, in the absence of opportunities for youth to gather knowledge about their community, cannot be expected to create youth that adapt to changes around them and to make better uses of their resources. Without these opportunities, we cannot expect resilience needed to persist when faced with resistance.

If connecting care to the environment helps students to see relevance in environmental issues and provides them with pathways to persist through the challenges of this work, an ethic care can be nurtured through care-based scientific work. However, this is not the same as indoctrination of belief systems. It involves arming students with the knowledge to see how what they care about is connected to the environment. As every community is unique, the problems they are facing and what they care about is also unique. Environmental educators need to give students opportunities to spend time in their environment and see how the problems of their communities are connected to their own actions and the environment. When students learn to care about each other and nonhuman nature, they begin to develop their own personal environmental ethic that will guide their actions rather than just knowledge about environmental ethics. Spending time personally experiencing nature will provide students the opportunity to emotionally connect with environmental issues and deepen their involvement (Goralnik et al. 2012).

Teaching while operating under an ethic of care involves modeling, dialogue, practice, and confirmation (Noddings 1988). For example, teachers *model* an ethic of care when they encourage accountable self-affirmation in their students (Noddings 1988). In the care-based scientific work, we modeled the scientific inquiry process during the mosquito investigation but also emphasized the human relationships to this investigation by linking the investigation to the health concerns of the students' community. Teachers also serve as models through demonstrating to their students what a responsible citizenships looks like in practice (Fien 1997). Concern for the natural world is not only nurtured but can also be inspired (Palmer and Suggate 1996). As we reflect back on Noddings' notions of an ethic of care, recall that the source of the felt obligation to act, the "I must," is the remembrance of caring relations. If the student does not naturally care, then the students' obligation to care will be drawn from times in their own life that they were cared for or observed others exhibiting caring relations. If students do not have strong remembrances of being cared for themselves or of seeing others model care, both care for other humans and for the environment, then they cannot feel a strong obligation to care. Therefore, the first way a teacher can model an ethic of care for their students is by caring for their students. Respectfully reach out to your students and meet them as one cared-for. Establishing caring relations with your students is foundational to nurturing within them an ethic of care for the environment (Noddings 1995).

Love and caring relations are necessary in the arena of education. In environmental science classrooms, as in any class, caring and expressing this care to your students is a "pedagogical necessity" (Gay 2002). The final way teachers model caring is by demonstrating desirable ways of interacting with the environment. If students observe their teacher actively engaged in caring for the environment, when they encounter situations in which the "I want" to act is not there, the teacher's demonstration of caring for the environment will increase the student's felt obligation to care and to act. Both in and out of the classroom, teachers need to let "qualities of care, empathy, concern, and understanding" guide our practices (Jickling 2003, p. 22).

Noddings mentions *dialogue* as the second component teaching care. In her approach, the dialogue is open, and the conclusions are drawn from multiple sources and through this dialogue. In our care-based scientific work, these discussions happened throughout the inquiry and provided opportunities for learning and further inquiry. An open dialogue is one in which the teacher does not hold the conclusion. Instead teacher and students together search for a fuller understanding and for what the responsible choice may be (Noddings 1995). Building in indefiniteness into the projects undertaken by our classes and the conversations we have will allow students to move beyond our goals and develop their own opinions (Jickling 2003). Conversations that challenge students to reflect and critically apply learned values in order to make decisions about what actions they "ought" to take will help develop a critical environmental consciousness (Fien 1997). As their teacher, Cassie never told the students what action to take; rather through discussions, the class decided to investigate alternatives to chemicals, alternatives that would still control mosquito populations but not at the risk of their communities' health.

Students should also be given opportunities to *practice* caring and working for the well-being of the environment (Noddings 1995). In this way, the care-based approaches are provided and nurtured so that students have opportunities to engage in this work. In the mosquito project, the students cared for the other students in the school and despite not being able to change the school-wide policy. The students advocated for the school to stop spraying chemicals out of concern for their and other students' health. In this way, they had the opportunity to practice caring for their school community. An ethic of care is not a set of principles to be learned. It is a way of living that must be practiced. Experience and emotion are allies in the process of understanding (Johnson and Fredrickson, 2000). Teachers need to be proactive in planning experiences that promote caring for the environment. As noted in the mosquito project, creating spaces for social action are useful opportunities to develop an ethic of care for the environment (Fien 1997).

Similarly, according to Noddings, students need *confirmation*—confirmation that they are acting ethically. Upon reflection of teaching care-based EE, the students needed more support was confirmation. As noted previously, I was viewing confirmation as indoctrination instead of as a component of teaching an ethic of care. Confirmation involves presenting students with the ideal image of themselves, giving students opportunities to see themselves as ethical persons devoted to caring (Noddings 1995). "In education, what we reveal to a student about himself as an ethical and intellectual being has the power to nurture the ethical ideal or to destroy it" (Noddings 1984, p. 193). Giving the students an opportunity to advocate for a change in the amount of chemicals used allowed students to see themselves as being part of the effort to conserve nature and care for the well-being of those around them.

Confirmation is such a key component of developing an ethic of care as it provides students with a picture of what care looks like and they are able to see themselves as being a part of that picture.

13.6 Implications and Complications of Teaching Care for the Environment

In the preceding sections, we conceptualized an ethic of care in EE, described what it looks like to nurture care in an environmental science class, and described components of an ethic of care. Teaching an ethic of care in EE allows students to develop a personal, intimate relationship with the environment, which in turn can act as a basis for protecting and preserving it. According to Martin (2007, p. 62):

> *Caring as environmental education demands that students work at getting to know nature. Caring demands that a sense of proximity be created by having students engaged in experiencing, learning and sharing time with nature in the same sorts of ways we might get to know a new friend. In such caring for nature, students need to understand their relatedness to the* environment *as a subjective relationship, individual to individual.*

We agree with Martin and believe traditional, indigenous approaches to knowledge offer valuable insight on how to foster such a personal, emotional connection with nature. A colleague, who partners with tribal communities in scientific research, shared the words of advice she received from a member of a local tribal community:

> *We don't put a big premium on everybody knowing everything about everything. Because in our spiritual approach to nature, if you go very specifically into learning a lot about a particular being, then you're taking on a spiritual responsibility to that being.* (Personal communication with Leslie Allee, 2015 about her experiences with Kateri tribal community members)

This approach to caring for nature is in opposition to the view of nature as an external, unknowable object. Care-based EE has the goal of teaching a rational understanding of nature but of equal importance is the goal of nurturing within students an emotional understanding and relationship with the nonhuman individuals of our environment (Martin 2007). Before we ask people to act for the good of the environment, we need to nurture a deep care and responsibility for nature. Love and the emotional connection to nature is a precursor to attentiveness to nature; love fosters a responsibility for nature (Elder et al. 1998).

Care-based environmental education provides another pathway. An ethic of care inherently utilizes positive emotions (care and love) as part of developing ethical thinking and provides affirmative educational experiences for students (Nazir 2014). However, this work is not easy. There are challenges to the care-based scientific work. To truly teach an ethic of care, teachers need to understand students' interests and values to connect students to the environment. In describing the components of an ethic of care (modeling, practice, dialogue, and confirmation), we found confirmation to be the most difficult to incorporate in our teaching. But as Noddings (1988) notes, confirmation needs an examination of the context of the learning.

To provide confirmation of care-based work to their students, teachers need to have clear expectations, and those expectations should be related to understanding our students. In this way, we must provide teachers with more opportunities to learn about their students. These types of opportunities would provide our teachers with time to get to know their students' hopes, fears, and dreams—and to do this, teachers need time. Despite its importance, this type of time is rarely afforded to our teachers.

The vignette highlights another challenge of care-based work, the foundational support needed to teach an ethic of care. During the mosquito project, the leadership (school board) did not support the work, and when this happens, students can become disengaged. Even though in our situation, this did not happen, there are many other cases that describe the disengagement of students because of inability to induce change. Therefore, an ethic of care is most effective if the school has the leadership to support its work. When this happens, EE emerges from these foundations, so that in the care-based scientific work activities, care is not an add-on but rather a place from which their practice continues (Mitra and Serriere 2012).

13.7 Future Directions for an Ethic of Care in Environmental Education

Care-based EE well aligns with standards-based reforms that emphasize the development of critical thinking and scientific inquiry skills while learning scientific content. For example, the Next Generation Science Standards (NGSS) in the United States move away from a focus on facts, skills, and efficiency of operations and move toward a deeper understanding of issues. NGSS emphasize the learning of science content integrated with real-world problems, problems relevant to students' lives. This emphasis on real-world context creates a deliberate educational experience embedded in the lived experience of students, which is a vital component to nurturing within students an ethic of care (Martin 2007). In addition, NGSS place great value on developing within students the skills to pose questions, define problems, plan and carry out investigations, and analyze and interpret data concerning sustainability and the environment. These standards support teachers collaborating with students to design scientific investigations of environmental issues in their own communities with the goal of developing citizens who are able to act responsibly in a complex society. In accordance with care-based EE, students are provided opportunities to design solutions, engage in arguments from evidence, and obtain, evaluate, and communicate information. Standards-based reforms emphasizing critical thinking and real-world problem solving can provide a pathway to developing within students a critical environmental consciousness.

Our future world does not need citizens who can spew off facts about the environment but have little motivation to act. Our future world needs problem solvers who can think critically and creatively and who care enough to do so (Holt 2002). The environmental problems students will solve are not scripted and will be context bound. Critical thinking will be vital as will be the ability to consider alternative lines of action and how one should act. Current major environmental issues and their associated consequences are complex, with many conflicting interests and politics involved. The persistence and strength it takes to work toward preserving the environment require more than a cognitive knowledge of environmental concerns. These concerns for the well-being of the earth must be felt by the heart and will be developed by nurturing a deeply felt care.

References

Allee L (2015) Personal communication on February 12, 2015.

Bandura, A. (1997). *Self-efficacy: The exercise of control*. New York: Freeman.

Cook-Sather, A. (2002). Authorizing students' perspectives: Toward trust, dialogue, and change in education. *Educational Researcher*, *31*(4), 3–14. http://dx.doi.org/10.3102/0013189X031004003.

Dirkx, J. M. (2001). The power of feelings: Emotion, imagination, and the construction of meaning in adult learning. *New Directions for Adult and Continuing Education, 2001*(89), 63–72.

Elder, J., Basnage, M., Caswell, K., Danish, J., Dankert, B., & Kay, J. (1998). *Stories in the land: A place-based environmental education anthology*. Great Barrington: The Orion Society.

Fien, J. (1997). Learning to care: A focus for values in health and environmental education. *Health Education Research, 12*(4), 437–447.

Fien, J. (2003). Learning to care: Education and compassion. *Australian Journal of Environmental Education, 19*, 1.

Gay, G. (2002). Preparing for culturally responsive teaching. *Journal of Teacher Education, 53*(2), 106–116. http://dx.doi.org/10.1177/0022487102053002003.

Goralnik, L., Millenbah, K. F., Nelson, M. P., & Thorp, L. (2012). An environmental pedagogy of care: Emotion, relationships, and experience in higher education ethics learning. *Journal of Experiential Education, 35*(3), 412–428.

Gough, A. (2002). Mutualism: A different agenda for environmental and science education. *International Journal of Science Education, 24*(11), 1201–1215.

Gruenewald, D. A. (2004). A Foucauldian analysis of environmental education: Toward the socio-ecological challenge of the earth charter. *Curriculum Inquiry, 34*(1), 71–107.

Gruenewald, D. A., & Smith, G. A. (Eds.). (2014). *Place-based education in the global age: Local diversity*. New York: Routledge.

Holt, M. (2002). It's time to start the slow school movement. *Phi Delta Kappan, 84*(4), 264–271.

Ingold, T. (2000). *The perception of the environment: Essays on livelihood, dwelling and skill*. London: Psychology Press.

Jickling, B. (2003). Environmental education and environmental advocacy: Revisited. *The Journal of Environmental Education, 34*(2), 20–27. http://dx.doi.org/10.1080/00958960309603496.

Jickling, B., & Spork, H. (1998). Education for the environment: A critique. *Environmental Education Research, 4*(3), 309–327.

Johnson, B. L., & Fredrickson, L. M. (2000). "What's in a good life?" Searching for ethical wisdom in the wilderness. *Journal of Experiential Education, 23*(1), 43–50.

Johnson, E. A., & Mappin, M. J. (2005). *Environmental education and advocacy: Changing perspectives of ecology and education*. Cambridge: Cambridge University Press.

Kelly, G. J., Carlsen, W. S., & Cunningham, C. M. (1993). Science education in sociocultural context: Perspectives from the sociology of science. *Science Education, 77*(2), 207–220. http://dx.doi.org/10.1002/sce.3730770208.

Kollmuss, A., & Agyeman, J. (2002). Mind the gap: why do people act environmentally and what are the barriers to pro-environmental behavior? *Environmental Education Research, 8*(3), 239–260.

Larson, R., Walker, K., & Pearce, N. (2005). A comparison of youth-driven and adult-driven youth programs: Balancing inputs from youth and adults. *Journal of Community Psychology, 33*(1), 57–74.

Li, J., & Ernst, J. (2014). Exploring value orientations toward the human–nature relationship: A comparison of urban youth in Minnesota, USA and Guangdong, China. *Environmental Education Research, 21*(4), 556–585.

Longino, H. E. (1990). *Science as social knowledge: Values and objectivity in scientific inquiry*. Princeton: Princeton University Press.

Loughland, T., Reid, A., Walker, K., & Petocz, P. (2003). Factors influencing young people's conceptions of environment. *Environmental Education Research, 9*(1), 3–19.

Lucas, A. M. (1980). The role of science education in education for the environment. *The Journal of Environmental Education, 12*(2), 33–37.

Macnaghten, P. (2003). Embodying the environment in everyday life practices. *The Sociological Review, 51*(1), 63–84.

Martin, P. (2007). Caring for the environment: Challenges from notions of caring. *Australian Journal of Environmental Education, 23*, 57.

Mayeroff, M. (1971). *On caring*. New York: Harper and Row.

Mitra, D. L., & Serriere, S. C. (2012). Student voice in elementary school reform examining youth development in fifth graders. *American Educational Research Journal, 49*(4), 743–774. http://dx.doi.org/10.3102/0002831212443079.

Namenwirth, M. (1986). Science seen through a feminist prism. In R. Bleier (Ed.), *Feminist approaches to science* (pp. 18–41). New York: Pergamon.

Nazir, J. (2014, June). Applying an ethic of care to environmental education: Insight from a study of outdoor educators. In *2014 conference of the Canadian Society for the study of education*. St Catherines.

Newfield, J. W., & McElyea, U. B. (1984). Affective outcomes, indoctrination and the use of case rhetoric in curriculum guides. *Journal of Curriculum Studies, 16*(1), 100–102.

Noddings, N. (1984). *Caring: a feminine approach to ethics and moral education* (2nd ed.). Berkeley: University of California Press.

Noddings, N. (1988). An ethic of caring and its implications for instructional arrangements. *American Journal of Education, 96*, 215–230. http://dx.doi.org/10.1086/443894.

Noddings, N. (1992). *The challenge to care in schools: An alternative approach to education. advances in contemporary educational thought* (ERIC, Vol. 8). New York: Teachers College Press.

Noddings, N. (1995). Teaching themes of care. *Phi Delta Kappan, 76*(9), 675.

Noddings, N. (2002). *Educating moral people: A caring alternative to character education.* Williston: Teachers College Press.

Noddings, N. (2006). *Critical lessons: What our schools should teach.* Cambridge: University Press.

Noddings, N. (2012). The language of care ethics. *Knowledge Quest, 40*(5), 52–56.

Noddings, N. (2013). Caring: A relational approach to ethics and moral education. In R. Shafer-Landau (Ed.), *Ethical Theory: An Anthology* (2nd ed., pp. 699–712). West Sussex: John Wiley and Sons, Inc.

Oreskes, N. (2004). Science and public policy: what's proof got to do with it? *Environmental Science & Policy, 7*(5), 369–383.

Palmer, J. A., & Suggate, J. (1996). Influences and experiences affecting the pro-environmental behaviour of educators. *Environmental Education Research, 2*(1), 109–121. http://dx.doi.org/10.1080/1350462960020110.

Sadler, T. D. (2009). Situated learning in science education: Socio-scientific issues as contexts for practice. *Studies in Science Education, 45*(1), 1–42. http://dx.doi.org/10.1080/03057260802681839.

Sampson, R. J., Morenoff, J. D., & Earls, F. (1999). Beyond social capital: Spatial dynamics of collective efficacy for children. *American Sociological Review*, 633–660. http://dx.doi.org/10.2307/2657367.

Shuck, B., Albornoz, C., & Winberg, M. (2013). Emotions and their effect on adult learning: A constructivist perspective.

Siegel, H. (2013). *Educating reason*. New York: Routledge.

Sobel, D. (1996). Beyond ecophobia: Reclaiming the heart in nature education. *Clearing, 91*(16), 20.

Tilbury, D. (1995). Environmental education for sustainability: Defining the new focus of environmental education in the 1990s. *Environmental Education Research, 1*(2), 195–212.

Venter, E., & Ferreira, J. G. (2014). A plea for environmental education that focuses on learning to care. *The Journal of Human Ecology, 46*(1), 33–38.

Weiss, R. P. (2000). Emotion and learning. *Training and Development, 54*(11), 44–48.

Willem Postma, D., & Smeyers, P. (2012). Like a swallow, moving forward in circles: On the future dimension of environmental care and education. *Journal of Moral Education, 41*(3), 399–412. http://dx.doi.org/10.1080/03057240.2012.691637.

Cassie F. Quigley is an associate professor of science education at Clemson University. Her research works to expand the participation in and perspectives of science. In doing so, her work is community- and place-based. To date, she has explored these questions in a variety of settings, focusing attention on specific places to potentially broaden the scope of whose knowledge garners value in scientific and science education communities.

Renée Lyons is a doctoral candidate in curriculum and instruction at Clemson University. Her research explores how science projects and educational experiences can become third spaces merging the discourse, practices, goals and values of the world of science with the world a person experiences outside of science. The goal of her research is to broaden participation in science by creating new forms of participation in science, forms which present participants with a vision of how participating in science fits in the larger context of their lives.

Chapter 14
Afterword: Science Education and Promises of Aesthetics, Emotion and Wellbeing

Steve Alsop

I sense a shared imaginary at the heart of this publication. It is science education in its fullest, most variegated and vibrant sense: science education as an expression and celebration of the promises, hopes and frailties of humanity's cultures. The arguments assembled are not interested in some reductionist and functionalised science training: a type of education unduly constricted by tick boxes and jurisdictional standards. After all, science education is not some anodyne process of thought control. It can never be reduced to a pre-packaged and predetermined technical exercise—downgraded to the status of a so-called curriculum 'deliverable'. Perhaps science education might be more hopefully conceived as an opening accompanied with an ethical invitation into a world of insight, beauty and delight: a world of coherence and opportunity. This is the science that has occupied many hearts and minds for such a long time. It is a science education that many of us strive for in our practices of teaching and research. When we think of this science education, we actively embrace the allure, the desires, the mysteries, the structures and the suspense of seeing and acting differently in dynamic, ever-emerging relationships with the world.

This science education is a roller coaster of adventure. Turning to W. B. Yeats' famous aphorism, this 'education is not the filling of a pail, but the lighting of a fire'. For Virginia Wolff (1972), it is an education that pushes back the 'cotton wool of daily life', enabling those involved to escape the mundane and the drab, in pursuit of hidden treasures, of patterns of connectivity within *Moments of Being* (also the title of her book). Such moments are, perhaps, never fully predictable. They are necessarily intelligible, often they are salient and germane, and very occasionally they are sublime.

S. Alsop (✉)
Faculty of Education and Department of Science and Technology Studies, York University, Toronto, ON, Canada
e-mail: salsop@edu.yorku.ca

A. Bellocchi et al. (eds.), *Exploring Emotions, Aesthetics and Wellbeing in Science Education Research*, Cultural Studies of Science Education 13,
DOI 10.1007/978-3-319-43353-0_14

The intensities of these moments give birth to wondering and to sensations and experiences that can shape both science and who we are, as well as what we might hope to achieve both now and in the future. Science education viewed in such a way is integral to personal and social development. It is a basis to join with others, releasing our collective and collaborative conceptual, emotional and ethical energies in pursuit of a world that is increasingly responsive, more caring, more equitable and just.

This is the science education that I want. It appears when I read Primo Levi's (1984) *Periodic Table*; or Stephen Jay Gould's (2000) *Wonderful Life*; or Evelyn Fox Keller's (1984) *Feeling for the Organism: the life and work of Barbara McClintock*; or Steve Jones' (2007) *Coral*. Perhaps, it is best conceived as an event horizon, as something I can aspire to and continuously struggle towards. By the way, I don't believe that I am alone with this imaginary. I sense I share it with multiple others—with many teachers, teacher educators, students and scientists whose daily practices aspire to help learners dream a little longer. This science education, by definition, is never normal, but always fluid and overspilling. It is never static and monochrome, but endlessly shifting colours that retain an enticing and provocative shine. This science education offers multiple opportunities for all those involved to change, to break with the 'taken for granted' or the 'natural attitude' and to 'impose different orders upon experience' (Greene 2001, p. 5).

In pursuit of this imaginary, the preceding essays are a testament to the generative possibilities of emotions, aesthetics and wellbeing. They extend previous research, offering new avenues for study by casting light on sciences and educations in all their subjective intensities, shimmers and nascent forms. By doing so, they open up new vistas and political possibilities for science education. I would like to thank the editors for this opportunity to add my words to this collection. I am delighted to be part of this book and greatly enjoyed an associated role as discussant in an ESERA research seminar in Helsinki, where earlier drafts of chapters were made public.

14.1 An Angle of Arrival

Everything, the affect scholar Kathleen Stewart (2010, p. 339) writes, 'depends on the feel of an atmosphere and the angle of arrival'. My interest in affect—my angle of arrival, if you like—is longstanding. Much of my earlier work emerged out of a desire to better understand living and learning within the heated emotional context of radioactivity and nuclear fear. I was a very young physics teacher in London at the time of the Chernobyl nuclear disaster and felt at a complete loss for how my professional practices could/should respond. Chernobyl changed my science education and fuelled my subsequent doctorate study that explored education within a seemingly idyllic rural village in Somerset, England, which was afflicted with very high levels of background radiation (Alsop 1999). I followed this with a collaborative study responding to a tragic nuclear accident in Brazil involving children and an abandoned radiotherapy machine (Watts et al. 2007). In these studies, what became abundantly clear was that the dominant psychological and sociological models of learning circulating in science education at the time—to which I had been dutifully introduced—were ill-positioned for thinking about science and education in these (and other) contexts of intense concern. Inspired by Mike Watts and others, I followed this work with a series of studies of affect in school contexts, including a special issue of a journal (Alsop and Watts 2003) and an edited collection of essays on the theme of Cartesian dualism (Alsop 2005).

On reflection, the promise of affect within these studies was partly epistemological: a desire to develop a more holistic and unabridged representation of education in particular science contexts. It was also pedagogical. I became convinced that affect offers a basis for renewed and enhanced practices of pedagogy. *Affective science education* needs, above all else, to better understand and work with *affect in science education*. This is a touchstone that I still grasp tightly today.

Of late, my thinking has been drawn to the 'affective turn' (Clough 2007), and this has resulted in questioning many of the assumptions of my physics degree and my early science education research (see Alsop 2014, 2015). I suspect that I am not alone, when I now feel more inclined to focus on the possibilities of rethinking ontologies in science education with affect, rather than adding to conceptually orientated epistemologies. My present preoccupation (Alsop 2011) is thinking about 'an ability to affect and be affected' (following a tradition starting with Baruch Spinoza and William James) (Deleuze 1988). This seems so simple and innocent—just seven words—and yet when I think with it, everything seems to change. In what follows, I offer some reflections on the collection, exploring ways in which arguments herein affect (and are affected by) my science educations.

14.2 Assembling the Collection

There is no doubt that this collection tackles questions that are of fundamental importance in contemporary science education. Few students, teachers and researchers, I believe, would disagree. What promises and prospects might aesthetics, emotion and wellbeing bring our sciences and educations? How might they affect (and be affected) by our theories and pedagogies? What *should* we understand as aesthetics, emotions and wellbeing science education? What could be possible if researchers, teachers and learners were more attuned to the aesthetics, emotions and wellbeing that they encounter in science classrooms? What are the aesthetics, emotions and wellbeings that might bring about the changes that we urgently desire at this moment of time? There are so many powerful questions percolating within this collection.

As teachers and researchers, perhaps our best starting points for deep-rooted change are to create a heightened sensitivity and awareness both within ourselves and within our practices of research and teaching. In order to lead others, we need to develop sensitised and more discerning capacities with aesthetics, emotion and wellbeing (A-E-W from now on) *in* science and education (I actually prefer *as* science education, because it is more provocative). In this way, our theoretical and empirical narratives of A-E-W can amplify our awareness and direct our attentions to new vistas of experience, such that we see, feel, touch, move, taste, hear, act and react in science education in different ways. This is the bigger picture—it is about tuning-in and shifting the sensory dial with prospects of realising and releasing different pedagogies, sciences and educations.

Affect theorists often write about such awakenings as orientations—being moved towards or away from something, by allowing something to come closer or to fade into distance. Others prefer the term 'intensities' (Betelsen and Murphie 2010) or 'bloom spaces' (Stewart 2010) or 'shimmers' (Barthes 2005). Sarah Ahmed (2010, p. 29) prefers 'sticky'. Affect, she writes, 'is what sticks, or what sustains or preserves the connection between ideas, values and objects'. In this manner, perhaps, some of the promises of A-E-W (as outlined in this collection) are in their capacities to direct us towards something different and generative, whether it is knowledge, feelings, practices, objects or new ways of being in science education. Some things, some movements, some utterances and some feelings now standout; they gain intensity, shimmer or become stickier in some sense. Our capacities thereby become more discerning; we notice (and can invite others to notice) what was previously eclipsed, or perhaps only partly in-shade. This is changing classrooms in subtle and very occasionally rather more dramatic ways. Surely we are better positioned to respond if we can walk into classrooms with enhanced capabilities to notice, feel and react to A-E-W?

This is how I wish to approach this text, by focusing on the promises of affecting and being affected in some way. I am immediately confronted with multiple discourses with possibilities. The chapters herein assemble under contrasting stories, assumptions and goals. Indeed, throughout this process, A-E-W becomes rather slippery and difficult to pin down: are they precepts or concepts? Mental schema? Experiences? Rituals? Openings? Outcomes? Are they inner states of the mind, or sensations of the body? Are they outer expressions of social identity? Are they all or some of these, or something else entirely? Whatever 'they' are, they just can't keep still! Perhaps in this way, they invite encounters that are less structured, more open and more exploratory. This is my desire, at least.

For some of the authors, A-E-Ws are better understood within the context of Western history and narratives of philosophy (following a 'synthetic' line of flight); for others, they reside in the social fabric of classrooms (within/as rituals, gestures or ruptures). For others, they are grounded within our immersive, embodied experiences of place and environments, or within our and others' evolving identities as teachers. For yet others, Western philosophical traditions seem unduly limiting, and A-E-W is altogether better understood as applied Eastern philosophy. Each chapter seeks to structure and create coherence, ordering and leading our thinking with A-E-W in different directions with varying intensities, emphases and 'stickiness'.

So what might bring these discussions together? There is a shared belief that theory and research can bring about desirable changes that inquiries with A-E-W might assist in some ways in moving closer to shared science education event horizons—the science educations that we truly want and desire more of. Moreover, there is also a type of 'negative solidarity' (Arendt 2003), holding these arguments together. This is solidarity across discourses that presently share a lack of attention and are often overlooked and thereby marginalised.

This edited collection is an assembly of different perspectives that venture outside of the normalised Cartesian knowledge project and its familiar dichotomies of mind/body, reason/emotion and affect/cognition. As so many of the authors note, this project is very much the 'bread and butter' of academic institutions (especially, I suspect, in the disciplines of mathematics and the sciences). The Cartesian project has indelibly scarred our educational institutions, shaping their technologies, languages and sociocultural practices. Of course, it makes little sense to exclude A-E-W from science education. Most will agree. And yet somehow we do, quite routinely within our curriculum, policies and research. Why? Critical theorists will remind us that our schools reproduce the subjects that they come to represent. These subjects, for the most part, are made in the mind's eye of Cartesian dualism—'the brain in a vat'. This is, perhaps, a key part of our A-E-W story.

So, what A-E-W offers in science education are subjugated standpoints. I embrace this as a key part of their purchase and appeal. Donna Haraway (1991, p. 191) joins with many other 'standpoint theorists' expressing a preference for such positions because 'in principle they are least likely to allow denial of the critical and interpretative core of all knowledge'. The preceding discussions of A-E-W, above all else, provoke us to reconsider our *subjectivities* and the roles they continue to play within our sciences, research and pedagogical practices. I return to discuss this further in the concluding section.

14.3 Venturing Beyond Cartesian's Grasp

If science education is so complex, then how should we sense it, how should we know it, and how should we perform it? If we are to escape our Cartesian blinkers, we need to teach ourselves new ways of thinking, of feeling, of sensing, of relating and, perhaps most importantly of all, of acting. This is my assertion.

Michel Foucault (1988) is widely cited on critique:

> Critique is not a matter of saying that things are not right as they are. It is a matter of pointing out on what kinds of assumptions, what kinds of familiar, unchallenged, unconsidered modes of thought the practices we accept rest. (p. 154)

What is so appealing about this quotation is the way in which it doesn't seek to settle debates in terms of being wrong or right, driven by a promise of getting that illusive 'right answer' or 'scientific truth' (see Adams St. Pierre 2014). Instead, it encourages us to reflect on underpinning assumptions, the practices, or modes of thought underwriting ourselves as science educators and researchers. Once these ontological assumptions become more visible, they invite talking and doing things differently—of rewriting and reforming ourselves, and thereby nurturing new capacities in science education.

In what follows, I offer some personalised reflections on the chapters. This reading is not offered as definitive—as an objectively crafted précis of the authors' work. Their work, of course, is so much more than my abridged reflections can ever hope to offer. But, within my interactions with the text, some pieces reached out to me with greater intensity and appeal. In line with my angle of arrival, at this particular moment in time, they were more 'sticky', as Sara Ahmed (2010) would say. This is not intended as comment on the author's work, as much as it is a comment on my own *subjectivities* as a science educator continuously in the making. When I return to reread the chapters, I will, of course, notice (and become affected by) so many other things.

The collection starts with aesthetics.

14.3.1 What Might Aesthetics Offer Science Education?

Most of us have had experiences that help us recognise that our encounters with scientific concepts, theories and experiments are never purely a detached rational affair. The opening chapters in the collection turn our attentions to aesthetics. Per-Olof Wickman (Chap. 2) explores a familiar historical dichotomy between 'synthetic' and 'analytic' philosophical traditions. This type of analysis encourages a foregrounding of history in the nature of what is taken to exist within our ever-emerging worlds.

So much science education, it seems, is grounded on naturalised analytic assumptions, centring an a priori knowledge subject. In contrast, Wickman invites me to explore emotion and aesthetics (and wellbeing) within synthetic traditions, mirroring a shift in Ludwig Wittgenstein's thinking, from his earlier Tractarian views to his later work as outlined in *Philosophical Investigations* (Wittgenstein 1958). Wickman directs us to attend much more closely to what scientists, teachers and students *do* (in terms of their unfolding aesthetic experiences, choices and actions) rather than what they *have*. For those with synthetic allegiances, meaning happens dynamically in cultures (broadly understood). This modus operandi contrasts with analytic ways of doing and talking that are predominately about *having* (*having* representations, *having* true propositions) and the associated mechanism of *having* (or acquiring) these—the cognitive mechanisms allowing for translation into mental categories and processes.

My choice of language, of course, steers my openings and thereby can betray historical alignments. To conceive of learning as a construction—social or otherwise—is to potentially slide into a familiar view of education as first *acquiring* and then *having* knowledge. Wittgenstein in his latter work was much more interested in how we 'do' language: how we know how to go on with language, in a given setting and time. Per Anderhag's (Chap. 3) enticing analysis of action-orientated methodologies pushes these discussions a little further with Pierre Bourdieu's concept of *taste*. This offers a new metaphor (with narrative associations) to think with. Again, it is about doing (habits of action or increasing participation) rather than having in the sense of internalised mental representations. A promise of Anderhag's analysis of taste is that it offers different ways of working with all-too-familiar science education concepts (attitudes/interest/motivation) that are largely conceived and measured as individualised attributes.

In Wickman and Anderhag's analyses, aesthetics (and emotion and wellbeing) are placed where many argue that they should be within emerging patterns of situated experiences, actions and activity. I am excited by the prospect of looking at classrooms with taste and intrigued by how this might redirect my attentions. What might stand out now that was previously unnoticeable? How could/should I respond differently? What can I do pedagogically with taste? I start thinking about this metaphor and differentiated tastes—the rich, bland, savoury and 'palatable' (Alsop 1999); the bitter, sharp, sweet and sour; and many others. I wonder, what might differentiated tastes subtract and add to my experiences and openings in science education?

Linda Hobbs and Leissa Kelly's study (Chap. 4) moves aesthetics in slightly different ways, under the guidance, or reins, of an empirical heuristic 'KIP' (knowledge-identity-passion). This directs my attention to 'particular moments' identified by primary science teachers as part of their evolving professional identities. Hobbs and Kelly's analysis encourages new reflections on professional growth and practices of teacher education. How might I more effectively create opportunities for pre-service teachers to articulate their KIPs and explore their effects? What might a teacher education curriculum look like that holds onto complicated relationships between knowledge, identities and passions? It is likely to be quite different to curriculum that I have been involved in.

14.3.2 What Might Emotion and Affect Offer Science Education?

The chapters in the second section offer new approaches to theorise emotions in science education. In the earlier chapters, our attention returns, once more, to what emotions do, their dynamic social traces, if you like. Alberto Bellocchi (Chap. 5) places emotion (and affect) firmly within social theorising, thinking with Durkheim, Collins and Goffman. He argues for a microsociology of learning based on Durkheim's social epistemology of knowledge that considers emotional energy as central. Bellocchi's invitation is to focus my gaze on emotional energy within interaction rituals, or focus on classroom-based moment-by-moment encounters.

Stephen Ritchie and Jen Beers-Newlands (Chap. 6) direct my attention slightly differently, to particular 'emotional events', centring on the significance or saliency that they hold for those involved, as revealed within post-event narratives. Science education now dissolves into a form of temporally and contingently threaded 'segments' that in many ways add up to more than the sum of their individual parts. I am left with a series of questions: How might I more purposefully punctuate my classroom space and time emotionally? When and how might I isolate and reflect upon segments of my classroom practices, exploring different ways in which they might be narratively threaded together? How might these narratives shape my science education imaginaries, as well as what might be lost and gained within these emotional distinctions?

James Davis (Chap. 7) skilfully guides us through an impressive array of theorists (including Kant/Hume/Durkheim) in building a persuasive case for social ontology and accompanying ethnomethods. A discussion of 'undramatic emotional energy' stands out, encouraging me to reflect on my empirical objectivities and ways in which 'dramatic' becomes 'undramatic' within my research and pedagogy (I return to this later). Kristin Cook and Gayle Buck (Chap. 8) offer an invitation to think about place-inspired pedagogy also with the construct of 'emotional energy' (EE). I was drawn to their positioning of EE within dynamically situated interactions, common moods and flowing group memberships. Their desires for emotion are normative, as a basis to reflect on the efficacy of (self-identifying) place-based pedagogies. This is clearly far-reaching.

Amongst many other things, these studies invite approaching emotion as relevant causal factors in acts of social pedagogical and temporal formations. What now emerges is dynamically situated, enfolding emotional sociality. When my teaching embraces 'group work' and 'communities of practice', these chapters remind me to embrace the dynamism and associated micro-processes of social assemblage (cf. Latour 2005). In a broader sense, communities are no longer rendered 'static objects' to be connected with but dynamic assemblages in which I (and others) might play a part in shaping. The authors encourage me to think with emerging, shifting social configurations, constituted by a greater number of factors, including emotionality and affect. In this manner, my gaze has been invited into vibrant social-affective agencies. Perhaps a more general point is that these studies push back on the rhetoric and self-evident desirability of 'group-work' and 'community-referenced' practices. They invite teachers and researchers to critically reflect on what social formations might be more desirable than others—as well as when and how (Alsop and Ibrahim 2014). Verbs and adjectives now seem more appropriate than nouns—grouping work, communing practices and others. As practitioners, these discussions invite us to experiment with how teaching might more purposefully support particular microsociologies and emotional events, those we deem to as eminently more desirable and efficacious for particular reasons at particular times.

We now turn to the body. What is the body of science education? What can it do? Spinoza famously asked: 'do we know what a body can do?' Brian Massumi's (2015, p. 177) response is 'simply no!' He continues, quite provocatively: 'there are powers of improvisation, powers of invention in a body that we have only begun to plumb'.

Liv Kondrup Kristensen and Kathrin Otrel-Cass (Chap. 9) direct our embodied thoughts to Maurice Merleau-Ponty's writing with desires to unpack relational and interactive dimensions within what are clearly visceral embodied encounters in physics education. These dynamic discussions seek to write a body back into science education where oddly it seems strangely unnecessary (Alsop 2014). I am reminded of the opening paragraph of Brian Massumi's (2002, p. 1) *Parables for the Virtual:*

> When I think of my body and ask what it does to earn that name, two things stand out. It *moves*. It *feels*. In fact, it does both at the same time. It moves as it feels, and it feels itself moving. Can we think a body without this: an intrinsic connection between movement and sensation whereby each immediately summons the other?

Kristensen and Otrel-Cass's study offers dynamic connections between movement and sensation in Doppler physics. The authors mention Merleau-Ponty's unfinished text (*The Visible and the Invisible*), which introduces the evocative notion of 'flesh'. I am now 'thinking' about—'wriggling' might be a more appropriate word here—an intermingling flesh as science education, as well as the rich possibilities for phenomenological analyses in science education. As Ostergaard et al. (2008) argue, phenomenology seems underrepresented in the field.

Morten Rask Petersen and Niels Bonderup Dohn (Chap. 10) search for relationships that are too often denied in the field of psychology. Whilst there is a plethora of studies of *interest* as a facilitator for learning, acknowledging the reverse, that learning is a facilitator for *interest*, is strangely—some might say bizarrely—overlooked. Many would read this asymmetry, once more, as a legacy of Cartesian dualism, which centres our gaze squarely upon the project of cognition. I am struck by the ways in which interest shifts perceptual time. The 'Aha moment' in biology education that the authors outline shifts an inner flow of time. Here, I am reminded of a heated debate over the politics of temporality in Paris in 1922 between Henry Bergson and Albert Einstein. Bergson insisted on talking about time in relationship to consciousness. Einstein saw no role for philosophers in considerations of time. Time, he argued, should be left for scientific study (see discussions by Canales [2009]). I am also reminded of Philip Payne and Brian Wattchow's (2009) work on slow pedagogy. Some questions emerge: How might my pedagogies and research rearrange temporalities? How could these practices reinforce and/or disrupt dominating temporal politics? How might my notions of time serve to engender and/or marginalise A-E-W? How might I more purposefully play with time in my professional practices?

Tatiana Chemi (Chap. 11) grounds our thinking with emotions and affect in curriculum. She invites us to juxtapose cultural, embodied educational practices of art and science, offering ways of reflecting on our insatiable desires to reintegrate curriculum (as either STEM or STEAM). Chemi leaves me fascinated by tinkering pedagogies as dynamically fluctuating meeting points of curriculum and child. I wonder about curricula and its relationship with experiences. What does curriculum mean at the level of conscious emotional awakenings, for instance? How might curricula serve to discipline and normalise our bodies in art and in science? What might I hope for my subjectivities within my acts of curriculum integration, commonly labelled as STEM or STEAM?

14.3.3 *What Might Wellness and Wellbeing Offer Science Education?*

In the third and concluding section, my attentions are moved to contemplate wellness and wellbeing. This offers the collection of a normative shift, inviting discussion of what might be more (or less) desirable within science education. For Ken Tobin, Konstantinos Alexakos, Anna Malyukova and Al-Karim H. Gangji (Chap. 12), wellness and wellbeing find their places in embodiment, guided by Eastern philosophy. These authors explore a body as circuits that pass through 26 pairs of safety energy locks (SELs). Their detailed empirical analysis raises my consciousness of particular body movements, orientations, touches, holds and flows occurring during teaching activities. This offers a very different way to order my experiences. I am afforded capacities to attend to science education practices in different ways. I start to think about the Western body and ways that it underwrites so much of my work.

For Cassie Quigley and Renée Lyons (Chap. 13), wellness is entwined with environmentalism. Niel Everden (1999, p. 4) once famously described environmentalists as those who publically 'confess a concern for the non-human'. Quigley and Lyons' arguments are framed by the need to care. Navigating subtle differences between 'education' and 'indoctrination', they offer helpful distinctions between the emotional caring, natural caring and ethical caring. These demarcations are especially generative, inviting me to reconsider assumptions of 'the natural' within my imaginaries of (environmental) science education. How might care be evidenced in science education practices? How might I orientate my research to care more?

I am reminded here of Maria Puig de la Bellacasa's (2011) invitation to move beyond Bruno Latour's 'matters of facts' and 'matters of concerns' and reflect on 'matters of care', asking 'who cares? 'why ought they care?' 'how ought they care?'. The topic of care, as Quigley and Lyons eloquently note, has been a topic of sustained interest in education, although it has yet to be a pressing theme in science education. It should be.

The notion of the 'natural' remains a hot debate in environmental studies. For instance, David Abram (1996) and Paul Shepard (1982) both suggest (in slightly different ways) that the modern 'development of self' results in compromised relationships with the world that surrounds us. They direct us to pay attention to our 'natural', sensorial awakenings as rudders to steer us towards more ethical and moral sensory-immersed existences within the world in stormy times.

Wellness and wellbeing, of course, have a long and established history of human desire. They are something that we all strive towards that gives us meaning and thereby ordering and disciplining our lives, in profound ways. Indeed, wellness and wellbeing are fast becoming ultimate educational indicators, surpassing even literacy and numeracy in Canada (and many other places) as major organising themes for national school reforms. Who could possibly argue against increasing wellbeing? Indeed, such reforms—more often than not in my experiences—have a self-evident desirability about them that escapes sustained critical reflection.

The chapters in this collection encourage me to locate my desires and promises for wellbeing and wellness somewhere: within a body and environmental care or, bringing in the other chapters, within aesthetics and emotion. This type of grounded analysis escapes what Haraway (1991) famously calls 'the God trick'—the view from nowhere to everywhere. The politics of demarcation, what gets to count as 'well' and 'being' (and not, of course), deserve our continuing attentions as science educators. I am immediately reminded here of a multiplicity of studies of wellbeing drawn from the margins, stories of wellness that are somehow silenced by dominant traditions. These include, for instance, Arthur Frank's (1995) wounded storyteller, Susan Leigh Star's (1991) allergies to onions and many others of course.

14.4 The Politics and Promises of A-E-W in Science Education

So, where to now? What do we need as science educators and researchers in pursuit of transformations and deep-rooted change? Martin Heidegger (1968, p. 3) once famously commented that, 'we learn to think by giving our mind to what there is to think about'. As I have previously noted, the processes of change with science education might start with our consciousness, our perceptions and our bodies, by bringing new and unexpected patterns into our openings and experiencing. We need to nurture our capacities to sense, to think and to feel science education in different ways, enabling us to challenge the static and the taken for granted. It is at that point we can seek to enlist others into the promises of our research narratives by nurturing others' capacities to share our encounters and experiences with A-E-W.

In this light, the promises of the chapters assembled herein reside within their invitations to different vantage points and vistas, different ways of knowing and evoking different political possibilities and actions. A central message—of course—is that we need to develop a heightened sensitivity to the roles and possibilities of aesthetics, emotion and wellbeing within our teaching and research methods. The authors encourage us to venture beyond our deep-rooted traditions of Cartesian dualism and embrace subjective, embodied, socially situated dimensions of our artistry and craft. Such dimensions are so easily discouraged by powerful voices of tradition forcing attentions towards standardised and universalised knowledge and representative practices.

John Law (2004, p. 2) writes so delightfully about different ways of knowing within research methods. This passage of text is provocative, definitely worth quoting in length:

> Perhaps we will need to know them through the hungers, tastes, discomforts, or pains of our bodies. These would be forms of knowing as embodiment. Perhaps we will need to known them through 'private' emotions that open us to the worlds of sensibilities, as emotionality or apprehension. Perhaps we will need to rethink our ideas about clarity and rigour, and find ways of knowing the indistinct and the slippery without trying to grasp and hold them tight. Here knowing would become possible through techniques of deliberate imprecision.

> Perhaps we will need to rethink how far whatever it is that we know travels and whether it still makes sense in other locations, and if so how. This would be knowing as situated inquiry. Almost certainly we will need to think hard about our relations with whatever it is we know, and ask how far the process of knowing it also brings it into being.

A key theme of this text is that science educators can better understand science education by attending more closely to social interactions and their underpinning processes of formation. Here, A-E-W take up their meanings in dynamically forming social worlds (Bellocchi 2015). The arguments offer explanatory frameworks to account for (and hopefully predict to some degree) social happenings. Other authors draw attention to different themes, or perspectives—to windows with views given life through narratives of embodiment from Merleau-Ponty, critiques of asymmetrical psychological orientations, Eastern philosophies and pedagogies of care and places.

Throughout all these discussions, perhaps what stands out most are the ways in which the 'what' slowly gives way to the 'how'. To think with A-E-W is an invitation to follow them through their effects. It is less about what A-E-W is, as essence or substance, but more about their manner, about what they *do* (and what they *might* do) in science education. It is more about their agencies, as Wickman eloquently notes in Chap. 2. Put slightly differently, this is a text with a prominent *process* theme. Sarah Ahmed (2004, p. 2) writes of emotions in these terms:

> Emotions do things, they align individuals with communities—or bodily space with social space [...] Rather than seeing emotions as psychological dispositions, we need to consider how they work, in concrete and particular ways, to mediate the relationships between the psychic and the social.

Aesthetics, emotions and wellbeing, viewed in this way, share ontologies consonant with William James's famous phrase, 'always in making'. This, of course, contrasts sharply with a world that is 'out there' in a static, objectively determined sort of way. For me, A-E-W come alive within their capacities to mediate their abilities—returning once more to my opening conceptual guide—to 'affect and be affected'. This is an understanding of science education that is always ongoing, influx and enfolding both meanings and contexts. It is also an understanding of science education that is deeply relational. Indeed, the processual nature of these discussions distinguishes them from much traditional disciplinary knowledge.

Now, I can imagine walking into a classroom with my attentions redirected to bodily postures, energy locks, prosody, emotional events, micro-social rituals, emotional energy (undramatic or otherwise), tastes, expressions of care and encounters with place. I can take the authors with me in my educational journeys. Their analyses offer me a structure and coherence within so much complexity. They invite crafting pedagogies that encourage my students to find their own voices and feelings with/in A-E-W.

However, throughout these discussions, my attentions have been largely focused on others' educational practices and theories. In the traditions of research, my gaze has been steered outward. I now want to move it inward, in a critical reflexive fashion. I have some questions: What roles might I allow A-E-W to have within my theoretical and empirical practices? What roles might the authors' A-E-Ws be allowed to play within their analyses? More generally, what roles should our subjectivities be allowed to have within our science education research practices?

There is such an impressive array of different research methods on display in this edited collection. These are so far in excess of the usual 'speaking subject' underwriting so much education research. Authors innovatively grapple with social, corporeal, exteroceptional, kinaesthetic and other phenomena.

In this light, I pause to reflect on this endnote and the ways in which it somehow disguises its embodied, sensory, emotional, social, material origins. It is another all-too-familiar case of an embodied, sensory, socially and materially situated person representing their thoughts in a disembodied way. Consonant with my previous citation from Foucault on critique, I'm not arguing that this is either right or wrong. But I wonder about the desirability of more openly acknowledging A-E-W within scholarly practices, including my own research and writing. How might my science education scholarship and research be serving to prop up worldviews that I am seeking to problematise? How might I be implicated in maintaining Cartesian worldviews, whilst also seeking for deep-routed change? What might I do differently?

Counter discourses (such as A-E-W) can carry prospects of radical transformations. And yet, by definition they are marked by origins of their departures, tied to the discourse that they are questioning and seeking to move beyond. Part of the allure of this collection is venturing beyond a narrowed cognitive and rationalistic constraint. The authors make this clear. However, an associated complexity is that this constraint structures so much of what we do as/in/with science education. This makes this text potentially far-reaching, but it also, perhaps, makes the ideas extremely vulnerable and fragile.

Elizabeth Adam St. Pierre (2014) makes a powerful point about education research, demarcating what she calls 'post-qualitative inquiry' or 'post inquiry'. She is quite critical of much qualitative research in education, claiming that it 'participates in and continues the centuries-old history of Cartesian knowledge projects—ferreting out what is known only on not being known—yet' (Adams St. Pierre 2014, p. 6). She continues noting that:

> Even texts that claimed to be 'interpretive' or 'critical' retained positivist structuring concepts like *objectivity, bias, data, coding data, grounded theory, saturation, audit trails, inter-rater reliability, triangulation* and systematicity, even as they introduced phenomenological concepts like *voice, lived experiences, narrative and/or critical concepts like authenticity, agency, emancipation, transformation, social justice and oppression*. The structure, indeed, deconstructs itself.

My general point is that this text opens up questions of subjectivity not only within practices of science teaching but also within practices of research and scholarship. Even though I like to conceive of myself as a progressive science educator, like many others, I can become caught up in 'analytic' traditions of separating the researcher from the researched—employing methods, which obviate my 'being there' (Adam St. Pierre, Ibid. p. 7) with research subjects in the field.

In this manner, these rich analyses encourage me to continue to grapple with new research traditions or 'new empiricisms' to use Patricia Clough's (2009) term. I am now thinking about how my tastes, or my bodily locks, or my passion, microsociologies, or my identities and places shape my empirical practices. How might A-E-Ws be more fully acknowledged in/as research? In what ways might my research and writing more firmly decentre the mind's eye?

It should be recognised that dominant cultural traditions in our field still require particular notions of empirical 'rigour', 'validity', and 'reliability', largely built on a Cartesian ontology (Alsop 2005). This leaves me asking, how might I extend what gets to count as 'good data' and 'good writing' by explicitly recognising A-E-W? When might my research data, for instance, be publically allowed to have 'emotional energy', 'taste', 'intensity', 'passion' or 'care'? James Davis's (Chap. 7) 'undramatic emotional energy' seems important here. However, distinguishing the dramatic from undramatic doesn't alleviate such considerations, even if it does helpfully lower their intensities.

Now this is radical territory: especially given that mainstream science education journals have their own particular orientations, largely, it seems to me, for 'conventional humanist methodologies' (Adams St. Pierre 2014). Such methodologies often follow 'analytic' traditions of rigour and objectivity. Such tastes, of course, seem to have no taste at all—the result of established cultural traditions that have become so naturalised that they have faded from our sensory gaze. It is never easy to work against hegemonic traditions, particularly as publishing in mainstream journals is so fundamental to successful academic careers.

Again, this is not a statement of what is right or wrong, but an invitation to reflect on ontologies as an ethos and/or emotionality, way of being in the world and the far-reaching nature of the arguments that this text embraces. What sense does it make to separate our research objects from our subjectivities? This is the tradition. I am left considering what these studies of A-E-W might offer my research? What might it mean to allow A-E-W greater agency within my empirical practices? What do I lose (and gain) when I turn to 'analytic' empirical methods within studies with 'synthetic' traditions (or vice versa)? What could A-E-W research-based methods entail and not?

There is so much to explore now and yet so little space. I can only leave these questions as provocations for future work.

14.5 Partialities and Future Possibilities

In any book project, there are lots of unexplored questions. I am reminded of some other ways of approaching A-E-W. These might include post-phenomenology and eco-phenomenology; psychoanalytic inquiries; feminist, queer theorist, disability, activist and subaltern inquiries; cultural geography and anthropology; critical discourses of emotion; and science and technology studies. Many of these perspectives are associated with the 'affective turn', as previously mentioned (see discussions in Gregg and Seigworth 2010). In the sciences, you could probably add other perspectives, including neuroscience, cognitive science and evolutionary biology. These perspectives, I suggest, warrant more attention in A-E-W research in science education.

Each of these perspectives, of course, brings differing questions, concepts and assumptions. Affect theorists, for instance, often mark distinctions between 'feelings', 'emotions' and 'affect' (see, for instance, Gregg and Seigworth 2010). The chapters in this collection, they might argue, dissolve these differences, thereby, perhaps, unnecessarily flattening affect—to their liking at least. In this text, there is an emphasis on the social body, and perhaps less attention is granted to 'the psyche'. The text, following well-established academic traditions (that some may trace back to Immanuel Kant), rests heavily upon the 'knowable rational actor'—for some affect theorists, such as Brian Massumi (2002) and Eve Kosofsky Sedwick (2003), affect is taken up as pre-personal and innate, operating largely outside of conscious intentions. Acknowledging unconscious dimensions of A-E-W raises some really fundamental questions. What are limits to our knowing and sensing?

I offer all these comments in the spirit of celebrating a multiplicity of possible worlds, each brought into being through different underpinning assumptions. My comments here are not intended as judgements on the preceding arguments. They are not premised on a notion of getting *the* right answer in a single world. In contrast to 'a closure', they are intended as an invitation to explore multiple openings.

In the same spirit, there are also, perhaps, lingering questions of how the authors and editors separate aesthetics, emotion and wellbeing. In this text, they occupy different sections. This provokes consideration of their separation: what is aesthetics without emotions? Or wellbeing without aesthetics? There are other permutations, of course. Perhaps some significant questions here are: what might each of these concepts distinctively bring? What might they enable us to *do* differently (and not)? When and how? What are some of the associated politics and ethics involved? Where might they be helpful, desirable and not?

In my experience, the history of A-E-W scholarship can become swept up in pursuit of deep-rooted transformation. Throughout this collection, it is repeatedly noted that our dominant ways of thinking are missing something essential, something foundational. This certainly drives and animates my work. But giving creative attention to A-E-W often carries with it our thirsts (building on a taste theme) for unlocking and realising truly subliminal forces—forces fundamentally shaping science and education in all its various expressions and disguises. My, and perhaps others, stories of A-E-W in science education can become propelled by such revelatory desires. Indeed, at times they might, somewhat regrettably perhaps, come across as mini French revolutions (see Ritchie and Beers-Newlands; Chap. 6). Arguably, this text is marked by these sentiments. I see this as a good thing. But having said that, A-E-W in all their disguises can, perhaps, never satisfy all my (and perhaps others) utopian cravings. I leave this as another open question.

Bringing things to a close, I offer this endnote within the context of my evolving knowledge of the field and with a series of commitments and wishes for science and education. I hope for more diverse and productive ways to talk about affect theory and subjectivities in science education. Above all else, what I want to celebrate in this endnote is a truly fabulous collection that provides so much for science educators to think about and to think with. This should never be underplayed, especially in education.

I would like to again thank all those involved.

References

Abram, D. (1996). *The spell of the sensuous*. New York: Vintage Books.

Adams St. Pierre, E. (2014). A brief and personal history of post qualitative research: Toward "post inquiry". *Journal of Curriculum Theorizing, 30*, 2–19.

Ahmed, S. (2004). *The cultural politics of emotion*. New York: Routledge.

Ahmed, S. (2010). Happy objects. In M. Gregg & S. Seigworth (Eds.), *The affect theory reader*. London: Duke University Press.

Alsop, S. (1999). Understanding understanding: A model for the public learning of radioactivity. *Public Understanding of Science, 8*, 267–284. doi:10.1088/0963-6625/8/4/301.

Alsop, S. (Ed.). (2005). *Beyond cartesian dualism: Encountering affect in the teaching and learning of science* (2nd ed.). Dordrecht: KluwerSpringer Academic Press.

Alsop, S. (2011). The body bites back. *Cultural Studies in Science Education, 6*, 611–623. doi:10.1007/s11422-011-9328-4.

Alsop, S. (2014). The body and the laboratory. In M. Watts (Ed.), *Dilemmas and debates in science education* (pp. 205–219). London: Routledge.

Alsop, S. (2016). Encountering science education's capacities to affect and be affected. *Cultural Studies of Science Education, 11*, 551–565. doi:10.1007/s11422-015-9692-6.

Alsop, S., & Ibrahim, S. (2014). School community projects. In D. Gunstone (Ed.), *Encyclopedia of science education*. Dordrecht: Springer.

Alsop, S., & Watts, M. (2003). Science education and affect. *International Journal of Science Education, 25*(9), 1043–1047. doi:10.1080/0950069032000052180.

Arendt, H. (2003). *Responsibility and judgement*. New York: Schocken.

Barthes, R. (2005). *The neutral*. New York: Columbia University Press.

Bellocchi, A. (2015). Methods for sociological inquiry on emotions in educational settings. *Emotion Review, 7*(2), 151–156. doi:10.1177/1754073914554775.

Betelsen, L., & Murphie, A. (2010). An ethics of everyday infinities and powers: Felix Guattari on affect and the refrain. In M. Gregg & S. Seigworth (Eds.), *The affect theory reader* (pp. 138–161). London: Duke University Press.

Canales, J. (2009). *A tenth of a second: A history*. Chicago: University of Chicago Press.

Clough, P. (2009). The new empiricism: Affect and sociological method. *European Journal of Social Theory, 12*, 43–61.

Clough, P., with Halley, J. (Ed.). (2007). *The affective turn: Theorizing the social*. London: Duke University Press.

Deleuze, G. (1988). *Spinoza: Practical philosophy*. San Francisco: Continuum.

Everden, N. (1999). *The natural alien*. Toronto: University of Toronto Press.

Foucault, M. (1988). Technologies of the self. In L. Martin, H. Gutman, & P. Hutton (Eds.), *Technologies of the self: A seminar with Michel Foucault* (pp. 16–49). Amherst: University of Massachusetts Press.

Fox Keller, E. (1984). *A feeling for the organism: The life and work of Barbara McClintock*. San Francisco: W.H. Freeman.

Frank, A. (1995). *The wounded storyteller: Body, illness & ethics*. Chicago: University of Chicago.

Gould, S. J. (2000). *Wonderful life: The burgess shale and the nature of history*. London: Vintage books.

Greene, M. (2001). *Variations of a blue guitar: The Lincoln Center Institute Lectures on Aesthetic Education*. New York: Teachers College.

Gregg, M., & Seigworth, G. (Eds.). (2010). *The affect theory reader*. London: Duke University Press.
Haraway, D. (1991). *Simians, cyborgs, and women: The reinvention of nature*. New York: Routledge.
Heidegger, M. (1968). *What is called thinking*. London: HarperCollins.
Jones, S. (2007). *Coral: A pessimist in paradise*. London: Little Brown Book Group.
Latour, B. (2005). *Reassembling the social: An introduction to A-N-T*. New York: Oxford University Press.
Law, J. (2004). *After method: Mess in social science research*. New York: Routledge.
Levi, P. (1984). *The periodic table*. New York: Schocken Books.
Massumi, B. (2002). *Parables for the virtual: Movement, affect, sensation*. Durham: Duke University Press.
Massumi, B. (2015). *The politics of affect*. Oxford: Polity.
Ostergaard, E., Dahlin, B., & Hugo, A. (2008). Doing phenomenology in science education: A research review. *Studies in Science Education, 44*(2), 93–121. doi:10.1080/03057260802264081.
Payne, P., & Wattchow, B. (2009). Phenomenological deconstruction, slow pedagogy and the corporeal turn in wild environmental. Outdoor education. *Canadian Journal of Environmental Education, 14*, 15–32.
Puig de la Bellacasa, M. (2011). Matters of care in technoscience: Assembling neglected things. *Social Studies of Science, 41*(1), 85–106. doi:10.1177/0306312710380301.
Sedgwick, E. (2003). *Touching feeling: Affect, pedagogy, performativity*. Durham: Duke University Press.
Shepard, P. (1982). *Nature and madness*. San Francisco: Sierra Book Club.
Star, S. L. (1991). Power, technologies and the phenomenology of conventions: On being allergic to onions. In J. Law (Ed.), *A sociology of monsters: Essays on power, technology and domination* (pp. 26–56). London: Routledge.
Stewart, K. (2010). Afterword: Worlding refrains. In S. Seigworth & M. Gregg (Eds.), *The affect theory reader*. London: Duke University Press.
Watts, M., Alsop, S., Zylbersztajn, A., & Maria de Silva, S. (2007). Event-centred-learning: An approach to teaching science technology and societal issues in two countries. *International Journal of Science Education, 19*(3), 341–351. doi:10.1080/0950069970190306.
Wittgenstein, L. (1958). *Philosophical investigations* (2nd ed.) (trans. G. Anscombe). Oxford: Basil Blackwell.
Wolff, V. (1972). *Moments of being*. London: Pimlico.

Steve Alsop is a Professor and Director of the Graduate diploma in Environmental Sustainability in the Faculty of Education. He teaches courses and supervises graduate students in the fields of education, science and technology studies, environmental sustainability and interdisciplinary studies. Steve has held a series of administrative appointments including Associate Dean (Research and Professional Development—York University); Teaching Studies Area coordinator (Science—Roehampton Institute, University of Surrey, UK), and Head of Physics (Haverstock School, London, UK); and a series of honorary positions at the Froebel Institute, Roehampton University and the Universidad Baja California, Mexico.

Index

A. Bellocchi et al. (eds.), *Exploring Emotions, Aesthetics and Wellbeing in Science Education Research*, Cultural Studies of Science Education 13,
DOI 10.1007/978-3-319-43353-0

CPSIA information can be obtained
at www.ICGtesting.com
Printed in the USA
LVOW04*1656231216
518588LV00009B/151/P

9 783319 433516